KU-535-390

SOCIETAL SYSTEMS

NORTH-HOLLAND SERIES IN

SYSTEM SCIENCE AND ENGINEERING

Andrew P. Sage, *Editor*

Wismer and Chattergy	Introduction to Nonlinear Optimization: A Problem Solving Approach
Sutherland	Societal Systems: Methodology, Modeling, and Management
Šiljak	Large Scale Dynamic Systems

SOCIETAL SYSTEMS
Methodology, Modeling, and Management

John W. Sutherland

NORTH-HOLLAND · NEW YORK
NEW YORK • AMSTERDAM • OXFORD

Elsevier North-Holland, Inc.
52 Vanderbilt Avenue, New York, New York 10017

Distributors outside the United States and Canada:

Thomond Books
(A Division of Elsevier/North-Holland Scientific Publishers, Ltd.)
P.O. Box 85
Limerick, Ireland

Library of Congress Cataloging in Publication Data
Sutherland, John W
Societal Systems: Methodology, Modeling and Management

(North-Holland series in system science and engineering)
Includes bibliographical references and index.
1. Social sciences—Methodology. 2. Social systems.
I. Title.
H61.S97 300'.1'8 77-16645
ISBN 0-444-00239-1

Manufactured in the United States of America

To Scatophagus Argus,
who has since moved on to be the muse
of a more elegant master.

Contents

Preface xi

PART 1: METHODOLOGY

1: *A Syncretic Paradigm: The Bases of Behavior* 4

Criteria for Social-Science Constructs 5
Steps in the Syncretic Process 11

The Senses and Cybernetics 14
The Id-Level Behaviors 20
Trial-and-Error Behavior 23

The Inductive Modality 25

The Deductive Processes 33
The Idiosyncratic and Exegetical Variants 36
The Discursive Variants 39

The Crucial Concept: from Individual to Collective Behaviors 45

Notes 47

2: *The Dialectical Challenge to Complexity* 50

The Dialectical Engine 51
Properties of the Paradigm 59

Dialectical Configurations 62

Dynamic Implications of the Dialectic 67
The Cognitive Corollaries 74

The Dialectical Disputations 77

Notes 85

PART II: MODELING

3: *The Determinants of Societal Structure* 90

The Case for Qualitative Discipline 91

The Cultural Constraints 97

Axiological Prophets and Moral Order 101

Axiomatic Prophecy and Rationalization 109

The Dimensions of Culture 117
The Cultural Set Concepts 124

Appendix: Ideologics (by Stephen E. Seadler) 140

Notes 150

4: *Ideal-Types and Societal Dynamics* 152

A Synoptic Ideal-Type: The Modalities of Organizational Behavior 153
Action Implications of the Synoptic Ideal-Type 157
Socio-psychological Implications of the Synoptic Ideal-Type 162

The Several Societal Referents 170
The Acculturated Ideal-Types 172
The Adaptive and Idiosyncratic Ideal-Types 181

Dynamic Implications of the Ideal-Types 188
Linking the Qualitative and Quantitative Domains 197

Notes 204

PART III: MANAGEMENT

5: *The Interdisciplinary Imperative* 210

The Societal Correlatives 211
The Units of Social Significance 213

The Economic Engines 216
The Several Socio-Economic Calculi 229

The Guises of Government 242

The Interdisciplinary Interfaces:
A Typological Construct 249

Notes 252

6: *The Criteria of Societal Quality* 254

The Mandate of the Masses:
The Search for Isomorphisms 255
A Schedule of Societal Benefits 260

Terms of the Societal Tradeoff: The Corridor Concept 266
Systems of Spiritual Significance 267
Systems of Social (Associational) Significance 272
Systems of Materialistic Significance 277

The Managerial Methology: Analogy and Prescription 283
An Analogic Construct: A Generic Calculus of Satisfaction 289
The Managerial Challenge: Optimizing at the Margin 294

Notes 303

Glossary 305

Index 315

The problem is not how to produce great men,
but how to produce great societies.

[Alfred North Whitehead]

Preface

In these pages, we are in pursuit of the substance and implications of societal systems. A societal system is a collection of individuals, bound together by complex and possibly amorphous sets of interests, pursuing an always complex and possibly indeterminate set of ambitions. It is the presence of human beings that lends societal systems their interest for the sociologist. The cultural, cognitive and *a priori* (e.g., axiological, axiomatic, ideological) roots of societal systems makes them the concern of the anthropologist and social psychologist. The manifest purposes of societal systems—their material dimension—commands the attention of the economist and management scientist. Finally, because societal systems represent the highest level of analytical challenge, they are a proper subject for the system scientist or for others whose concern is primarily with the extension of our methodological capabilities. To some extent, then, the societal system is everyman's child, claimed by all but cared for by none. There are at least three reasons for this.

First, societal systems—as composites of social, cultural, economic, behavioral and political properties—frustrate the neat academic demarcations we have laid down. The sociologist tends to define a society in terms of the structure of personal relations or class references. For the economist, a societal system becomes comprehensible primarily in terms of the factors of production and the method of their organization and control. To the anthropologist, particularly if he is of the functionalist persuasion, societal systems are intelligible primarily as complex adaptations to environmental properties. The political scientist sees societal systems primarily as sets of power relations and tends to define them mainly in terms of various matrices of authority. Finally, the behavioral scientist views societal systems mainly as constraints on individual prerogatives, and perhaps more as a source of repression than as an arena of opportunity.

From the system-science standpoint, virtually all of these perspectives are useful and necessary. But none, of itself, completely exhausts the properties of the societal system. Therefore, he recommends an explicitly interdisciplinary mode of inquiry, one where the variables from the several traditional social science disciplines are all entered endogenously, and without *a priori* prejudice as to their relative importance. Moreover, his interdisciplinary approach must exhibit an explicitly syncretic sympathy, for there is some disagreement within the individual disciplines about the truths of their subjects. Thus, before a meaningful interdisciplinary approach can be evolved and put to work, the conflicts internal to sociology, economics, anthropology, etc., must be resolved. To do this, the hypothetico-deductive method must be put to work. For to resolve *intra*disciplinary arguments—and to construct the *inter*disciplinary interfaces necessary for a coherent treatment of societal complexes—issues must generally be raised to a higher level of abstraction. Moreover, the hypothetico-deductive method gives credence to the long-standing principle that as the process or system we are studying becomes more and more complex, the inductive and experimental procedures popular with modern social scientists become less and less useful.

This leads us directly to the second reason why societal systems have received such scant attention from formal science: they do not, except in the most trivial cases, permit us to use the instruments, methods and paradigms that are evolved from—or respond to—the empiricist epistemology that dominates our disciplines. In comparison with formal organizations or most functional institutions, societal systems always have very complex origins and purposes. In particular, at least part of their structure and substance owes its definition to essentially *a priori* predicates—to axiomatic, axiological or ideological bases. Thus societal systems cannot be fully explained as functional adaptations to the environment in which they happen to reside. In short, they are usually not intelligible as homeostatic, cybernetic, conditioned or functionalistic systems. We are thus denied the use of these powerful concepts when treating societal complexes, though they are of undeniable benefit in the treatment of most lower-order phenomena. Models of societal systems must, then, be based on constructs more sophisticated than the first-order feedback mechanisms that we find in corporate or institutional models. Societal models cannot, as a rule, be deterministic in any sense at all; rather, they will tend to take a distinctly dialectical form, and this is a form with which the social scientist has little familiarity. In developing dialectical constructs, we shift our attention from neat causal inferences to force vectors, and from well-behaved stimulus and response sets to often very complicated reticular configurations or other network referents. In summary, then, societal systems have been ill studied largely because of their basic incompatability with the preferred procedures of social science—because (1) societal systems are not always empirically accessible in all their components, (2) their critical properties are not always amenable to precise measurement, either directly or through surrogation, and (3) being essentially "organic" entities, societal systems cannot be taken apart and studied analytically in a laboratory environment.

The third reason for neglect goes beyond the parochial structure of the social sciences and the dominance of the empirical-inductive epistemology. Rather, it is concerned with technological limitations that are to some extent inescapable for the present. It is an uncomfortable fact of modern scientific enterprise that as a system becomes more complex and indeterminate, it becomes more and more immune to the mathematical and statistical instruments in which we justifiably take pride. Our mathematical modeling techniques fail, beyond a point, to accomodate the complexity of all but the simplest societal systems. The number of relationships that can be considered—the number of state variables we can manipulate—places very severe limits on our ability to treat societal systems as integral "wholes", as organic entities. Even our normal statistical procedures are vulnerable to awesome confusion when turned on societal systems, even those that are fairly well bounded in time and space. The problem of handling enormous numbers of state variables and relationships is amplified when we consider the nature of the relationships we must accommodate. For, as was just suggested, deterministic causal chains and neat stimulus-response configurations are scarce indeed in the societal domain. This means that any model we might try to develop, were it to seriously attempt to reflect societal realities, would be so riddled with contingencies and stochastic branches as to be unusable by the policy-maker (even if it were manageable by the scientist). In all, then, scientists have shown a great deal of discretion and good sense in avoiding a confrontation with societal systems and in concentrating their attention on the parts rather than on the whole.

But the problem is this: societal systems cannot be understood merely as some product of their separate parts. That is, in moving from the individual to the organization, from the organization to the institution and from the institution to the societal system, something new is added at each stage. Any higher-order system is, as a rule, something more than the sum of its parts. And unless we are able to capture this "something more" that is added, we cannot adequately comprehend societal phenomena or rationally manage them. Now I am at least partially in sympathy with those who suggest that we might do well not to try to manipulate societal systems. But I suspect that men will continue to try to control the structures that involve them, and that more knowledge implies a potential for better management. For this reason, science has always thought it important to challenge complexity, even while much of the rest of the world was at work trying to accommodate it. And with respect to societal systems, we may take some comfort from the suggestion that none of the barriers to the study of societal systems we just discussed is insurmountable. A solution to the structural problem posed by the historically parochialized social sciences is already gaining effect in the movements toward interdisciplinary teams and transdisciplinary research procedures. To these forces already at work, we can add a bit by considering the implications of syncretic methods for forcing linkages among social, cultural, economic and political variables. But parochialism is on its way out, and will likely disappear with no help from this volume.

So too with those empiricist predilections that have so long frustrated serious inquiry at the societal level. In virtually every one of our disciplines, serious challenges have arisen to the suggestion that the *sine qua non* of science is observation, measurement and manipulation. It is simply no longer sophisticated to propose that the modern social scientist should emulate the nineteenth-century physicist or restrict his analytical procedures to those that might be applauded by the organic chemist or electrical engineer. Rather, the leading edges of social-science methodology are pushing toward new appreciations of the role to be played by conceptual, theoretical and deductive instruments, and

I hope that what we do in these pages will contribute to this push in some way.

Finally, even the technological limitations we mentioned can be compensated for, though not immediately overcome. Particularly, much of the complexity that quantitative models cannnot accommodate may be absorbed by prefixing them with qualitative models—hypothetico-deductive constructs. More specifically, the strategy is to get around instrumental inefficiencies by using normative or deductive models as a flexible "front end", and then linking them through logical devices to the lower-order quantitative constructs. In this way, the two major arms of scientific inquiry—deductive and inductive inference—are joined, and conception and perceptions once again become complements rather than competitors. Of necessity, this will involve a shift away from analytic procedures and toward a heuristic platform, a shift away from laboratory experimentation and toward action-research. And before these deductive-inductive linkages become a practical artifice for the social sciences, much further work needs to be done in the area of qualitative analysis. It is here that I hope these pages will be of most direct assistance.

In summary then, the analysis and management of societal systems implies some minimal innovation, but relies primarily on an amalgamation of skills, data and techniques that already have a long and honored history. We attack societal phenomena by providing syncretic and interdisciplinary interfaces among existing disciplines; we restore the hypothetico-deductive method to parity with inductive procedures, and pay constant attention to the empirical dictate that all constructs should eventually be validated through some type of experimentation; finally, we establish critical linkages between qualitative and quantitative constructs, hopefully producing levels of insight and discipline that neither could obtain in isolation. These, at least, are my ambitions in this book. I cannot pretend to have achieved them all in equal or even adequate measure; limitations of both wit and energy have forced constant compromise. But while this work may be short on conclusions and unequivocal laws, it is long on enthusiasm and liberally laced with the sure knowledge that more elegant and ingenious scholars are at work to repair my deficiencies.

Piscataway, N.J.
April 1977

PART I
METHODOLOGY

Most social scientists agree that our methodological base is in need of some repair. So our first two chapters will concern themselves more or less directly with methodological matters. In Chapter One we shall show the great diversity of opinion and presumption about societal phenomena, and suggest that at least some of this diversity is due to the operation of parochial perspectives that social scientists sometimes adopt. The scientist, like the common man, occasionally lets his prejudices and predilections color his inquiries. The problem for us is to make these prejudices explicit, and thereby compensate for them. One technique for doing this is to translate parochial paradigms into higher-order constructs, and in the first chapter we try to show something of the technology that is available to us for this purpose. This technology takes the form of syncretic analysis, and attempts to produce paradigms that are less tendentious than the individual perspectives from which they are constructed. When generalized, the syncretic process holds great promise for generating the kind of interdisciplinary linkages we need if we are going to successfully attack societal systems. The second chapter takes a new look at an old subject: the dialectical method. In the form in which it appears in these pages, the dialectical paradigm presents us with a procedure for inquiry that is particularly well suited to challenge the complexity that societal systems exhibit. Our comprehension of the dialectical procedure is still somewhat immature, but it is offered here as an indication of the direction in which social-science methodology is probably heading. The only qualification I would add is that we still have a long way to go.

1
A Syncretic Paradigm: the Bases of Behavior

We have two concerns in this initial chapter, one methodological and the other substantive. Methodologically, we want to show the importance of syncretic constructs as a vehicle for countering social-science parochialism and for developing proper paradigms. In a much later section, we shall return to the concept of the syncretic model and show how, when generalized, it can be used to provide interfaces to join entire disciplines, not just intradisciplinary "schools". In essence, a syncretic construct seeks to accommodate several different parochial models, and join them in a working whole that has greater research potential than any of the competitive components from which it was formed. The syncretic paradigm, then, is generally at a somewhat higher level of abstraction than that of its components. As for the substantive considerations, we are going to focus our syncretic instruments on a single discipline: psychology. We could as easily have chosen economics or anthropology. But because the human being is the lowest-order component of the societal system, we must gain some comprehension of his various behavioral modalities—and his possible cognitive referents—if we are eventually going to be able to link individuals together into societal collectivites. Even more directly, the different behavioral modalities that we develop within the confines of our syncretic paradigm in this chapter will be the primary dimensions on which we define our societal ideal-types in later chapters.

CRITERIA FOR SOCIAL-SCIENCE CONSTRUCTS

From the standpoint of the modern system theorist, the criteria for proper social science constructs would be at least the following:

1. Constructs should be apodictical—that is, they should be susceptible to validation or invalidation through the offices of normal empirical science.
2. Constructs should—internally—be constructed according to the principles of deductive inference, so that lower-order components become comprehensible as *conclusions* derived from higher-order components as *premises*.
3. Connections between components should be non-elliptical, and terminology should be universalistic.
4. Qualitative constructs must be recognized as being primarily of use only in conjunction with quantitative models. The primary rationale behind a qualitative construct is thus not to displace quantitative models, but to provide a tentative, hypothetical "envelope of certainty" within which quantitative techniques may operate. That is, a qualitative construct becomes significant as a *front end* to an objective construct.

It is the last criterion that leads us into some comments about the model-building process as it would be conducted under a system platform, as sketched in figure 1.1.

A labored and abstract defense of this construct is not really useful to us here, so I shall be very brief [1]. Initially, it is important to understand that any proper model will involve development on four levels of analysis: state-variable, relational, coefficient and parametric. The state-variable level informs the model about the major structural determinants of the problem or phenomenon at hand; it thus serves to define the domain of the model in terms of the factors to be operated on or considered. At the relational level, we are interested in defining the interface conditions among the state variables (i.e., what state variables are related to which others) and the *direction of influence*. Now, in a

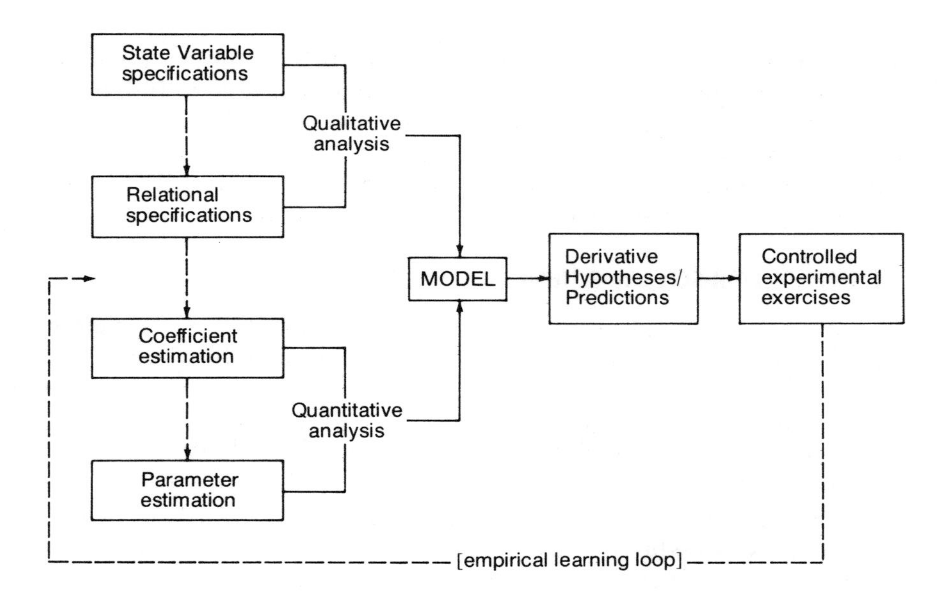

FIGURE 1.1: Aspects of the model-building process.

quantitative model, the relational aspect is usually taken care of through the offices of the normal mathematical operators (+, −, log, powers, etc.). In the statistical variation on the relational theme—and in the case of a qualitative construct—we are usually interested in assigning each interface a correlational factor: positive, negative, indeterminate or uncorrelated. In other words, this level of analysis seeks to provide us with a "mapping" of relationships and the broadest level of detail about the nature of the interrelationships. And in terms of model technology, we may be indifferent whether the quantitative (mathematical operators) or correlative connectors are used. In practice, however, a complex subject is most likely to be treated in terms of correlative specifications at the initial stages of inquiry, simply because the mathematical operators usually imply a greater level of predictive specificity (and hence knowledge of the subject) than do the *qualitative connectors*. In either case, the result of operations at the relational level of analysis is either an explicit or an implicit *network* of some kind. Implicit networks usually leave the relationships expressed in tabular or equational form (as with decision tables or traditional "systems" of algebraic or difference equations). Explicit network formulations usually involve an attempt to demonstrate the structure in direct or surrogate graphic form. In simpler variations, these demonstrations may provide us with flow diagrams or some other block-and-linkage mapping. In more sophisticated network analyses, we are interested in developing topological or set-space constructs (the most interesting of which are the lattice or t_{ϕ} spatial transforms that can inform us about the extent to which relations are hierarchical or reticular, etc.).

Operations at the next level of analysis—the coefficient level—are responsive to the type of relational network that is employed. And it is here that the distinctions between qualitative and quantitative analysis become most evident. In general, all quantitative constructs will employ coefficients which seek to express a direct (dimensional) *magnitude of relationship* for all interfaces defined above, as a complement to the directional or configurational specifications emerging from the relational operations. Where we have employed the mathematical operators, this means assigning a numerical quantity (or a range estimator) in place of any general attributes. In short, we are interested in the specific strength of any relational operators. The usual situation in the social or behavioral sciences, however, is that we have employed the correlative connectors. In this case, for our quantitative models we usually seek to replace the static operators (positive, negative, indeterminate, uncorrelated) with specific numerical quantities typically regression coefficients expressing the degree of interdependence. Of course, these numerical coefficients may be either point or interval estimates, and either probabilities or deterministic quantities; if probabilities, that may be either objective (empirically predicated) or subjective (judgmental). Of course, when the relational analysis shows a system or subject to be additive (or effectively so), the coefficient level of analysis becomes gratuitous. But for non-trivial societal phenomena, the system structure may be very complex (stochastic-state or even equifinal), so that coefficient estimation is always required.

Now, what we have suggested above refers to coefficient operations carried out under some sort of quantitative discipline, by those equipped or willing to develop the statistico-mathematical properties of a problem or phenomenon. But there is also a qualitative-analysis aspect to coefficient estimation. In this case, we are not concerned with deriving or imposing numerical values, but will rather develop qualitative specifications. For example, we might establish a set of crude intervals: strong, moderate, weak. We would then try to assign one of these modifiers to each correlative connection, suggesting whether a relationship were "strongly positive", "moderately negative", etc. And of course, as the continua of qualitative modifiers become better defined, so that there are more and more intervals from which to elect, then it becomes possible to translate the qualitative operators into *surrogate* form. This simply means that we replace cardinal values with *ordinal* values, and assign a ranking quantity to each of the qualitative modifiers. The various forms of surrogation are not generally very interesting, though such a technique as "fuzzy set theory" may be of help when it has matured a bit [2].

Moving to the final level of analysis—the parametric—we are here interested in assigning point-in-time values to the state variables, much as the evaluation of an equation or system of equations demands that each of the independent variables be replaced by a manipulative value or specification. And as with the coefficient operations, these point-in-time values may be either quantities or qualitative attributes. In either case, the model-building tasks are now complete at the first pass. But our little diagram now shows some procedural logic which may be explained very simply. Once the state-variable, relational, coefficient and parametric specifications have been made, any proper model will have some sort of predictive implications or, for a static model, projective implications (predictive implications are time-dependent, while projections refer to spatial or structural extrapolations, as in cross-sectional analysis). These predictive or projective implications are easily transformed into *derivative hypotheses* by using the model as a set of articulated premises. Now, the testing of hypotheses is the business of normal empirical-experimental science, and it is thus that every proper model will be *apodictical*: be developed in a way such that testable derivative hypotheses are made available to the experimental operations. When the above-mentioned criterion of "deductive discipline" is met, so that the various components of the model are interelated according to the laws of deductive inference, then the evaluation of the derivative hypotheses is effectively (but not logically) equivalent to a testing of the validity of the *a priori* logic and estimative properties of the model [3]. Particularly, to the extent that the predictions or projections derived from the model are not borne out under experimental (empirical) analysis, the points of modification become apparent. This, essentially, is the utility of the *empirical learning loop* shown in figure 1.1. In short, the deductive (*a priori*) components of a complex model become susceptible to field validation through the processes of normal experimental science—given that the model has the apodictical dictate in mind. And, as many readers will readily recognize, quantitative coefficient and parametric values considerably ease the demands of empirical validation. Therefore, it is distinctly in the interest of socio-behavioral science to attempt—to the extent consonant with the nature of the phenomenon under study—to generate a set of quantitative coefficients and parameters to complement the generally qualitative and deductive "front end" consisting of the state-variable selections and relational specifications in any proper model.

At this point it is well to suggest some propositions which we are not yet in a position to defend in any detail:

1. In general, the validity of any model will be affected most seriously by errors at the state-variable level, and least seriously by errors at the parametric level. That is, it is a more serious analytical breach to mis-specify a state variable (or perhaps ignore a determinant altogether) than to make an error of estimation on the lower levels of the model.
2. As the subject being modeled becomes more complex (so that projective or predictive implications become less deterministic), direct estimation of quantitative parameters becomes less a matter of experimental manipulation or unmediated statistical measurement and more a matter of "learning" best described by the type of learning logic that is embodied in the many interesting variations of Bayesian *convergence* [4].
3. Thus many sociobehavioral phenomena cannot be tested in the normal laboratory environment, but must rather be tested under some sort of *action-research* scheme. The key to this sort of process is that this allows us to treat societal systems as organic entities, and escape the reductionism associated with analytic experimental designs.
4. Finally, as the subject at hand becomes more complex, we are less likely to be able to exhaust the relational, coefficient and parametric properties using only quantitative specifications, and therefore more deductive (qualitative) substance will remain in the finished model. With this increasing residue we should also expect some increasing likelihood of projective or predictive error due to the lack of *descriptive precision*.

Now I am not too concerned if these technical points remain a bit obscure as yet, for we shall be amplifying them in later sections (particularly in Chapter Two). But what should be clear is this: most sociobehavioral constructs will tend to be unbalanced. Those developed by the empiricist-quantitative functionaries will have a great deal of specificity and detail at the coefficient and parametric level, but have sorely constrained and perhaps deliberately "artificial" state-variable and relational sets. That is, limits on the ability of mathematical instruments to accommodate large sets of state variables—in any simultaneous optimization or even simulation effort—may tend to provoke the response we earlier mentioned: the tendency to transform determinants into *exogenous* variables, and thus give credence to the concept that many quantitative model-builders sacrifice analytical validity for expediency. It is thus that many models that are promoted as being system models—or having a more or less "organic" quality—really tend to be instances of *analytic* construction (e.g., implicitly operating under the *ceteris paribus* assumption). At the relational level, analytical expediency shows itself in many subtle ways, the most usual of which are the tendencies to use linear operators where non-linear ones are indicated by the nature of the problem, and the almost universal tendency to avoid using partial derivatives in favor of simpler difference formulations in non-linear constructs. The general explanation that might be offered for these shortcomings in quantitative constructs is the obvious one: the quantitative socio-behavioral scientists—emulating "hard" science—are often more interested in technique than in content. We may add to this the explanation we also offered: that the empirical traditions—both epistemological and instrumental constraints—can really only be successful in the face of essentially simple problems, with the result that inherent complexities may often be wished away in the eagerness to get to grips with the empirically accessible, measurable and manipulative components.

At the other extreme, rhetorical models are usually characterized by enormous detail and complication at the state-variable and relational levels, little of any attention being paid to specifications at the coefficient or parametric levels. Thus rhetorical models tend to restrict themselves to affective or *a priori* significance only, and can seldom offer any but the most gross prescriptions in the real world. In short, from what is perhaps an extreme (but convenient) perspective, the socio-behavioral sciences tend to consistently provide only partial models, only incomplete allegories. It is thus that we find models of great discipline and precision but of no normative or theoretical significance, or models of great normative importance but of no operational significance.

Now, a majority of my colleagues in both the social and behavioral sciences are well aware of these defects and embarrassments; to belabor them any further would be gratuitous. And, of course, many outstanding scholars are at work trying to develop proper constructs. Our contribution to this effort will be restricted to two rather well-defined questions: (1) how the hypothetico-deductive method can be used to produce properly apodictical qualitative constructs, and (2) how quantitative coefficient and parametric specifications can be used to complement the usually qualitative "front end" of most socio-behavioral constructs (i.e., the state-variable and relational specifications, which are in general deductively predicated). As was suggested earlier, we shall be trying to make these points by concentrating on the development of theories that we may hope will actually describe societal structure and dynamics.

Steps in the Syncretic Process

The best place to begin is with the development of a syncretic paradigm to treat the subject that is central to societal systems: human nature and its behavioral implications (and predicates). Most social scientists will readily see that we have a problem here, for there are as many conceptions of human nature as there are psychological "schools". Each of these schools (behaviorism, *Gestalt* psychology, etc.) are efficient at explaining some particular aspect of human behavior, but do not comprehend well the issue as a whole. If we consider a school to be the product of some parochial paradigm—as many epistemologists indeed do—then we can see the various schools of psychology mainly in terms of their *a priori* restriction of the boundaries of inquiry, and the preselection of certain research and explanatory paths from among all alternatives that might be followed. Each school treasures its own paradigmatic assumptions, and moves its adherents along trajectories that do not intersect with adherents of other schools. So, to treat human nature as a phenomenon in its own right, we need to develop some sort of syncretic construct—a supraparadigm—under which the implications of the several schools may be joined and considered as an articulated set. This supraparadigm would not pretend to be *the* true explanation of human behavior, but would be expected to contain the true explanation somewhere within its boundaries.

The construction of such a supraparadigm must treat the subject of human nature broadly enough such that virtually all logical probabilities are encompassed somewhere within its components and structure. This task is somewhat facilitated by trying to group behaviors into ideal-type categories. Fortunately, an analysis of existing behavioral platforms indicates that such a grouping is indeed possible. For when we reduce all the platforms to their most abstract properties, we find that there are really only three bases of behavior: (1) deductively predicated, (2) inductively predicated, and (3) data-driven (sense-driven). This provides the base we need for the synthetic supraparadigm, whose structure is presented graphically in figure 1.2.

As a general rule, the construction of a syncretic paradigm involves the following [5]:

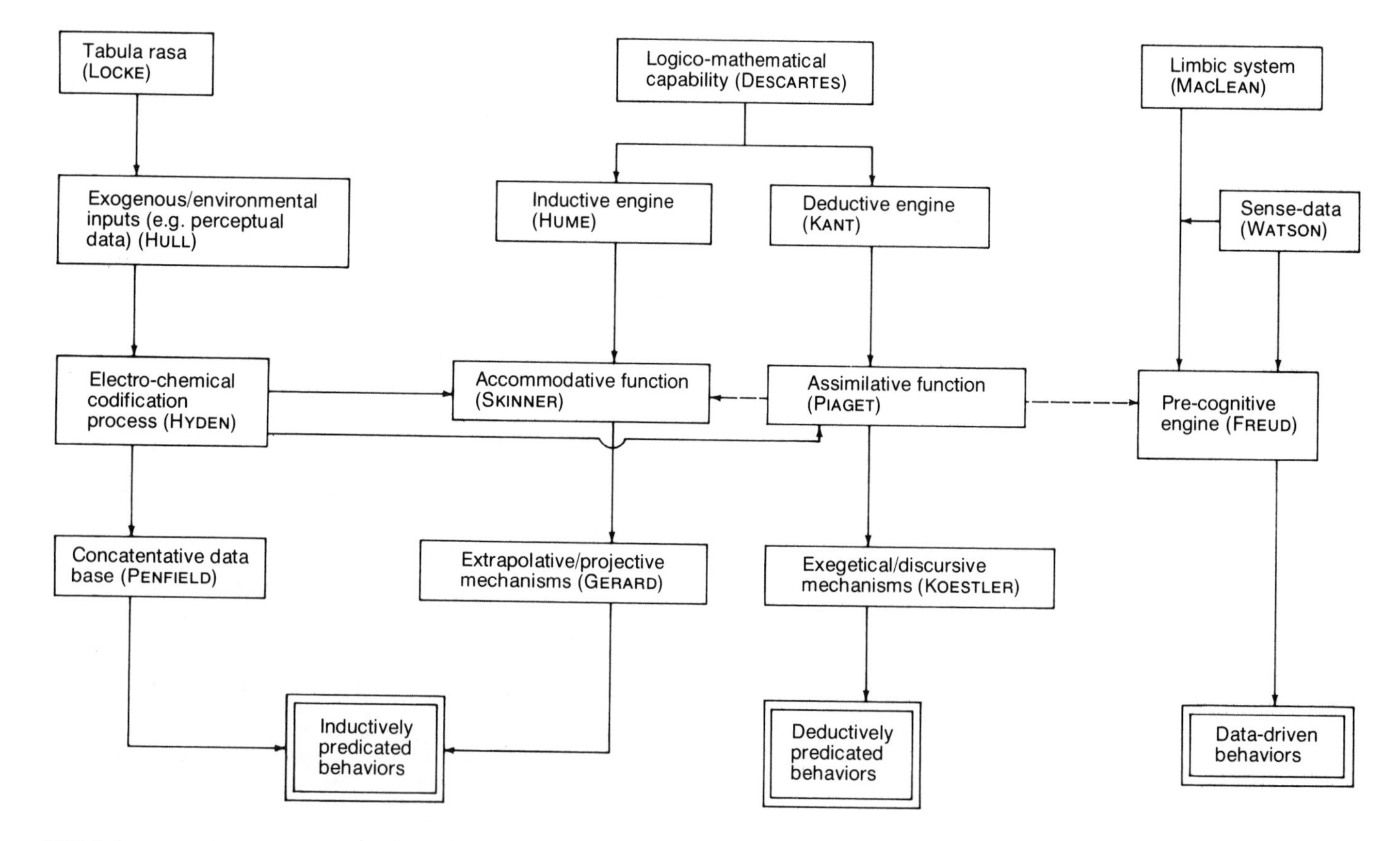

FIGURE 1.2: A Syncretic Paradigm for Disciplining Inquiry into Human Behavior. (Note: As this schematic is intended to portray structure rather than process, the interconnections among the various components are only incompletely specified.)

1. Develop unique causal trajectories for each of the various parochial paradigms involved in a discipline (in the present case, we thus include the physico-chemical, the behaviorist, the structuralist and the psychoanalytic school—via Hyden, Skinner, Piaget and Freud, respectively).
2. Make explicit the usually tacit (transparent) first premises underlying each of the competitive paradigmatic trajectories [6]. We usually do this by raising the level of abstraction and looking for fundamental epistemological predicates. In the present case, we accomplish this by incorporating the Lockean, Humesean, and Kantian positions as hypothetico-deductive alternatives.
3. Establish the linkage between methodologically disparate perspectives. For the syncretic paradigm at hand, this is done by providing an interface between the empirical-positivistic platform (due to MacLean and Hyden and to Watson and Hull) and the rhetorical positions of Koestler, Skinner, Piaget, Freud, etc.
4. The test of sufficiency for a syncretic construct is the emergence of a categorical ideal-type *output* for each of the *a priori* input trajectories (e.g., the deductive, inductive and data-driven behaviors serve as the outputs in our model, given the several paradigmatic and methodological perspectives).
5. Finally, the syncretic solution should be such that the categorical outputs have a significant probability of fully exhausting the range of phenomena with which the discipline is concerned (in this case, the various behavioral types should be viewed as referents against which all empirically and logically accessible behavioral phenomena may be compared) [7, 8].

Although we shall treat each type of behavior in some detail in subsequent sections, and in the process explain the components of the model in terms of both their processual and structural significance, one thing must be noted immediately: this model is not simply a restatement of the Freudian position.

There is certainly some correspondence. Freud also postulated three levels of human behavior: (1) the id, the sense-based behaviors originating in the primitive part of the brain, (2) the ego, essentially the cognitive level of behavior, and (3) the superego, which houses broad sanctions or values which constrain behavioral trajectories. The data-driven behaviors, stemming from the limbic system, are indeed essentially the same as the id-driven behaviors defined by Freud. But the ego level here is divided into two separate dimensions: the *inductive* and the *de-*

ductive. Both these dimensions reflect cognitively mediated behavior, but carry implications which Freud's ego-level behaviors do not. The superego-level behaviors will in later chapters emerge as what we shall call "culturally" conditioned behaviors, those driven by sets of *a priori* determinants. These externalized behavioral constraints must be assimilated through the internal mechanisms involved in our model. To say, for example, that someone does something because he has been "taught" to do it, or because society approves, does not take us a long way toward understanding the process by which these external behavioral constraints become internalized. Moreover, the methodology for examining external behavioral engines—operating at the superego level—is very different than that which must be employed to comprehend intrinsic behavior. Especially for reasons of analytical efficiency, then, some modification to the Freudian position has to be made. For it, along with the behaviorist paradigm, represents the "null hypotheses" of modern psychology, and hence carries some measure of guilt for the confusion which currently exists in the field.

At any rate, we may now somewhat briefly describe each of the three major types of behavior, and examine the three process algorithms which lead to each. For simplicity, we shall begin with the data-driven behaviors and their origin in the limbic system. These are the simplest forms of behavior, and therefore occupy the area of greatest consensus.

THE SENSES AND CYBERNETICS

Data-driven behaviors can be identified as those which occur in the absence of any cognitive mediation. They arise from the level of the brain which is generally referred to as the "primitive" or animal-level brain [9]. These functional mechanisms, which we appear to share with the lower animals, are collectively referred to as the *limbic system*. Now, in the generally accepted version of the Freudian model, these behaviors could be graphed as in figure 1.3.

The interpretation is that some stimuli arises in the subject's environment, and energizes some component of the limbic system. In the limbic system there are a set of pre-programmed (automatic) response routines, which then elect a behavior and cause its implementation. Such behaviors are tantamount to the Freudian "drives", or the basic physiological processes (e.g., fear, sexual arousal, hunger). But in the Freudian system (and properly so), we must also consider another aspect of this level of behavior, one which arises internally. For there are some behavioral stimuli that have their origin within the body or the

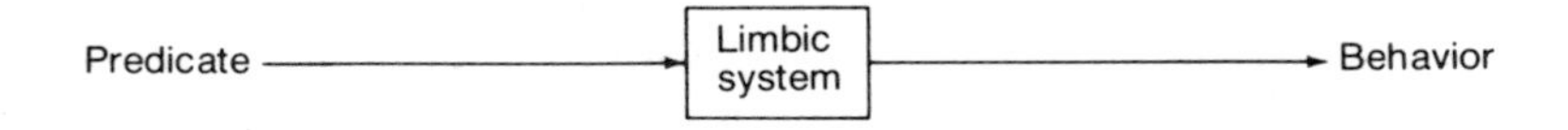

FIGURE 1.3: The black-box model

primitive brain. For example, it is possible for individuals to make themselves afraid on the basis of "nameless terrors". By imagining a sexual object, one may arouse oneself. Moreoever, it is possible for some individuals to exert a determination, from the cortical or cognitive level, on physiological behaviors once thought to be beyond control. For example, much recent literature has been devoted to the process by which certain oriental mystics can control their respiration rate and blood pressure, etc. [10]. Finally, there are the homeostatic behaviors, which have a source within the body (e.g., the arousal of hunger on the basis of internally determined nutritional needs; the internally determined need for sexual release; the automatic mechanisms which act to maintain the body temperature and respiration rate).

The simple model which we presented above may now be complicated somewhat. We will add a *feedback* loop. This feedback capability provides us with the ability to judge how far a drive has been fulfilled—how completely the behavior has adapted to the predicate. The model is sketched in figure 1.4. This is what is known as a closed-loop system; with the feedback provision, we can eat until we are hungry no longer, drink until our thirst is slaked, etc. In short, behavior becomes a sequential process, with two fundamental forms, shown in figure 1.5.

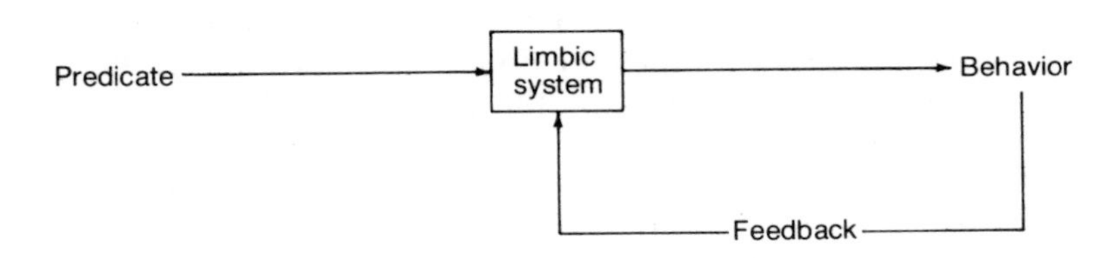

FIGURE 1.4: The feedback-based model

Satisfaction
Adequacy
Food intake

FIGURE 1.5: (a) Convergence, where a goal is sought and pursued until adequately attained.

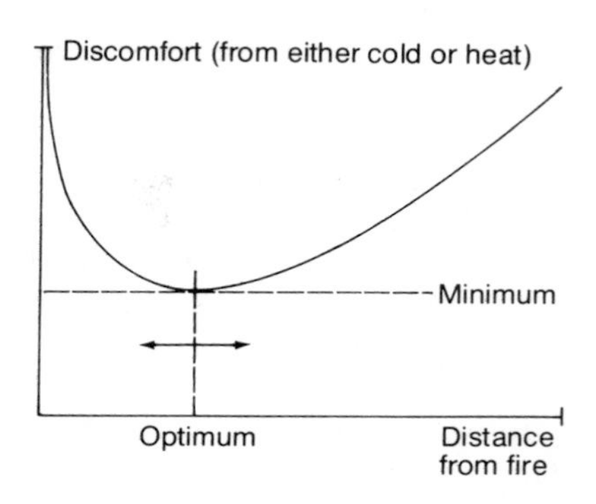

(b) Compromise, where an adequate balance between competitive forces is sought through trial and error.

Now, both these mechanisms are instances of simple cybernetic processing, and its provisions adequately explain the way in which low order behaviors may be generated and controlled. In figure 1.5(a), the vertical axis measures the degree of satisfaction of the hunger drive, which the horizontal axis measures the total intake of food. A curve (typically concave downward) is followed until the drive is satiated. Figure 1.5(b) deals with another simple problem: how close an individual—seeking warmth—will get to a fire. Distance is measured along the horizontal axis, while the pain level is indicated on the vertical. The motion here is oscillatory; the person gets burned and first overreacts, moving too far away. He next moves in less closely than at the first trial, but still gets burned and so overreacts again (but does not move quite as far away as he did the first time). As he learns more about the properties of the fire through trial and error, the oscillations gradually dampen and he converges on an optimal distance from the fire, where the pains of heat and cold are effectively balanced at the margins. Note that were we to restate the objective of figure 1.5(a) in negative terms—e.g., the avoidance of pain from hunger pangs—the nature of the process would not really change; the direction of the curve would merely be reversed. In the realm of human behavior (though not in the domain of all systems), the simple feedback processes are distinguished primarily by the nature of the *learning curves* (plots of behavior against time). The convergence curves are usually cumulative and monotonic; the compromise-seeking curves are usually oscillatory and represent the trial-and-error base of so much simple human behavior. Thus, the behaviors of classical (Pavlovian) conditioning and the basic motor skills may generally be thought of as instances of negative trial and error, whereas most of the basic physiological and Freudian "drives" may be thought to be pursued through a cumulative, amplifying convergence process. While these distinctions are not of critical importance for us here, they shall be when we discuss certain types of higher-order behavior.

Now, while the feedback-based models seem to provide an adequate explanation for low-level behaviors, there is a problem. For the behavior to be strictly data-driven, it is necessary for the predicate and the reaction (the behavior) to be effected below the level of consciousness—that is, at the "id" or precortical level. In short, data-driven behaviors are a more or less direct response to a stimulus, with no "thinking" or speculative process intervening. The nature of the sensory input itself (the datum or stimulus) determines completely the associated response or behavior. With this model at hand, the tendency is to eliminate *a priori* the cognitive component from certain behaviors, without equivocation. Under normal circumstances, this is a very appealing model. But let's consider a rather simple variation on a theme presented earlier. Suppose that, as in figure 1.5(b), we have a man who wants to find the minimal distance from a fire which will provide him with maximum warmth but only tolerable discomfort. As we saw, the compromise process, which can take place largely below the level of directed cognition ("thinking"), can adequately explain the dynamics of behavior. The tendency is to generalize, and suggest that all men, everywhere, would follow essentially the same process. In short, the tendency is to suggest that the behavior of men with respect to fire is isomorphic among all individuals, in terms of process at least. That is, the same optimal distance will not pertain for each individual, but the method by which he solves the problem will be the same. Much the same thing could be said about the other ostensibly data-driven behaviors: hunger, sex, fear, motor manipulations, respiration rates, etc. And, indeed, this is the basis for the classification of data-driven behaviors in the first place. The classification is merely a normative one, and not a *law*. For some men deliberately burn themselves, for their own reasons. Some men eat way beyond the point where hunger is satisfied, for reasons which have nothing at all to do with nutrition. Some individuals have sexual appetites which bear no relation to the physiological impulses or adequate level of release from sexual stimulation. In such instances (and as Freud saw, they are by no means rare), hunger and sex and avoidance of pain no longer become data-driven behaviors, but are driven by motives in the higher-level brain, in the cortical region. Thus, to suggest that there is a set of simple behaviors which are exclusively and invariably pre-cognitive (direct) responses to sensory inputs is an oversimplification. What we can suggest, however, is the following: were we to postulate a criterion of *efficiency* for all behaviors, such that the mechanisms which mediated between a given predicate and the resultant behavior were as simple as possible (e.g., maximally direct), then the data-driven model would ideally apply to the wide range of simple behaviors noted: motor skills, homeostatic physiological function, basic drives and emotional responses of limbic origin or mediation. That is, were an engineer to design the human behavioral mechanism as an optimally efficient one, he would not allow these behaviors to be determined other than at the limbic level. Thus, such behaviors would *normatively* be strictly data-determined—direct reactions to stimuli with the nature of the reaction entirely dependent on the nature of the sensory input (the datum itself). Because human behavior is apparently not constrained by any efficiency criteria (at least of which we are aware), we still have to account for instances of *pathological* behavior, where the mechanism by which the behavior is determined is not that expectedly congruent with the nature of the input. In short, we have to allow for mediation at the cognitive level even for behaviors which would seem not to warrant this level of attention or determination. Thus, the simple models previously introduced for data-driven behaviors must be modified as shown in figure 1.6.

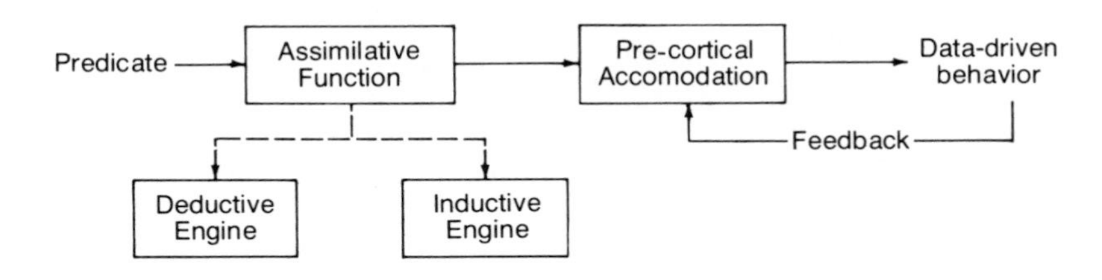

FIGURE 1.6: The generic data-driven model.

Under this model, the assimilative function acts as a kind of switching device, which can take any predicate (e.g., stimulus, event, circumstance) and route it to any of the three behavioral engines for processing. Now the assimilative function may be either conscious or pre-conscious in any instance (where the pre-conscious function would be performed according to traditional Freudian theory of unconscious substitution, displacement, etc). [11]. But the emergence of the assimilative function as a mediator for all behavior means that human behavior has two potential properties which introduce enormous complications:

1. Any predicate of behavior is equifinal in potential—that is, there need not be any specific response associated invariably with a given predicate. For example, a sexual stimulus may arouse the normative response in one individual (e.g., physiological arousal) but may cause another individual to have a rage reaction or still another to feel hunger.
2. This means that for any particular behavior, there may be more than one cause: some set of possible predicates rather than a single deterministic predicate. In particular, there is an *equipotential* aspect of human behavior. Thus, were we to observe an individual eating (the behavior), we could not be entirely sure that the predicate was hunger; as suggested above, the real predicate could have been a sexual stimulus, with the assimilative mechanism intervening to convert the normative response to a pathological one.

In summary, then, the determination of whether or not a particular behavior is data-driven or has a cognitive origin is not always simple and straightforward. All we can suggest is that the criterion for normality is available to us in terms of efficiency: the normative set of data-driven behaviors are those which do not demand cognitive mediation. Thus, when inductive or deductive engines can be shown to drive a behavior which could be adequately taken care of at the pre-cortical level, we have an anomalous or pathological situation. A theoretical anomaly would be, for example, the case earlier mentioned: the occasional mystic's ability to control physiological functions long thought to rest beneath the level of volitional control. A pathological instance would be the emergence of a behavior incongruent with the nature of the predicate—the eating response to a sexual stimulus, for example. The assimilative function or faculty thus must be interposed between predicate and behavior if the entire range of behaviors is to be understood. Thus, in no case can behavior be assumed deterministic. All we can say is that there are some responses to a given stimulus or predicate which are deemed *most probable*; and the assimilative process may cause even probabilistic laws to be violated. Thus, we are forced to synthesize the Freudian and the Structuralist school.

The Id-Level Behaviors

In general, the mechanisms we have thus far discussed are adequate to explain the range of id-oriented (e.g., emotional) behaviors as well as the homeostatic processes of the physiological system. The former are presumed—following Freud—to be stimuli that reflect the basic instinctual emotional urges or drives with which the human species is endowed. That is, it is possible to think of fundamental biophysical stimuli, arising internally, giving rise to empirical behaviors. These behaviors are not really well understood as yet, but the general proposition is this: as part of his fundamental genetic endowment (probably resting somewhere within the limbic system), man has inherited certain fundamental *needs* which are more or less equally distributed across the entire species. The needs become the innate predicates for what appear to be pre-programmed behaviors. In short, these predicates are the basis for *Sentient man*—the slave of his senses. *Sentient man* has a significant proportion of his behavior (or in pathological cases, virtually his entire behavioral repertoire) determined by the exercise of these internal drives.

There is some dispute as to what precise set of activities these drives constitute, but there are at least two types which must be considered. The first are what are generally referred to as emotions, presumably derived from man's emergence as a social animal; these are, as it were, the innate, inherent bases for social association (e.g., love, the need for reflective identification, jealousy, the tendency toward self-sacrifice). The other type of drives are really context-independent referring to the object as such rather than to its social aspects (e.g., the search for individuality and for alleviation of guilt; the territorial imperative; the desire to minimize uncertainty). We shall have more to say about these intrinsic predicates later. Here all I want to suggest is that they are probably bases for behavior which has no apparent external stimulus or predication [12]. That is, they are initiators of behavior rather than mere reactions to exogenous stimuli. In short, in certain conditions, human behavior will be impossible to relate to any particular cause, unless we are equipped with a knowledge of these internal drives and the forms in which they are reified in the real world.

For example, we are often faced with what might be described as an *identification* problem. In economics, this quandary arises very often: we cannot be sure of the relationship between supply and demand in certain instances. Is there a demand because there is a supply, or is supply the response to demand? In fact, depending on conditions, either answer can be right. Something like the same problem can arise in psychological inquiry. For example, is some innate drive for social expression the basis for the spontaneous and intense associations which sometimes spring up between strangers? Does one fall in love because of some external predicate (e.g., the emergence of a love-object)? Or is the prerequisite for such a relationship an internal stimulus derived from the id level? I suspect that, in many instances, intrinsically oriented explanations are necessary for a valid appreciation of certain behaviors. The temptation, given the empirical preoccupation of normal psychology, is to seek exogenous explanations for everything. But here the assimilative theory of Piaget takes on a special meaning for us. It suggests, in effect, that no exogenous stimulus will be accommodated unless the individual is *ab intra* prepared to assimilate it. Carrying things one step further, however, I want to suggest that under some conditions, the individual may seek an exogenous entity to accommodate—himself initiating the search in response to an innate drive. In short, the individual is potentially an active, and not merely a passive, agent in his own interests. Thus any synthetic paradigm for psychology must consider the possibility of internal factors engineering behaviors, behaviors which have the net effect of causing the individual to structure his own environment.

There is another significant aspect of such behaviors that we must at least raise for speculation: the possibilty that they are actually conducted in an open-loop process, largely without regard to the responses elicited from the environment or other individuals. That is, they may be instances of compulsion. In mild forms, the "driven" individual will merely be reluctant to assess or accommodate external responses to his behaviors (as, for example, under the influence of alcohol or certain de-cognating drugs). Then the behavior of the driven individual may be anomalous with respect to his milieu. He is, as it were, out of phase with reality. The critically important point here is this: *driven or compulsive behaviors need not be reinforced by exogenous responses in order to persist*. The individual simply shuts down the normal feedback mechanism, and hence becomes the arbiter of his own reality. It is thus that driven behaviors are generated largely below the cognitive level, and can persist for some time before the incongruence between behavior and environmental response become evident to the conscious mind—if it ever does. We can elicit such behaviors in the laboratory by administering certain drugs or by other techniques. But in the real world as well, they have a long history, and man's search for agents to retard the cognitive mechanism has been an eternal and ingenious one. Thus, the normally rational individual may shift his behavior artifically to a data-driven modality, and there are satisfactions (utility) found in that domain which are more or less reflective of the satisfactions one finds on the rational dimension (the ego level).

One final note should be made here. There is a popular thesis that there is constant conflict between the conscious (ego-level) and unconscious (id-level) mechanisms for controlling behavior. The presumption is that this conflict is due to the mental immaturity of the human species—the behavior mechanism has not fully evolved. A further assumption is that this postulated conflict is dysfunctional (harmful). But the introduction of the assimilative mechanism as a mediator between all predicates and all behaviors offers an alternative. It suggests that behavioral responses expected or deemed probable from the standpoint of a supposedly objective outside observer are not necessarily "positive" from the standpoint of the individual, equipped, as he is supposed to be, with highly idiosyncratic and potentially complicated utility functions. Thus, the *equipotentiality* and *equifinality* we note in even simple behaviors may not be a sign of the immaturity of our behavioral apparatus, but of its enormous ingenuity and sophistication. What would be a sign of immaturity is the potential for only a single behavior in response to any given predicate, the "mechanization of the mind". Indeed, a deterministic mental apparatus is more likely to have been a primitive endowment of man than to be waiting for him further down the evolutionary path [13].

But this relativism can be tolerated only within certain limits. As we have seen, not every improbable behavior is an innocent anomaly. Some are pathological. For example, the man who responds to every situation of stress with a craving for a drink—even when a rational, rather than visceral response is clearly indicated—is impaired. If such an evaluation cannot be made with the individual's own interests in focus, the welfare of society at large may be introduced. As we shall see, the shift from an individual to a societal focus is rather well developed in existing social psychology. But there is another and perhaps more directly compelling reason for avoiding the sophistic tendency to shun any moral or evaluative judgments of behavior. It is particularly important to propose that the "programming" of the assimilative mechanism is not entirely beyond the control of the individual. This, in essence, is the important contribution of the structuralist platform—that the assimilative functions that are performed during any interval are in part products of the previous assimilations, and that not all the factors determining the assimilative structure are beyond the individual's control [14]. In traditional language, some pathological behaviors are as much faults of character as faults of mind, and hence are reparable by the individual himself.

Trial-and-Error Behavior

Although we cannot fully appreciate the facts of trial-and-error behavior until we have adequately explored the inductively predicated behaviors, it is a variant of data-driven behavior and so must be introduced here (though with almost constant reference to arguments which will really not be defended until the next section).

There is an unusual circumstance in which trial-and-error behavior becomes the preferred (congruent) behavioral modality: the *maze* context. A proper maze is one which holds virtually no possibility for exercising any strategic or rational options. The maze is constructed (or occurs naturally) so that no clues are available to the subject about the proper path to follow, either to achieve some objective within the maze or simply to remove himself from it. Lord Russell once noted that not even Isaac Newton could devise a better algorithm for running a maze than trial and error.

Raw trial-and-error behavior is indicated by the absence of any cognitive mediation. The individual abandons reliance on any intellectual discipline and "plunges in." It is generally a primitive behavioral response because, as mentioned, true maze situations seldom exist in the real world. In more technical terms, this means that natural situations where a true indeterminate distribution occurs are negligibly rare. Thus, trial-and-error behavior is generally dysfunctional, except in the following circumstances:

1. Where the possible event alternatives are extremely few, so that trial and error promises to exhaust them quickly.
2. Where the expected loss (or cost) of an error is very low.

Let's take a somewhat crude example which will point up the sense of the above assertions. A man is taken into a room with a single light fixture and five switches. He is told to find the switch that works the fixture. Clearly, an undisciplined trial-and-error process is called for. Now, suppose the experiment is repeated, but this time he is told that if he touches a wrong switch, he will get a mild electric shock, but when he finds the right switch he will receive $100. Now, a strategic decision of some sort is called for, one which will probably involve the following: the determination of the relative utility of $100 against the expected disutility of a mild electric shock. It is conceivable that an individual would decline to participate in the experiment for any of several reasons: First, he might consider such experiments beneath his dignity, and resent the suggestion. Or he might be so wealthy that another $100 holds virtually no utility for him. Or he might want the $100 but be mortally afraid of electricity, perhaps having once had a bad experience with it. There are other logical alternatives, but let us suppose that this will represent a situation where the majority of unimpaired individuals will indeed participate. Again, the preferred method will be trial and error, with some hope that one will get the right switch first.

To develop the example still further, let us introduce another condition. In the experiment, the individual is told that if he hits the wrong switch, he will be electrocuted, but the right switch will earn him $1000. It is possible, under certain circumstances, that the individual will participate, but acceptance of the terms would probably indicate some impairment (e.g., a "death wish", or an inordinate fondness for money, or simply an incapacity for dealing with probabilities and expected values). The normative reaction, in such a case, would be simply to refuse to make a decision. The expected value of being right simply is not as great as the expected disutility of being wrong. But to make our point somewhat clearer, let us suppose that the individual was forced to make a decision (e.g., that the refusal to participate meant certain death). The question is: would trial and error still be the preferred method? The answer could be yes for the problem as given. The normative response, however, would be to attempt the development of a data base, such that the decision could be made under the inductive modality. The call would go out for predicates—the wiring diagram, for example. In short, the individual would exhaust every opportunity to bring some discipline to the situation, to increase the odds in his favor through analysis. Thus, the trial-and-error modality would be abandoned, if possible, because the second condition would not be met, the expected disutility being too great. This would be done despite the fact that the first condition still holds: there are only five, clearly identified alternatives (the several switches). In general, then, trial and error becomes more likely to be the normative (appropriate) behavioral modality as the number of alternative events diminishes, and as the expected value of an error declines.

Aside from such artificial contexts, it is difficult to imagine a real-world situation that would make the trial and error the rational response. The example most often used is that of the "lost soul". An individual is in the middle of a desert, with no notion of where he is. He is without food or water. He cannot stay where he is, but must elect a direction in which to travel. Now the probability of his using trial and error is by no means complete. However, it will be large to the extent that:

1. The individual's experience is small.
2. He lacks intellectual discipline.

In short, trial and error becomes probable to the extent that an individual lacks information, or lacks a tolerance for anxiety (which formal analysis of alternatives implies). In such a situation, panic is likely: the desire to reach a conclusion— irrespective its nature—may be more powerful than the need to increase the odds for success in a stressful condition. For the ignorant, or for those with little analytical experience, many significant events appear random, and so trial-and-error behavior is especially likely among the intellectually unprepared. Under such conditions, trial and error has the appeal which is so often voiced as ". . . it's better to do something, anything, than just sit!" Of course, as we have seen, this is an appropriate dictum only under the two conditions mentioned earlier: an extremely limited event set and low disutility expectations. At any rate, an appreciation of the trial-and-error process is a prerequisite for an understanding of the inductively-predicated behaviors to which we now turn. For trial-and-error behavior is postulated to be the process by which individuals acquire the experience and information base necessary for the generation of inductively predicated behaviors.

THE INDUCTIVE MODALITY

For the concept of this second set of behaviors, we have to concern ourselves with phenomena such as prejudice, rote performances, operant conditioning, and one aspect of rationality: statistical or probabilistic reasoning.

To appreciate the range of inductively predicated behaviors, we have to introduce two genetic endowments (whereas the limbic system was the only such mechanism required for the range of data-driven behaviors, at least in their normative guise):

1. The *tabula rasa*, a blank slate on which life's experiences will write and which will gradually become a residence for the empirical-experiential data base of each individual. For those familiar with computer jargon, the *tabula rasa* is an initially empty information set—a blank memory.
2. A logical engine capable of performing inductive inferences. In short, there must be a mechanism for reasoning from the particular

to the general, and for ampliative manipulations (of which extrapolation and projection are the most common).

Equipped with these, man has the necessary and sufficient bases for the passage through life as Locke, Hume, the behaviorists and many others would envisage it. For, from their perspectives, the basis for all learning, and hence the basis for all behavior, rests essentially with cumulative experience, gained largely through successive trial and error. Hence, the convergence and compromise processes described in the last section for data-driven behaviors are the link from the pre-cortical and primitive behaviors to the more sophisticated behaviors involving a cognitive mediation. Initially, just after birth, the predominant behaviors are data-driven, raw experience taking place largely below the level of cognition. However, these primitive experiences gradually provide the individual with a data base (the accumulation of experience), such that trial-and-error activities may gradually be displaced by more sophisticated processes—particularly, the process of using past experience and/or environmental clues as the basis for determining current action.

Now, inductive processes may be defined in either of two ways. From a formal perspective, inductive inference is the process of reasoning from the particular to the general, which implies an ability to generalize from past experience. But from a functional standpoint, inductive inference may be described in the following way: It consists in the exercise of models which amplify or extend the implications of a data or informational input. A simple graphic example is shown in figure 1.7.

The implication of an inductive process, given the above figures, is that in each of the three cases, the morphology (configurational attributes) of the inputs was maintained throughout the inductive transformation. What changed, was the magnitude of the figures. This is an example of ampliative inductive inference—the extrapolation or projection of properties of the input stream, without qualitative restructuring. (We shall later qualify this to suggest that, in some cases, quantitative magnitudinal changes do imply qualitative changes, when the former become sufficiently significant.) In short, there is a significant *qualitative correlation* between inputs and outputs when the inductive process is in force.

In plain language, the key to inductively predicated behaviors is learning through experience and observation. In general, a single predicate for behavior—as in the data-driven case—is insufficient. Every inductive behavior will have both a direct and an indirect predicate.

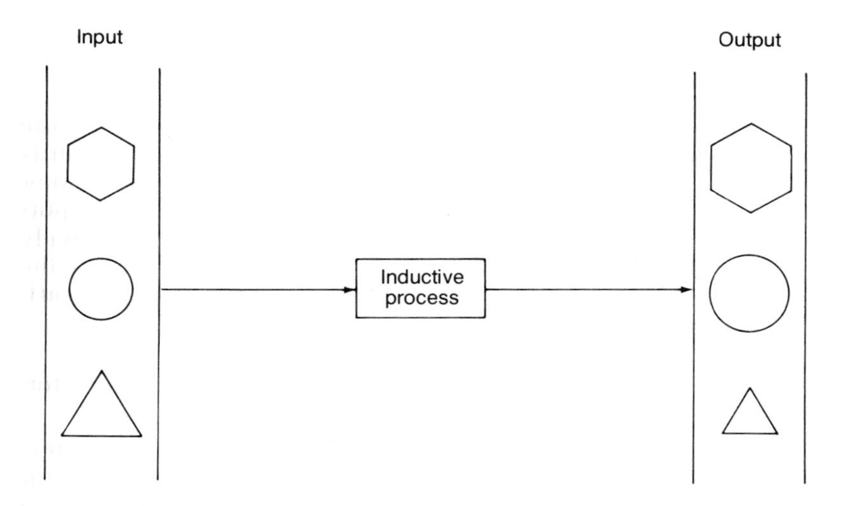

FIGURE 1.7: The ampliative inductive function.

The direct predicate is some stimulus, event or circumstance, usually arising outside the individual's behavioral apparatus. The indirect predicate is found in the cumulative experience base—the data base. Let's take a simple example. Let us suppose that the direct predicate is a storm cloud on the horizon. If the individual were facing this circumstance for the first time, we could expect nothing more than mild, undirected attention. However, if the individual has some previous experience with storm clouds, the indirect predicate (the experiential inventory) would suggest an appropriate behavior, the generation of a response of shelter or protection (e.g., getting indoors or taking an umbrella when going outside). The mediating process between the predicate (the storm cloud) and the behavior (seeking shelter or finding an umbrella) is clearly an inductive one: the generalization that, on the basis of past experience, a storm cloud indicates rain, and rain is uncomfortable. The situation may easily be converted into a utility issue. Suppose that, for some reason, the individual cannot find shelter (he must be outdoors) and that he finds an umbrella uncomfortable or embarrassing to carry (or must contemplate the purchase of one). His behavior could then be generated by probabilistic reasoning. In particular, he would weigh the probability of a storm cloud meaning rain (again going back to his experiential inventory), include the assumed discomfort of being wet, and then incorporate the expected utility of shelter or the purchase of an umbrella. If seeking shelter means the possibility of foregoing some profitable experience (expectedly), or the purchase of an umbrella means foregoing some other purchase that is expected to bring greater pleasure, he may be willing to tolerate a rather significant risk of being wet. On the other hand, if neither the opportunities of time nor those of money are significant, even a relatively slight probability of rain may cause him to seek shelter or purchase the umbrella. In sum, then, even this very simple example has three variables which carry significant potential for complexity:

1. The probability that a storm cloud means rain, predicated on historical (experientially predicated) probabilities.
2. The expected disutility of getting wet, which may vary with contextual circumstances (e.g., some people think getting wet is romantic).
3. The structure of opportunity costs, which involves the calculation of any other uses to which time or money might be put, were shelter or the purchase of an umbrella foregone.

The calculation of expected disutility and opportunity costs could again take place with respect to the inventory of experience. Ignorance or lack of experience in any of these dimensions would likely lead to a dysfunctional or suboptimal response. But even so, the fact of the experience—even an unpleasant one—would serve to elaborate the relevant experience base and hence reduce the probability of a dysfunctional response the second time the circumstance (a storm cloud) occurred. Hence, both good and bad experiences help remove man from the envelope of continual trial and error, and help gradually elevate him above data-driven, non-cognitive behavior. In short, experience is the prerequisite for sophistication, and sophistication means increasing emphasis on rational or cortically mediated behavior.

The normative process of generating inductively predicated behaviors is thus fairly simple, and may be abstracted from the general diagram of figure 1.2, as shown in figure 1.8.

Very simply, a predicate arises (e.g., a stimulus, event, or circumstance set) requiring some response. The predicate, in the form of a sense datum, is converted into information acceptable to the brain through the elctrochemical codification process which converts visual, oral etc. inputs into brain-readable form [15]. The dashed line between the electro-chemical codification and the data base indicates that some perceptions or sense data are likely to become part of permanent memory, and therefore go through a process which concretizes and stores the input. At any rate, preliminary to the response, the cumulative data

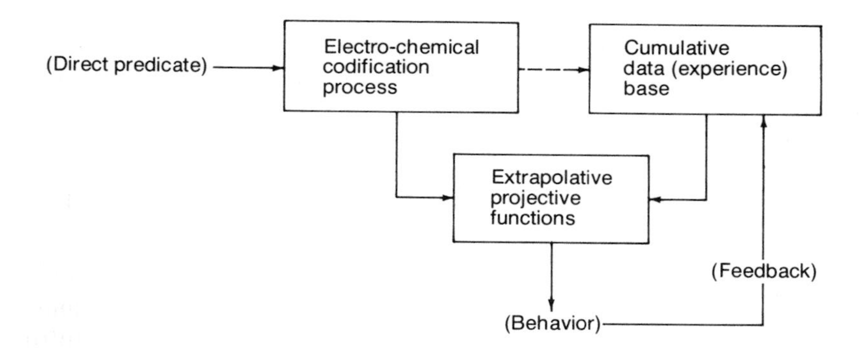

FIGURE 1.8: Inductive engination.

base is searched for appropriate stored information. If any is found, it constitutes the indirect predicate. The extrapolative/projective functions are performed (which actually generate the inductive inference), and the behavior is determined. In some cases, the results of the behavior will become a sense input and stored in the permanent data base. Hence the potential for feedback.

In slightly more mechanical terms, the sensory inputs which form the direct predicate are translated into terms correlative with those in which the permanent (or cumulative) experience base is stored in memory. Thus inputs and memory are both translated into the same language. This facilitates the search for a relevant information base, which becomes the indirect or secondary predicate. The initial function that is performed at the cognitive level is thus the *association* of a direct predicate (e.g., the storm cloud) with the companion predicate (e.g., rain). The response predicate, also derived from the experience base, would be wetness, which is associated with discomfort. At this point, then, we have the following set of associations:

$$\text{Cloud} \rightarrow \text{Rain} \rightarrow \text{Wetness} \rightarrow \text{Discomfort.}$$

Now assuming that there is an inherent criterion of avoidance of discomfort, a response (behavior) is called for. The comfort response would be conditional on the nature of the predicate. In this case, rain would imply either of the following (at least in terms of our example):

Rain → Shelter → Behavior
Rain → Umbrella → Behavior

Now, if one or another of these behavioral alternatives were exercised, and it did not rain, this would weaken the probability of such a response next time (as the association between cloud and rain would be statistically dampened). In short, the cumulative experience base tends to either reinforce or weaken existing associations (or perhaps introduce new ones if the result of the behavior was unprecedented).

The mechanics of this type of inductive inference process are essentially those that have come to be known as *Bayesian* decision making. In a Bayesian process, new information is constantly being sought to redefine, reinforce or modify prior information. Behavior (determined by sequences of decisions) is thus disciplined jointly by contemporary data (via observations) and by one's cumulative experience and information base. For our purposes here, the ampliative form of inductive inference thus finds the individual using empirical and historical data in combination, and making extrapolations of the type indicated in figure 1.7. Thus, the proper theoretical reference for ampliative behaviors is the Bayesian model, coupled with the more sophisticated aspects of statistical inference.

There is a simpler inductive modality, as we have suggested: the *associative*. The theoretical reference for associative behaviors is simple *statistical regression*. Under this scheme, the existence of some predicate (e.g., a storm cloud) implies virtually without equivocation some specific reaction (e.g., shelter). Here, then, past experience is the sole engine of behavior. The link between predicate and behavior thus becomes direct, and is effectively unmediated by cognitive processes of the type involved in ampliative behaviors. Specifically, the individual operating under the associative modality becomes intelligible in terms of *operant conditioning* [16], the popular engine of modern behaviorist theory.

Thus, the associative and ampliative modalities differ in their sophistication and in their cognitive demands. For associative (regressive) behaviors the probabilities become fixed and invariant, and each predicate will gradually become associated with an effectively automatic, unique response (behavior). In technical terminology, the individual operating under the associative modality thus exemplifies a *finite-state* program. And, indeed, for those who care to investigate this implication, the concept of operant conditioning can only lead to such a behavioral modality. The calculus is either minimization of pain or maximization of pleasure. The explanatory logic suggests, then, that each individual, for any given past stimulus or event demanding a response, has managed through trial and error to find a response that is adequate within the framework of the pleasure-pain calculus. Now, associative behaviors become *rational* when the circumstances to which one is exposed are precedented (which implies that the milieu within which one dwells is largely a simple and invariant one) [17]. Long periods of exposure to a set of similar stimuli (events) naturally tend to breed associative behaviors. Indeed, as we shall later see, primitive societal systems may in large measure be defined by the fact that associative behaviors dominate their activities.

There is great comfort in such a situation, for cognitive demands (and, consequently, anxiety) are effectively minimized. Much resistence to change—even when change promises to bring material advances—stems from the fact that change automatically implies the necessity of abandoning the associative modality. Under one of the postulates we shall be discussing—that which suggests that *individuals will normally tend to take the path of least analytical resistance*—it will be clear that some individuals will be reluctant to abandon the psychological comfort of association despite promises of greater economic benefits, etc. The long history of frustrations which many reform movements have suffered may perhaps be laid directly at the door of this proposition, as we shall later see.

Now, where the environment is not simple and invariant, but subject to limited change, the simple associative form of inductive behavior must be abandoned. In this case the *ampliative* modality arises. As explained earlier, ampliative inductive inference involves more than mere association or simple regression. One must make projections, extrapolations or any of the forms of generalization. In short, one must develop a capacity for *tactical* reasoning. Thus, as the milieu becomes more complicated, one moves from the comfortable and deterministic domain of simple association into the arena of probabilities and risk. Again, in terms of our simple example, the associative behavioral modality makes a direct link between a predicate (the storm cloud) and a specific behavior (shelter). But when we begin to assess the relative expected value (or probable relative utility) of different responses (e.g., shelter or umbrella or getting wet), we enter the ampliative domain. Simple associative processes tend to generate one specific response for any single predicate (behavior is deterministic or "finite-state"). Ampliative processes involve using one's experiential base and environmental clues to determine an expectedly optimal behavior from among a set of alternatives, not just the historically "adequate" behavior we find with simple association. For our purposes here, the difference between associative and ampliative behaviors is not very critical; in a later chapter we shall have to differentiate them more fully.

To reiterate: Inductive behaviors may be as simple as developing regression relatinships between a stimulus and a response, or between some predicate and a repercussion (e.g., storm cloud means rain). Or they may involve making probabilistic calculations under the ampliative modality (e.g., generating expected-value or expected-utility functions, which lend a magnitudinal aspect to associations, and permit the discrimination among alternative behaviors one may display in the face of a specific stimulus or event). But either type of inductively driven behavior may be either functional or dysfunctional. Specifically, the inductive inference modality becomes appropriate (functional) where there is a significant correlation between present and past contexts, such that the experience base becomes relevant. Where the event or circumstance (the direct predicate) demanding a response is effectively unprecedented, then one of two other behavioral modalities becomes appropriate: (1) raw trial and error, or (2) the deductive modality. Where correlation between present and past contexts is weak or absent, reliance on the inductive modality may be inappropriate. A good example of this is prejudice (a distinctly inductive behavior), where generalizations are too broad or procrustean in their implication. More severe forms exist, such as a pathological fear of darkness or crowds based on an isolated instance unlikely to repeat itself. And there are, of course, instances where errors of judgement are largely innocent of any neurotic or psychotic implication. These, most usually, are cases where we miscalculated the probabilities on which our ampliative inferences were based, a natural occupational hazard of individuals forced to live in a world that holds some potential for surprise.

There is an unfortunate tendency to try to extend the implications of the inductive modality to cover virtually all aspects of human behavior. In particular, there is the program of citing all human behavior as an instance of "operant conditioning", either spontaneous or induced. This is a central theme of behaviorist psychology, and for those not already schooled in the concept, it will still be fundamentally familiar through the above discussion of the regression-based association processes. When utilized by experimentalists in the laboratory, it essentially involves attempting to induce a relationship between a stimulus and response (a predicate and behavior) through an intermediate predicate. Schematically, it is the act of inducing a correlation between an event x and an event y. It has been suggested that only moderate success has been had in inducing such correlations, and only under essentially simple circumstances with essentially simple and tractable behavioral subjects. But the general notion exists that the trial-and-error process, coupled with the *tabula rasa*, is sufficient to explain all behaviors—that, in accordance with Lockean and Humean presumptions, all human behavior is a product of experience.

Both logic and common sense would, however, suggest that a lot of the appeal of the inductive modality is due to the fact that it happens to be the currently popular concept of the "scientific method" as well [18]. Thus, in trying to explain all behaviors as instances of inductive inference, we are at the same time justifying our particular form of science and restricting rationality to the process of generalizing from empirical experience. But there are some behavioral phenomena which the inductive modality cannot explain at all, and others which it explains only so inefficiently that it is unlikely to be the true explanation. It is for these reasons that we must develop a behavioral alternative to stand beside the inductive, other than the data-driven modality already introduced. This we do in the next section, examining the definition and implications of what we shall call *deductively predicated* behaviors. As we shall see, these deductive behaviors accommodate the heights of both man's genius and folly.

THE DEDUCTIVE PROCESSES

When we speak of the proper inductive behaviors, we are actually speaking about behavior which is deemed *rational* within the framework of objective (empirical) probabilities. The associative and ampliative processes involve the calculation of statistically significant relational and projective functions, respectively. The "objectivity" of the probabilities rests in their having a real-world character—a residence within the experience base of the individual, or within the perceptual domain. We can, at least theoretically, distinguish between successful and unsuccessful inductive behaviors largely on the basis of the apparent accuracy of the probability calculations made by the individual. That is, successful or unsuccessful behavior cannot be measured simply with respect to the extent that one's ends are achieved, or to the extent that behavior appears rational. The basis of inductive rationality is the *rightness of the procedure*, not the desirability of results. For, as was suggested, we live in a world that offers ample opportunity for surprise: the real events of life are not driven by the probabilities we calculate. Thus, the virtue of the inductive process is simply this: in the long run, over many iterations of behavioral and decision instances, he who employs proper probabilities and obeys their implications is likely to make fewer mistakes than he whose behavior is casual (e.g., trial-and-error) in determination.

This interpretation of rationality—its process orientation—leads to another issue. At what point should reliance on the inductive process be abandoned? Even behaviors predicated on the proper calculations of probablities drawn from an experience base—disciplined by inductive analysis—are not congruent or "functional" under all circumstances.

In fact, there are two distinct cases where the rational man will abandon the inductive modality for another:

1. Where the circumstance, stimulus or event to which one must respond is effectively unprecedented, so that a cumulative experience base is unlikely to contain any relevant predicates.
2. Where the engine driving the context in which one is resident is essentially random, so that no real structure exists which can be described by any disciplined probability distribution (e.g., every event has essentially the same probability of occurrence as any other).

The response in the face of situation 2 has already been discussed: it should be the trial-and-error behavior developed above (see pp. 23 ff.) as one of the data-driven variants. So it is situation 1 that sets the stage for the introduction of *deductive* behaviors.

The depth of misunderstanding about the role of deductive processes and the confusion about proper responses to complexity has roots in the history of the scientific enterprise. As I have explored these elsewhere in detail [19], here I shall be very brief. Since Francis Bacon stood at the threshold to the age of empiricism several centuries ago, science has become almost totally preoccupied with the inductive method. This is, in part, an understandable overreaction against the ultra-deductive and sometimes delusive science of the Middle Ages. At any rate, we have tended to emphasize the inductive modality—and tried to make it the exclusive form of human behavior—largely because it is the exclusive form of scientific behavior. This, more than any other factor, explains the failure of psychology to provide valid referents for the entire range of human behaviors, and the failure of science in general to secure success outside the technological or instrumental realm. Psychology, as I earlier suggested, under the banner of the inductive method, sought to make human behavior intelligible in the same terms as components of the physical world (also seeking, in a rather sophomoric way, to transport the instruments of classical physics into the behavioral sciences). In order to do this, it became necessary to deny man *a priori* any autonomy or volition, and to ignore the possibility of empirically transparent phenomena such as will, inherent values, or any behavioral predicate that was not external to the individual, or capable (at least theoretically) of being manipulated in a laboratory. But particularly, it was necessary (or perhaps simply natural) for any behavior not predicated on experience to be ignored or denied *a priori*. For such behaviors would have to be explained with reference to a deductive base that science itself chose to disparage at every turn as the instrument of mysticism, dogma and delusion.

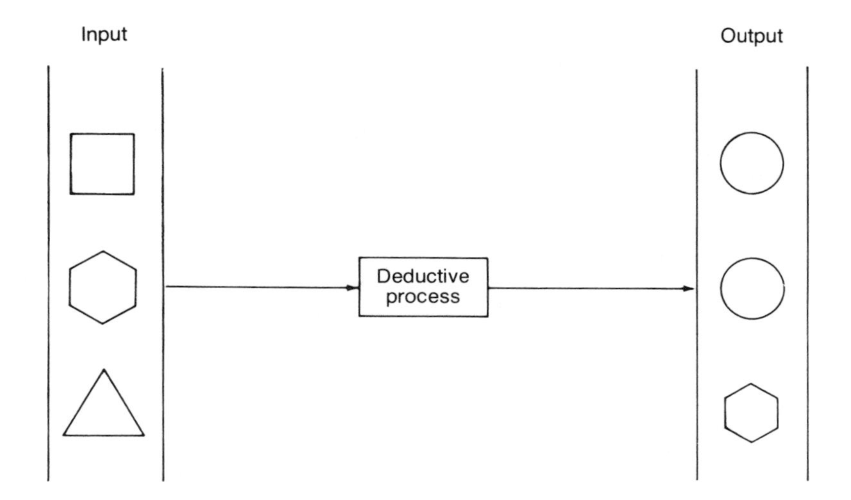

FIGURE 1.9: The deductive function.

Now, the deductive process, given our earlier illustration of direct inductive process (figure 1.7), can be contrasted as a case where the basic qualitative character of an input is altered, so that inputs and outputs no longer bear a configurational or qualitative correlation, as diagrammed in figure 1.9. Here we get a situation where the deductive process adds something fundamentally new. The input bears little qualitative resemblance to the output, and in fact it would be very difficult to trace the output back to an input unless we knew the structure of the specific deductive process intervening. Thus the deductive process is very different than the inductive, where the major alterations to inputs were magnitudinal rather than qualitative. The deductive process involves *morphological* change.

It is interesting to note that we can simulate such behaviors in the laboratory. Specifically, we can allow a relatively simple input to be manipulated by a complex set of equations or manipulative processes. We can either employ a computer for this, or simply use a set of mathematical models. In either case, we get the result we seek: the transformation of an output to the point where its apparent correlation with an input is very weak.

This introduction of morphological change between input and output is a characteristic of deductive behaviors. However, there is yet another form of deduction, derived from the formal interpretation of deductive logic *per se*: the act of imposing the general on the specific (i.e., reasoning from the abstract to the particular). We thus have the basis for three distinct forms of deductive behaviors. All involve an *aprioristic* (non-experimental, non-empirical) component mediating between predicate and behavior, but the precise nature of the aprioristic determination is different.

The Idiosyncratic and Exegetical Variants

As we have seen there are three distinct forms of deductive behavior. The first is what we shall refer to as *idiosyncratic* behavior, where the predicates arise within the individual, and are likely to be entirely unique (i.e., incapable of being generated in quite the same way by any other individual). Two variants of idiosyncratic behavior are especially important:

1. *Interpretive* behaviors are those engaged in by creative artists (for example). Here some qualitatively unique component is introduced, by which reality becomes reordered. For example, modern or abstract art is of this type, whereas the "photographic" art of the Renaissance was weakly inductive (i.e., it sought to portray the subject as it existed in reality) [20].
2. *Hypostatized* behaviors are those driven by predicates that arise within the individual, and for which there may not even be a direct (external) predicate. Much paranoid (e.g., hallucinatory) activity may be of this type, but so are certain forms of functional behavior [21]. We shall later see that compulsive behaviors usually become intelligible as instances of hypostatization.

As suggested, interpretive behaviors simply involve output that is morphologically distinct from the input. Here, then, interpretive behavior always has an external predicate (a subject) to which the individual adds qualitatively new components. In ordinary language such behaviors are called instances of creativity. Hypostatized behaviors, on the other hand, are instances of *reification*. That is, the individual's behavior may be thought of as driven by predicates that arise within, and the behavior may become independent of external realities. As with some data-driven behaviors, there is an "open loop" condition, such that no feedback between apriorism and reality exists. Thus, the individual displaying hypostatized behavior may seem to be marching to the sound of a "different drummer". His behaviors may run the gamut from relatively innocent eccentricities to deep-seated pathologies. These hypostatized behaviors differ from id-level behaviors in that the predicates of the former are conceptual, whereas the predicates of the latter are physiological or "instinctual". Finally, in those instances where there is a direct (external) predicate or stimulus for a hypostatized behavior, the individual's assimilative mechanism will have transmogrified it beyond all recognition, so that the empirical-experiential root is shorn almost completely. Only through deep psychoanalysis—and often not even then—can the direct predicate be located.

But, as suggested, there are also essentially functional behaviors in the repertoire of hypostatized activities. For example, two of the most rudimentary human achievements—language and mythology—may be the best evidence that hypostatized behaviors do in fact exist, and should be treated within the framework of general psychology. In this respect, there are two strong pieces of theoretical evidence: (1) the work of Levi-Strauss, which indicates strong similarities (isomorphisms) among the many human mythologies, isomorphisms so intense and pervasive that a genotypical origin is suggested [22], and (2) the work of Chomsky on "deep structures", instances of fundamental similarity among all human languages, again arguing for a genotypical origin. Because it is so very improbable that these isomorphisms among myth and language structures arose from shared experiences (or were accidents of accommodative evolution), the assumption of hypostatization gains plausability. There is a strong suggestions that myth and language are both products of the innate logical apparatus of man, a universal endowment of all our species. They are, thus, potentially explicable as hypostatizations—as reifications of some inherent logical capability that is innately the inheritence of all men, a spontaneously given substance [23].

There are at least two other kinds of evidence that hypostatizations are behavioral bases. First, there is the difficulty we have in explaining the prodigal genius in any other way, and the fact that prodigal geniuses appear only in strongly "deductive" fields: mathematics, music, chess and, to a lesser extent, the mastery of languages. All these disciplines are responses to a set of logical rules, which limit the permissible moves at any stage of the process, but which nevertheless allow great latitude in the different structures or variations which can emerge. In short, the significant basis for all four of these areas is *process*—a set of abstract rules. And the speed with which such rules may be absorbed and permuted by children is astounding, and makes one suspect that these skills are *educed* rather than imposed or acquired. The second phenomenon which I think deserves study in light of the possibility of hypostatizations is the *idiot savant*, a person showing incredible mathematical or calculative capabilities, but no mature or sufficient skills in other dimensions of behavior [24]. The normal

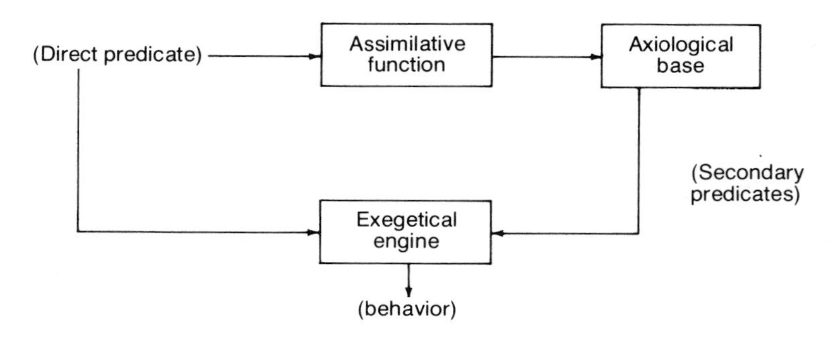

FIGURE 1.10: The exegetical process.

learning capabilities of these persons are highly impaired, so the probability that their knowledge of mathematical forms and behaviors were acquired by normal means (through accommodation or operant conditioning) is unbelievably low. Thus, until a better explanation is developed, the *idiot savant* stands as a strong suggestion that hypostatization occurs. Our synthetic paradigm, then, cannot afford to neglect hypostatization as a potential behavioral base.

The second major variant of the deductive modality we shall call *exegetical* behavior. Exegetical behaviors are driven by a set of aprioristic referents (usually axiomatic in nature) which set forth an array of more or less specific proscriptions or prescriptions. That is, the *a priori* base defines certain behaviors which are forbidden or undesirable (proscribed), or desired, recommended or imperative (prescribed). For the process diagram, we have figure 1.10.

As we shall see in Chapter Three, the provision of the axiological bases under which societal systems have labored has been the work of prophets. Thus, individuals operating under the exegetical modality have been those responding—with constancy and assiduity—to some ideological platform where the behavioral dictates are quite explicit: they are the proto-Christians, the proper Marxians, the "good old boys" of the Ku Klux Klan, or the devout and constant followers of any of the great "isms" inspiring collective behavior.

The equation of deductive behaviors with their exegetical form is the root of the general distrust of deductive exercises in general. The empiricist sees the exegetical engine as an oppressive one, one which has given the world much grief. And to the extent that exegetical processes become divorced from reality, and lead to fanatical behaviors, his point is well taken. Where the exegetical form rules science or society, we find the nadirs of civilization: the confinement of human intellect to dogmatic service (as with Lysenko's biology or the Catholic persecution of Galileo); the formulation of immutable codes of political behavior based on entirely unvalidated (and often pejorative) first premises, as with National Socialism in Germany; the abrogation of any attempt at contextual interpretation of activities (as with the ancient oriental forms of law or the absolutist religiopolitical platforms once predominant in Europe); the many coercive forms of government; and other activities where the revelatory or exegetical base reigns supreme (as with the Soviet mutation of Marxism). But what is often forgotten is this. If the intuitive behaviors represent some of the basest exercises of human nature and intellect, the other variant of deductive behavior has been the source of some of the most elegant and enlightening ones. For the third deductive variant is the discursive—very different, in both form and implication, from its exegetical and idiosyncratic counterparts (in the same way that the simpler associative behaviors differed from the more sophisticated ampliative forms in the inductive domain).

The Discursive Variants

The difference between the exegetical and discursive variants becomes apparent when we diagram the latter process, as shown in figure 1.11. Here, the discursive engine replaces the exegetical engine, and there is a direct feedback link between behavior and aprioristic predicates. The individual has the option to interpret reality in his own way (an idiographic license), but this time there is a rational mediation, for the discursive engine joins together the direct (empirical) and secondary (*a priori*) predicates in a hybrid set, and generates a behavior that is logically consistent with those predicates. In this sense, the discursive behavioral process becomes intelligible in terms of the ability to make disciplined deductive inferences, such that conclusions are congruent with the substance of premises.

In the real world, discursive behaviors appear in two forms, both complex and both of great significance for us. First, there are the range

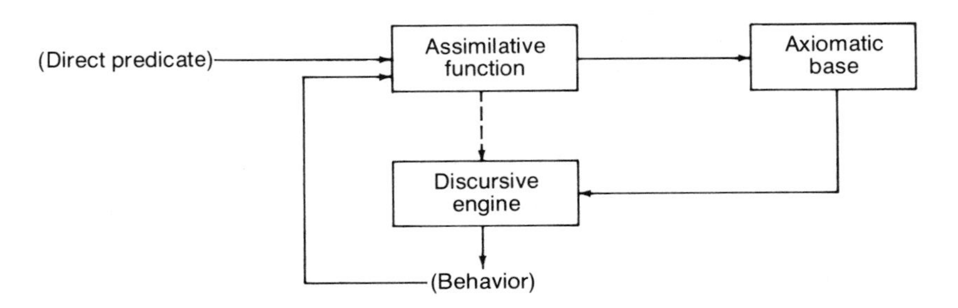

FIGURE 1.11: The discursive process.

of *principled* behaviors. Principled behaviors differ from simple exegetical behaviors in an important way. True, both rely on the generation of behaviors consistent with some aprioristic base, but principled behaviors make a special kind of cognitive demand on the individual. Namely, the individual is here forced to make proper deductive inferences, whereas in exegetical or intuitive behaviors the normative behavior was explicit within the frame of conceptual reference. The reason for this is that the apriorisms underlying principled behaviors are non-specific. That is, they may articulate a general or abstract rule of behavior (e.g., the Golden Rule; the thesis of self-actualization), but the individual is responsible for relating this rule to action within a specific context. Thus, behavior under this modality becomes a complex product of deductive inference and observation, the individual trying to *deduce* a reaction to a specific context that embodies as fully as possible the spirit of the *a priori* base. Thus we ask questions such as: What should I do in this situation as a good Christian? Now, in the exegetical world, such questions emerge only rarely, for the apriorisms there give specific rules or algorithms of behavior. But here, we have only normative (axiological or axiomatic) guidelines to follow, and are responsible for setting our own specific actions within the vague *a priori* constraints provided by the principles. We shall have much more to say about this in later chapters. For now, we may generally class principled behaviors as those that reflect some *ideology*, and where the cognitive demands on the individual consist in giving the abstract ideology substance in a succession of concrete contexts.

The second variant of discursive behavior is what is known as *heuristic* activity. This differs from the principled case in an important way. Here, the context in which one resides (or the direct predicates to which one must respond) is both complex and unprecedented. But there is no existing principle which promises to be productive as a discursive predicate. So, whereas principled behavior consists in the attempt to rationally apply abstract principles to a unique situation, the heuristic process demands the creation of *a priori* predicates before any attempt to discipline behavior can be made. In actual practice, heuristic behaviors become evident in two ways: (1) A *disciplined* trial-and-error process takes place, disciplined by deduced rules of procedure; (2) the initial trials are determined with reference to initially artificial presumptions about what might be the "logically" most productive search trajectory. There is constant feedback between the results of the trials and the conceptualization (e.g., heuristic-building) process [25].

Heuristic formulations have provided the great deflection points in science. They represent *a priori* "masks" which, when worn, bring to light patterns and purposes in the real world which remained transparent in their absence. The deductive nature of these masks is evident from the unlikelihood that they are derivatives of experience, or products of extrapolation or projection. Rather, they involve the postulation of some unique element—presumably derived from the innate idiographic base of the individual himself—*that restructures rather than reflects reality*. Thus, the fundamental engine that drove the Darwins, the Einsteins, the Harveys, the Newtons, was deductive-discursive rather than inductive-extrapolative. To a greater or lesser extent, most of the great historical intellects would be subject to a similar interpretation, including those who operated on the social or moral dimension—the phophets—rather than the scientific *per se*. In summary, then, the significance of the heuristic behaviors is the generation of rational, but original, responses to unprecedented situations. This thus becomes the mode of human behavior most appropriate for dealing with significant complexity. To the extent that heuristic behaviors are exercised, man rises above reality and in part *creates* it. For the great conceptual leaps of the heuristician reorder the reality that we perceive and make us all agents of deduction, no matter how indirectly or unwillingly. Indeed, as we shall see, it is our universal susceptibility to deductive predicates which gives the concept of "society" real substance for us.

To a certain extent, the ultimate ability to gain any real empirical or tangible grip on deductive processes—or any of the variants developed in this section—will depend on our skill in developing definitions of these processes which will permit some hard experimentation. At the very least, we must act to remove their consideration from the realm of rhetoric and speculation. As things stand today, we have little or no idea about the exact mechanisms by which the mind of man performs its deductive functions. All we do know—thanks to philosophers as diverse as Arthur Koestler, Maslow and Epicurus, and hard scientists such as Hyden, Gerard and Penfield [26]—is that human capabilities for both generating and manipulating conceptual constructs are enormously sophisticated, and truly ancient in origin and practice. We know, also, that conceptual constructs are intelligible as temporal *Gestalten*, where many diverse factors are ordered into complexes of multidimensional significance, where reality and speculation, projection and origination collide and mix in indeterminate ways. The heart of the process is probably connected with the ability to create, internally, those *a priori* "masks" we spoke about, which, when worn by the subject, make patterns visible in the real world which might be transparent in the absence of the mask. This, essentially, has been the contribution of the great theorists: the Darwins, the Harveys, the Einsteins, and the great moralists as well; they have offered us spectacles through which heretofore hidden possibilities appeared. To borrow one of Voltaire's thoughts, the creative deductive processes are like "metaphysical flashes of lightening on an otherwise dark night". In short, deductive processes not only amplify or reorder reality into new forms, but they are the mechanism by which man transcends reality—by which he rises above experience.

As was suggested, we know very little about the physio-mechanical or biochemical bases for deductive behaviors. There are two particularly appealing pieces of theory that are beginning to attract some attention. The first is work that indicates that there are essentially two structurally separate hemispheres in the brain, and that these two hemispheres evidently perform different tasks [27]. In terms of the dimensions we have been using, the left hemisphere houses the inductive capabilities, being concerned with essentially algorithmic tasks such as the processing of linear sense inputs and the implementation of normal language. The right hemisphere, on the other hand, presumably houses the deductive capabilities: these are put to use in broad pattern recognition (e.g., conceptual manipulation and ordering) and the other *Gestalt*-based human processes: the reduction and synthesis of either sensual or conceptual complexes, such as musical or artistic compositions. The second interesting piece of information is that memory tends to be stored in terms of *Gestalten*, not in terms of unrelated bits or pieces. This was brought to light through the researches of Penfield and his colleagues at the University of Montreal, where patients' memories emerged as "reliving, refeeling" experiences of entire scenarios [28]. If this is the case (and his evidence appears very strong indeed), then we perhaps have some basis in suggesting that the deductive process—the discursive capabilities—are rooted in intrinsic rules and vehicles for the manipulation of conceptual *Gestalten*. Presumably, the intuitive behaviors would have a similar basis, but with rules of transformation—and a conceptual calculus—of a much simpler order.

At any rate, it is not the internal mechanics of the deductive process which concerns us here. Rather, it is the historical tendency for so much societal error, waste, deprivation and dysfunction to be directly attributable to the predominance of the cruder deductive variants (e.g., intuition, exegesis) as a basis for both the design and conduct of the societal systems we have evolved. In practice, this means that a significant proportion of both the structure and the behavior of these sytems is based on simple exegesis performed on some crude and non-apodictical *a priori* base. Deductive predicates thus serve to fix the substance—or some measure of it—of the social, economic and political dimensions of societies, and are the source of the "irrationality" we shall find on so many fronts in later chapters. Now, our attempt to provide an apriorism-free paradigm for the analysis of human behavior—the syncretic paradigm of figure 1.2—may be viewed as a first crude attempt to show the way in which such predicates may be removed from the determination of activity. But as we shall later see, to remove them entirely is to condemn our social, economic and political activities to a most unsatisfactory envelope of realism and opportunism. Thus, it is not so important that we eliminate *a priori* predicates as that we learn to discipline and constrain our use of them, and that we come to understand the process by which they exert their control. This, essentially, is the task for the chapters that remain.

Finally, because we shall be referring to the various bases of human behavior so frequently, it is useful to summarize our arguments in the form of table 1.1, some of whose points will not really be clear until later in the book.

The hope is that we now have a more disciplined and precise grasp of the range of human capabilities, and the facilities we have both for accommodating ourselves to the world and for structuring it. But there must be a great sense of disatisfaction, for it appears that there is simply no way to adequately predict or explain behavior except with reference to specific instances in specific contexts. The potential—from the standpoint of our treatment of the bases of behavior—is for every individual to be effectively unique. In short, the bases of behavior suggest that human activities are *inherently indeterminate*—in and of themselves capable of achieving more forms and complexities than we can ever hope to comprehend.

This, indeed, seems to be true. But the real issue is this: How do we reconcile this innate potential for enormous complexity and variation with the fact that human behavior, in actuality, is not nearly so complex or protean as we might be led to believe? The answer, I think, rests in two specific propositions: (1) that, for the most part, only the simplest engines of human behavior are exercised by the great majority of human beings (the associative and intuitive-exegetical modalities); (2) that, therefore, the majority of human behavior consists in *accommodation* to either environmental predicates (contextual conditions) or axiological predicates (sets of *a priori* constraints about what constitutes "functional" or acceptable behavior in certain instances). In short, human behavior is essentially much simpler than our theoretical work would indicate *because the potential for complexity is seldom realized in practice*.

TABLE 1.1: Summary of Behavioral Bases

	PREDICATE	ENGINE	VARIANTS	MECHANICS	CATEGORICAL BEHAVIORS
1. DATA-DRIVEN	Sense data	Limbic system (pre-cortical)	a. Homeostatic	Closed-loop cybernetic	Maintenance of physiological parameters
			b. Id-level	Open-loop (cathectic)	Expression of instincts, innate drives
			c. Trial/error	Non-algorithmic	Random search processes
2. INDUCTIVELY PREDICATED	Perceptual data (sequential and linear)	Left brain[a] (algorithmic hemisphere)	a. Associative	Regression	Conditioned behaviors
			b. Ampliative	Inductive inference	Projective/extrapolative (instrumental) behaviors
3. DEDUCTIVELY PREDICATED	Conceptual data (Gestalten)	Right brain[a] (abstract hemisphere)	a. Idiosyncratic	Interpretation	Intuitive or interpretive behaviors
				Hypostatization	Language, myth, etc./ Hallucination, compulsion, etc.
			b. Discursive	Deductive inference	Principled (rationalized) behaviors
				Hypethetico-deductive	Heuristic behaviors
			c. Exegetical	Reflective association	Ideological (fanatical) behaviors

[a]Note that the left and right brain functions may be transferred in case of injury to one of the hemispheres; hence the distinctions are normative rather than axiomatic. However, the use of the distinction between right and left brains is strictly symbolic here; I am not aware that their existence has been completely determined by physiological psychologists, though the distinction between abstract and algorithmic bases is clear.

THE CRUCIAL CONCEPT: FROM INDIVIDUAL TO COLLECTIVE BEHAVIORS

We may now spend a very few lines describing the concept of societal systems that underlies virtually all of our work in this volume. This concept is not particularly radical, but neither is it without its subtleties. Specifically, it is my contention that the societal systems men form are primarily creatures of intellect and cognition, not unconscious and mechanical adaptations to environment. In time, admittedly, the cognitive or deductive origins of societal forms and practices may become obscured, but it is the different cognitive-behavioral modalities that best serve to distinguish one basic societal form from another. This, at least, is the hypothesis that we carry well and fully into our subsequent discussions; indeed, much of the argument of later pages is designed specifically to elaborate and defend this preliminary hypothesis.

This does not mean that we can force a direct correspondence—via typological constructs—between the contents of our table 1.1 and the societal ideal-types we eventually build in Chapter Four. But it does mean that generic societal systems may be distinguished largely by their predilection toward one or another of the cognitive (or more broadly, behavioral) engines just discussed. For example, some societies become comprehensible as collectivities employing mainly the simple associative cognition; others are predominantly characterized by their constant recourse to abstruse discursive referents; others, by the fact that their structure and dynamics appears to be a product of exegesis performed on some *a priori* axiological or axiomatic base. Moreover, using the behavioral modalities as the primary structural input for the definition of societal ideal-types, it is possible to show the ancestry of various social, economic and political subsystems in terms of cognitive predicates.

All this may seem a bit confusing at this point, but it is in the nature of fundamentally syncretic constructs that we get a form of allegorical reflexivity. This means that we could just as easily have begun by defining a logically exhaustive set of economic or social or political modalities. And to the extent that societal systems are indeed ordered sets of behavioral, social, economic and political properties, we should eventually have come around to the theoretical recognition that a unique specification on any of these four dimensions implies a unique specification on the other three. In short, societal systems (unless they are anomalous) present us with highly correlated social, economic, behavioral and political attributes. Therefore, to a great extent, a knowledge of the bases of individual behavior—the various cognitive alternatives—also provides us with a provisional knowledge of the several social, economic and political modalities that societal systems exhibit.

Now nobody likes to have to write paragraphs like those immediately above. They are, even for the charitable reader, obscure, and they smack of procrusteanism. But the concept itself is simple: If societal systems are primarily cognitive in origin, then each unique cognitive modality will be expected to force, with significant probability, a set of equally unique and highly complementary responses to that cognitive base on the other three system dimensions. This possibility is at the heart of the syncretic technique we have been urging. For it suggests that when we raise societal systems to a sufficiently high level of abstraction, we find causal relationships between societal properties that are simply not apparent from the perspective of any particular discipline operating at the institutional level.

Let's just assume, for the moment, that there is merit to this supposition. The immediate question is then this: How do cognitive and behavioral predicates, defined for individuals, become translated into behavioral predicates that pertain to collectivities? The tentative answer to this question rests with our next chapter, where we explore the dialectical perspective.

Notes

1. For a detailed exploration of the model-building logic summarized in this figure, see Chapter 5 of my *Administrative Decision Making* (New York: Van Nostrand Reinhold, 1977).
2. For more on this thesis, see L. A. Zadeh's "Outline of a New Approach to the Analysis of Complex Systems and Decision Processes" (*IEEE Transactions on systems, Man and Cybernetics* SMC-3, no. 1, January 1973).
3. As many readers are no doubt aware, it is virtually impossible to completely validate or invalidate any hypothesis (or set of hypotheses) (metahypothesis), i.e., through normal experimental methods. See I. Mitroff and M. Turoff's "On measuring Conceptual Errors in Large-Scale Social Experiments: The Future as Decision", (*Technological Forecasting and Social Change*, November 1974).
4. Of particular interest to those with a quantitative bent is the fine paper by K. S. Fu, "Learning System Theory", in *System Theory* (edited by Zadeh and Polak; New York: McGraw-Hill, 1969).
5. The most authoritative work on the phenomenon of paradigms is that done by Thomas Kuhn. See *The Structure of Scientific Revolutions* (University of Chicago Press, 1963).
6. I have attempted to show just how pervasive axiological predicates are in my "Axiological Predicates of Scientific Enterprise" (in *General Systems Yearbook*, vol. 19, 1974, pp. 3–13).
7. See chapter 5 of my *Systems: Analysis, Administration and Architecture* (New York: Van Nostrand Reinhold, 1975).
8. See Robert Merton's discussions of "theories of the middle range" in his *Social Theory and Social Structure* (New York: Free Press, 1968).
9. See Maclean's paper "Man and His Animal Brains" (*Modern Medicine*, 3, no. 2, 1964, pp. 95–106).
10. See the technical note by Edward Schafer and Marilyn Marcus, "Self-Stimulation Alters Human Sensory Brain Response" (*Science* 181, July 1973, pp. 175–177).
11. For a progenative treatment of such mechanisms, see Paul R. Miller's *Sense and Symbol: a Textbook of Human Behavioral Science* (New York: Harper and Row, 1967).
12. That even mediated loop-learning is an inherent and continuing human

function is asserted by Miller et al. in *Plans and the Structure of Behavior* (New York: Holt, Reinhart and Winston, 1960, pp. 103–116).

13. Emery and Ackoff, in describing goal-oriented behavior, make much the same point. See *On Purposeful Systems* (Chicago: Aldine-Atherton, 1972).
14. See, for example, Piaget's and Inhelder's remarks in "The Gaps in Empiricism" (in *Beyond Reductionism,* edited by Koestler and Smithies; New York: Macmillan, 1969).
15. For a broad, non-technical discussion of the mechanisms which are supposed to perform this codification-translation function, see D. S. Halacy's *Man and Memory* (New York: Harper and Row, 1970).
16. For a discussion of operant conditioning, see Chapter 2 of Skinner's *Beyond Freedom and Dignity* (New York: Alfred A. Knopf, 1971).
17. There are some obvious correlations between cognitive demands and the nature of the environment in which an individual (or societal system) resides. For a brief interpretation of the correlations, see the section on the field concept in my *Systems* [7].
18. That this is a somewhat stilted view is suggested in the brilliant work by Errol Harris, *Hypothesis and Perception* (London: George Allen and Unwin, 1970).
19. See my book, *A General Systems Philosophy for the Social and Behavioral Sciences*, or my "Attacking Organizational Complexity" (*Fields within Fields . . . Within Fields* 11, Spring, 1974).
20. Interpretation as data-reorganization process is analyzed by Bateson in his "Style & Communication in Primitive Art" (in *Steps to an Ecology of Mind* Chicago: Aldine-Atherton, 1972, pp. 128–152).
21. For example, Kekule's insight into the structure of benzene was apparently an *ab intra* exercise. Jung also ascribed a functional role to dreams in "The Importance of Dreams" in *Man and His Symbols* (New York: Dell, 1968, pp. 25–40).
22. Cf. Caysell, *The Masks of God: Primitive Mythology* (New York: Viking Press, 1959).
23. Levi-Strauss's theory is given substance in his recent *Introduction to a Science of Mythology* (New York: Harper and Row, 1973). Chomsky's thesis is explained in "Some Empirical Assumptions in Modern Philosophy of Language," (in *Philosophy, Science and Methodology*, edited by Morganbesser et al., New York: St. Martins Press, 1969).
24. For a discussion of this phenomenon, see M. Scheerer et al., "A Case of Idiot-Savant: An Experimental Study of Personality Organization" (*Psychological Monographs* 58, No. 4, 1945).
25. For a more complete and somewhat parallel discussion, see Miller et al. [12] (especially Chapter 12, pp. 159–174).
26. The most gripping and imaginative exploration of these creative conceptual exercises is Arthur Koestler's book *The Act of Creation* (New York: MacMillan, 1964).
27. See either of the following: D. Kimura's "The Asymmetry of the Human

Brain" (*Scientific American*, March 1973) or R. E. Orenstein's "Right and Left Thinking" (*Psychology Today*, May 1973).

28. See Penfield and Robert's *Speech and Brain Mechanisms* (London: Oxford University Press, 1959).

2
The Dialectical Challenge to Complexity

The work we did in the first chapter was in reference to the single individual, acting in effective isolation from contextual factors. But clearly the cognitive processes available to the individual are the mechanisms by which he becomes associated with—or responsive to—the components of the world around him. And this world around him includes both the empirical, environmental stimuli to which he is exposed, and the axiological or axiomatic engines—the cultural predicates.

Now, we have to develop a model of the exact process by which external factors, environmental or cultural, become internalized. Of all the possible paradigms at our disposal, that which appears most promising is the *dialectical* engine. From the dialectical perspective, collectively significant behavior becomes intelligible in terms of a force vector—a moment-to-moment resolution of complex and often competitive predicates and determinants. The individual becomes integrated more or less fully into a collectivity (or set of collectivities) through the dialectical process. In short, a proper dialectical paradigm may serve as the vehicle for our transfer from individual to collective behavior, and hence as the axial model of societal dynamics. It is through the offices of the dialectic that values and realities become intermixed, and here also that the individual becomes both an agent of individuality and a victim of circumstance.

As an ancillary benefit, the specification of the dialectical engine will allow us to introduce some of the more powerful instruments of modern qualitative analysis. Thus, as with our previous chapter, what we will be doing here should be of both substantive and methodological significance.

THE DIALECTICAL ENGINE

In the most basic sense, the dialectical position will be our link connecting the bases of behavior defined in the previous chapter with the social, economic and political contexts to be defined and elaborated in the remaining sections of this book. In short, the dialectical engine is a mediating behavioral engine, serving to connect the individual to his cultural context. If we take the broadest possible interpretation of the dialectical behavioral process, we might diagram it as in figure 2.1.

What this simple schematic tries to convey is this. Any instance of behavior is potentially a product of any or all of three predicate sets:

1. The *intrinsic* bases of behavior are those unique to individual (e.g., *ab intra*).
2. The *axiological* and/or *axiomatic predicates* are those which are used primarily as the secondary predicates for exegetical or principled behaviors: *a priori* referents which serve to prestructure reality and thereby predetermine, at least in part, culturally significant behavior.

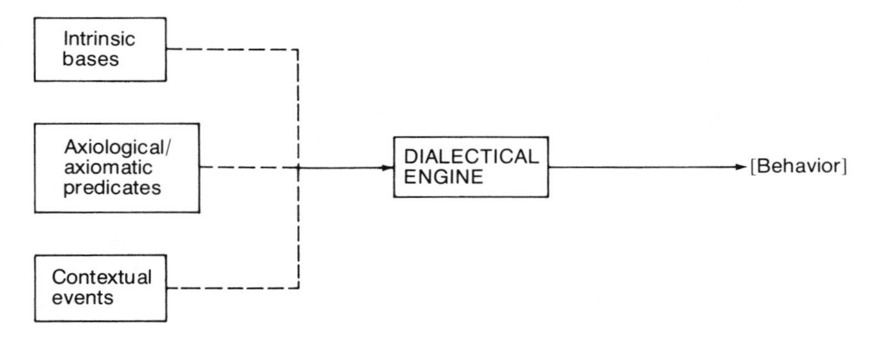

FIGURE 2.1: The dialectical process.

3. The *contextual* events are the reality components in the individual's predicate set, the tangible and empirical aspects of his milieu. In the simplest form, these contextual factors become intelligible as the bases of realistic (associative or ampliative) behaviors.

In great abbreviation, then, the intrinsic bases of behavior are the properties of the individual, the axiological-axiomatic predicates are the products of the prophets, and the contextual variables are the attributes of reality.

Because these distinctions are so critical to us in this volume, I want to briefly illustrate the kind of logic we shall be going through, even though the details will not become apparent until much later on. For the purposes of this preview, let us take the concept of the "afterlife"—of continued existence after death. This, among many other examples we have chosen, is of interest for several reasons:

1. The existence of the concept of an afterlife has distinct behavioral implications. It serves, in various ways, to control, limit or direct the behavior of those individuals (and hence those societal systems) susceptible to its suggestions.
2. In its historical forms, it has also served as a basis for the structuring of the social, economic and political sectors of societal systems.
3. But, perhaps most interestingly for us here, the origin of the concept of the afterlife may be explained—with more or less equal authority—as an instance of either empirical, axiomatic or intrinsic predication.

For example, to those who search for all aspects of societal structure and behavior in the domain of the empirical, the concept of the afterlife must have natural or "positive" roots. In particular, they suggest that the concept was a product of prehistoric man's dreams, in which dead friends and relatives reappeared. In other words, the concept of afterlife was a reification of a natural (empirical) function.

Now, we can also suppose that the concept of the afterlife has an axiomatic origin, which owes nothing to the world of natural or empirical phenomenon. The axiomatic base may be found in the literature of the neoplatonists and certain post-Aristotelean teleologists. Their argument is the following: Since there is no positive origin of the universe, there is no specific origin of the entities therein. Human existence is an entity of the universe. Therefore, it has no specific beginning; lacking a specific beginning, it can have no end. Therefore, it is logically necessary that there be an afterlife.

We may also specify the basis for the axiological predication of the concepts of the afterlife. This emerges in an entirely different framework of reasoning, and serves preliminarily to point out the difference between axiological and axiomatic bases of behavior. The axiological argument goes as follows: It has been revealed (rather than deduced) that God is benevolent. Yet there is evil and corruption in the world. No God that is good would permit his children to suffer so. The concept of the after life thus emerges rather neatly as a resolution of this apparent paradox. To wit, the mortal span represents just an iota of time in an eternity of individual consciousness and existence. The mortal life is lent us soley for the purpose of preparing us for an eternity in heaven. Moreover, the calculus of God is a compensatory one: the worse one's mortal condition, the higher the probability of a beneficent external existence (e.g., it is easier for a camel to pass through the eye of a needle than for the rich man to gain heaven).

Now the axiomatic and axiological positions, were we to reduce them to abstract logic, would have essentially the same generic configuration (a point we shall explore in a later section). For example, both derive from unvalidatable first premises. The axiomatic assertion is that the universe has no beginning; the axiological assumption is that God is good. But the distinction between the axiomatic and axiological positions is important to understand, and it evolves from this point: the first premise of the axiomatic position stemmed from an attempt to understand nature, and was consistent with both the theoretical and empirical knowledge of the day. The axiological position, on the other hand, was an assertion of affective or value significance: it attributed a quality to an abstract entity, with no attempt to justify that attribution. Moreover, and this is critically important, the subsequent elements in the axiomatic model are intelligible in terms of the laws of deductive inference, whereas the structure of the axiological model is elliptical, at least in the first analysis. As we shall later see, however, very few axiomatic positions (except those with no pretentions to reality, such as the bases of mathematics or chess, etc.) escape the elliptical and non-apodictical qualities which affect all aprioristic constructs. But we need not be concerned about this yet.

There is also the *intrinsic* origin to consider. And, as is true whenever we consider the intrinsic base of behavior, we shall have both a functional and a pathological variant. The latter is the easiest to dispose of in the case of the afterlife: it historically emerged as a personal and idiosyncratic revelation to a number of different prophets at different times, the creator speaking to them. Even more distinctly pathological are the thrusts toward conceived "immortality" which are the products of certain forms of schizophrenia or more distinctive pathologies, such as a power complex. This origin of the concept differs from the axiological in that the concept emerged *ab intra* and as an imperative, not as the result of any value judgments or normative assessments. It differs from the axiomatic base in having no logical structure—no attempt being made to deduce the afterlife from any logically probable premises.

Another possible intrinsic origin for the concept of the afterlife might be *instrumental* genius. The notion might simply be an original and brilliant solution to the problem of how to keep the lower classes in their place. If so, the promise of an afterlife and the notion of the compensatory calculus was struck upon as a solution to a pervasive political problem, and was thus utilitarian. Moreover, this origin of the concept of the afterlife is discursive in engination, in the sense that we earlier explained, and results in what is tantamount to a hypothesis (whose instrumental validity may be tested empirically—e.g., does the belief in an afterlife actually make the people more tractable?).

There is also a second variant of the intrinsic (functional) case: For those individuals who are primarily *purposive* (i.e., susceptible to teleological arguments), existence takes its significance from the postulation of somewhere to go when it is over. To this extent, they may find direction in the concept of an afterlife, irrespective of the way the concept originated. But, more realistically, those individuals preferring rational explanations will be most comforted by the axiomatic basis, while those individuals exercising a prior penchant for mysterious explanations will find the revelatory or axiological origins sufficient. I mention this here only to indicate that, in later sections, we shall have occasion to note that many behavioral predicates appear in different guises, with each of the guises particularly potent with respect to particular classes of individuals (stratified according to their cognitive susceptibilities). The key to any determinant of human behavior, and also to collective or system behavior, is thus first that the predicate must serve some need, and secondly that it must be phrased in an appealing way. These simple assertions will have to be elaborated later. In the meantime, however, consider table 2.1, summarizing what we now know about the bases of behavior. This table very briefly summarizes some of the points we have already made, and some that will occupy us for many pages to come. These, basically, are the components on which the dialectical engine operates. Our concern now is with how it operates.

The functions performed by the dialectical engine may be suggested only reflectively. The dialectical engine decides, at any point in time,

TABLE 2.1: The Bases of Behavior

PREDICATE CATEGORY	BEHAVIORAL MODALITY	BEHAVIORAL CATEGORY
Intrinsic/ idiosyncratic	(a) Hypothetico-deductive	(a) Heuristic (sapient) behaviors
	(b) Psychogenetic	(b) Compulsive/hypostatized/ intuitive behaviors, etc.
Axiological/ axiomatic	(a) Exegetical	(a) Ideological/affective behaviors
	(b) Deductive inference	(b) Rationalized/principled behaviors
Contextual/ realistic	(a) Association	(a) Conditioned (deterministic) behaviors
	(b) Ampliative inductive inference	(b) Instrumental (stochastic) behaviors

the proportion of influence which either of the predicate sets will exercise. Generally, the dialectical function is to effect a priority scheme between the various behavioral predicates which *could* be active in any given situation. The word "dialectical" connotes a situation of potential conflict, and indeed this is exactly what I am trying to convey. Specific behaviors are often products of a resolution of competitive predicates. That is, the intrinsic, axiological-axiomatic and contextual factors may all suggest different responses to a situation, and the dialectical engine is responsible for reconciling these competing suggestions and deciding the actual behavior. It is, then, an operational component of the assimilative mechanism we spoke of earlier.

Basically, to reiterate, intrinsic predicates may be thought to drive three types of behaviors: (1) the id-level responses, considered as a variant of the data-driven modality; (2) the heuristic (hypothetico-deductive) behaviors, which maximize sapience; and (3) the idosyncratic behaviors, including compulsion, hypostatization, etc. As we suggested in our discussions of these behavioral varieties, there will be no significant correlation between the direct predicate—the empirical or reality-based stimulus, condition or event—and the resultant behavior. Or the behavior may arise *ab intra*, without external cause. The axiological-axiomatic bases are the predicates which energize principled (rationalized) and ideological/affective behaviors, where *a priori* determination is strong. In such a case, the ability to predict behavior would depend on our prior knowledge of the aprioristic inventory of the individual (e.g., the particular philosophical, religious or ideological platform to which he subscribes). Finally, as should be clear from our discussion of the inductive modality, the contextual bases of behavior are the reality components, operated on by the associative and ampliative modalities.

If we now consider the predicate set of any specific behavior to be definite and known (the sum of all predicates), then the relationship between the three predicate subsets (the intrinsic, axiological-axiomatic and contextual) is definite and competitive. That is, they play a zero-sum game,* where whatever actual influence one subset exercises must be at the direct expense of the influence available to the others. We may therefore graphically describe the dialectical behavioral modality by means of the simple iconograph [1] in figure 2.2, composed of three individual axes (or dimensions), one for each of the predicate subsets. Positions on any of the axes closer to the origin indicate weaker influence than than positions farther out—i.e., a smaller influence of that subset on the behavior in question.

Thus the iconograph has the virtue of representing, in neat schematic form, the integrated effect of the several influences. The precise relationship between the several dimensions of behavior (which we approximate as linear) defines an artificial plane (as opposed to a mathematical point). The orientation of the plane corresponds to the general "state" of the several influences. And, as we shall shortly see, the iconograph permits us to portray a limited range of distinct behavioral "states".

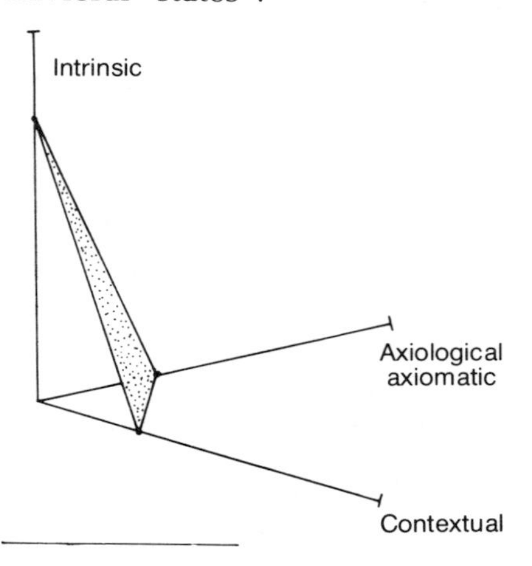

FIGURE 2.2: Essentially intrinsic behavior.

---

*qua Constant-sum.

Thus, the iconograph has the purpose of easily and quickly (though non-rigorously) conveying the broadly different behavioral states available to individuals, and their relative influence. In this regard, consider the iconographs in figure 2.3 and compare them with the configuration of figure 2.2.

In figure 2.3(a), the behavioral "state" is one characterized by significant influence of the axiological-axiomatic predicates, implying that the behavior was driven primarily by exegisis operating on some *a priori* base (as we described when treating the intuitive modality). In (b), describing the contextual case, the behavior was in large measure an associative or ampliative (i.e., inductive) response to the realities of the milieu (it was conditioned by environmental factors). Or the behavior might have been a product of inductive inference, using the environmental factors as the direct predicate and the experience base as the origin of the secondary predicates.

We must now briefly return to an issue we left behind in our previous chapter: the issue of behavioral congruence or "rationality". The essential implication for our current work is this: the dialectical engine, in its function as the author of the basic modality of behavior, can lead to either a congruent or an incongruent response. Specifically, for any given circumstance, event or stimulus (predicate), one or another of the basic modalities will be the congruent one, (i.e., the most effective *and* efficient modality for the case at hand). We already know, from our work in the first chapter, something about the criteria for congruence. The inductive, deductive and data-driven modalities all have their specific roles to play in different conditions. Thus, the dialectical engine operates functionally when it elects the proper predicates to drive

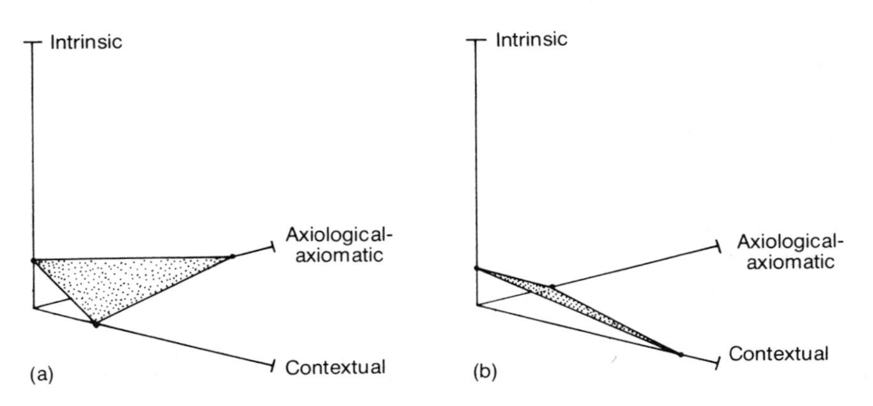

FIGURE 2.3: (a) The Axiological Axiomatic case. (b) The contextual case.

a particular behavior. When the wrong predicate subset is used, then the dialectical resolution is incongruent or *dysfunctional.*

In the rational individual, then, the election of a particular predicate subset is dependent on the attributes or properties of the circumstance at hand. In the impaired individual, two other conditions may be present: (1) the same predicate subset may be used continually, irrespective of the circumstance (e.g., an individual constantly generates responses predicated on intrinsic bases); or (2) in any circumstance, the predicate subsets are equiprobable (the individual is just as likely to employ intrinsic bases as axiological-axiomatic or contextual). To elaborate on this a bit—and to clear up the terms we are employing—a simple example should be of use.

We draw on a minor classic. The *circumstance* is this: a man is brought into an arena with two doors, both closed. He is then told that there is a tiger behind one, and a beautiful lady behind the other. He must open one, and either suffer or enjoy the consequences. Now, let us see what each of the three sub-predicate-bases might suggest. The intrinsic base might involve the employment of a psychogenetic mechanism (e.g., intuition). The hypothetico-deductive modality is irrelevant here, for there is no opportunity to "learn" anything, or approach the decision in stages. The intrinsic predicate might be the *feeling* that one door or another is the correct one, where the basis for the feeling cannot be articulated. It is, in short, a visceral ("gut") reaction. The axiological-axiomatic case would involve something like the following: The angels sit at the right hand of God; therefore, good things will be on the right, and bad things on the left; therefore, take the right door (this, of course, is an enthymeme, and not a syllogism, for the prisoner is leaving some premises tacit—for example, that God is facing in the same direction that he is). In short, the *a priori* response was exegetical, conditioned by the individual's *a priori* base. Now, if the circumstance permitted a contextual predicate to be used, it might take the following form: The prisoner might look to see if there were any physical evidence as to which entity might occupy which door. If, for example, he noted paw prints in front of the left, and high-heel shoe prints in front of the right, he would most probably elect the right door. However (and this is an important adjunct of contextual-based behavior), he must reach his experience base for the probability that the agents of this experiment are trying to fool him. If, in the past, the associative solution (e.g., the association of empirical evidence with a result) has been the wrong one, then he may very well choose the left door. This latter response would then be ampliative, the more complex counterpart of simple associative behaviors, depending as it does on the exercise of inductive inference rather than simple regression.

This, in its simplest form, is the logic behind the dialectical process. We shall return to the issue of congruent and incongruent instances in a later section, but must now explore the dialectical engine in a bit more detail, and especially the opportunities we have for employing a dialectical platform as a legitimate base for societal research and explanation.

Properties of the Paradigm

We must first try to convert the dialectical position into a proper paradigm, treating both its substantive and its methodological implications.

A proper paradigm is simultaneously of epistemelogical (*a priori*) and methodological significance. The epistemological predicates—as apriorisms directing our attention toward certain causal or structural patterns which might remain transparent in their absence—serve to bring hypothetical order to otherwise chaotic fields of inquiry (generally by tentatively constraining the permissible search and solution space). But unless the epistemological predicates are accompanied by specific methodological instruments that permit their experimental validation or invalidation, the substantive implications remain condemned to the realm of rhetoric and beyond the reach of the serious scientist. They are, however, fully at the disposal of the non-scientific community or those who sit at the academic fringes, and are there susceptible to manipulations which make them unattractive to legitimate scientists irrespective of the promise they hold. This, I think, has been the case with the implications available from the dialectical perspective—a coincidence that is particularly unfortunate for the behavioral sciences. For when removed from the unpalatable contexts in which it has historically been forced to labor, the dialectical epistemology directs our attention to certain descriptive or explanatory constructs that are simply not available from other perspectives; and our current knowledge of our subjects is not in such a state that we can afford to reject help out of hand from any source.

The dialectical epistemology acts as an *a priori* "mask" that the investigator can wear, enabling him to recognize certain patterns of structure or behavior that might otherwise pass beneath his level of comprehension. It does this by making tentative (i.e., heuristic) assertions about the reality of some phenomena, assertions that lead the investigator to favor particular search trajectories from among all those that might have been available. Now, from among the various epistemological predicates that have at various times been advanced in the name of the dialectical perspective, a set of strategic dictates may be developed that are particularly appropriate for treating societal phenomena, and that thus become the predicates of a methodology for the analysis of societal systems *per se:*

1. That investigators treat all societal subjects as *field* phenomena, as irreducible. This means that the fundamental and only proper unit of societal analysis is a subject defined in all four dimensions: social, economic, political and behavioral [2]. More specifically, no unit of analysis about which any policy or decision prescriptions are to be made may be allowed to remain isolated from the broader context of which it is part; however, where research is non-instrumental (will have no practical implications), then reduced phenomena, appropriate for traditional analytic procedures—may be used with impunity.
2. That not only the fundamental unit of analysis be structurally defined in field terms, but also longitudinal (historical; dynamic) specifications be made specific and endogenous. This means that the proper dialectical subject is simultaneously of structural and dynamic significance, either directly or by logical extension. By logical extension, this is meant: that there is an unambiguous and calculable correlation between structural properties and dynamic behavior, such that structure determines function or the reverse. Direct (rather than logically mediated) dialectics would allow incongruences between structure and function, or what will be referred to as anomalous relationships between the two dimensions.
3. That no *a a priori* "unilateral" causality be assigned any member of the phenomenal set. With this dictate, the proper unit of dialectical inquiry thus becomes a *field*, i.e., the nexus of (i) the focal subject itself, with its array of endogenous properties; (ii) the exogenous or static environmental determinants; (iii) the longitudinal or historical determinants. This is a field rather than a simple system because each of the components is allowed to exercise some influence on the others. That is, the subject (e.g., an individual entity or some social unit) is allowed to alter the environment, just as the environment is allowed to determine certain aspects of the subject's structure or behavior. In the dialectical field, then, the subject is not simply the passive slave of the environment in which it happens (or has elected) to reside. By extension, the future is not *a priori* condemned to be some simple extrapolation of the past [3].
4. Next, the dialectical epistemology demands that the investigator attempt to free himself from any vestige of the traditional predilection for stability, equilibrium and causal invariance derived from the

socio-behavioral sciences' roots in post-Revolutionary French social theory [4]. Rather, every social-science phenomenon should be treated (again until empirical investigation indicates otherwise) as "emergent" in potential, so that normatively desired states of conservation, homeostasis and adoption become merely referential rather than preferential conditions. But neither is the dialectical investigator allowed to hold any unaverred axiological preference for revolutionary as opposed to evolutionary dynamics [5].

5. Finally, as perhaps the most striking aspect of the dialectical epistemology, the investigator is asked to initially consider the possibility that the behavior of his subjects is determined under "force vector" conditions, as the constantly shifting product of possibly competitive forces, whose relative strengths are allowed to vary in potentially indeterminate ways.

It is this set of dictates, then, that serves to distinguish the dialectical perspective as it might be employed by the student of societal systems. The substantive assumptions serve the paradigmatic cause of narrowing an initially ill-bounded search space through a set of hyptheticodeductive assumptions. These initially constrain the investigator to certain paths of inquiry, and direct his attention to certain determinants among the universe available in the initial stages of an investigation. But the basic unit of analysis is more encompassing (less restricted *a priori*) than those employed by certain other schools of psychology—especially those that deny volitional or endogenous elements any role, or disallow significant discontinuities between past and future behaviors. In short, the constraints are largely procedural rather than substantive, as is certainly to be preferred in an epistemology with pretentions to scientific integrity.

Any set of methodological prescriptions will have to be fully responsive to the epistemological dictates just set out if a proper paradigm is to be the result. But, in addition, there are some other implications of the dialectical methodology that should be made explicit. First in developing a dialectical methodology, we have an opportunity to devise instruments such that structural and dynamic analysis can be conducted using the same unit of analysis. Secondly, it is possible, given the dialectical focus, to develop instruments that do not distinguish between individual and collective units of analysis, but are capable of handling both within a single frame of analytical reference. Thirdly, as an adjunct to the syncretic procedures discussed earlier, the dialectical engine can unite now isolated subdisciplines (e.g., link social psychology with personality and developmental psychology), and at a still higher level point out the terms of the tradeoffs between the major dimensions of any societal system: between sociological, economic, political and behavioral benefits of collective membership. These ancillary aspects of the dialectical method will, however, have their day later in our work. Here we turn to the matter of determining configurations.

DIALECTICAL CONFIGURATIONS

In the broadest sense, the dialectical system analyst sees all societal phenomena as collisions of separate and possibly competitive interests. When the collision is among essentially compatible forces, we get a synergistic system; when the collision leads to conflict, we get pathology or anomaly. In more operational terms, every societal phenomenon becomes comprehensible as a force vector—the moment-in-time resolution of a potentially large set of forces, with the resultant relationships running the gamut from competitive to complementary. In modeling language, societal subjects are defined as configurations in n-space, where n is the number of different factors (or state variables) affecting the system. Societal systems thus become, again, composites of cultural, material, spiritual agencies, etc., all of which must be considered simultaneously. The individuals of which societal systems are composed—the lowest-order components—become intelligible as driven, at any point in time, by a potentially complex set of cognitive or behavioral forces, with empirical behaviors ultimately defined as temporal resolutions of the force-vector components (with the dialectical engine setting the priorities that will determine the final behavioral configuration.

Procedurally the dialectical method is very simple, and has strong links to contemporary social-science practices. In particular, we are concerned always with the explanatory or heuristic potential of dichotomoies. The dichtomization of properties corresponds to the pseudo-axiomatic base of generalized dialectics, the concept of *thesis* and *antithesis* being defined as extreme positions on some dimensional continuum. The next step in the process is to consider the construction of some sort of iconograph, where each of these several continua represents the several forces on the system at hand. When we then assign numerical values or qualitative attributes to these continua —when we *scale* them—we have a measure for representing the force-vector configuration associated with the subject under study. We plot the system's properties within the confines of the iconograph, and hence define the configuration of a societal phenomena in more or less dimensioned terms. And it goes without saying that the iconograph, as a geometrical or topological construct, is usually easy to transform into symbolic or mathematical terms.

The process of arriving at values to be represented by discrete intervals on the continua is a difficult one. Here I can only mention that when the polar positions are properly defined—as thesis and antithesis—so that the properties of the one are entirely absent in the other (or are the logical inverses of the other's properties), then it is possible to use surrogate techniques to develop intermediate positions with some precision. For example, the polar positions can be converted to "fuzzy sets" [6], or other disciplined qualitative representations. Any movement along the continuum will then represent an increase in emphasis on one set of properties at the direct expense of the other: playing a zero-sum game on the continuum. Thus, every point of the continuum (we use as many as our level of precision will permit) becomes uniquely identifiable as a surrogate proportion or ratio—the presence of properties from the thesis relative to properties from the antithesis. In evaluating the propriety of our antithetical positions, we have at our disposal the various logics that can assist us in validating the fact that our two polar positions are true inverses (or perhaps complements, etc.), and in assigning logical surrogate values (e.g., binary strings representing ordered presences or absences of specific properties) to the intermediate positions. For example, an "or" between the property sets of the two polar positions should produce a string of ones; an "and", a string of zeros indicating a complete lack of correspondence between the two sets of properties.* Thus, while the instruments for qualitative analysis are not necessarily simple to operate—nor completely mature as yet—our exercises in partitioning qualitative continuua into adequately precise intervals are not completely beyond us. Indeed, when we later define certain societal ideal-types, we shall array their properties with these logical criteria in mind.

To lend some substance to this discussion, we can develop a highly simplified typology to establish a set of continua and poles, which will here be used for illustrative purposes only. In table 2.2 I employ five dimensions, which are not intended to be exhaustive or even particularly adequate. Beside each of the dimensions is a pair of terms which I hope will be accepted as reflecting antonymical positions—thesis and antithesis, repsectively. In a later section, when we try to define the concept of "culture" in a relatively formal way, we shall have occasion to talk about each of these terms in more detail. Here I am interested

*It should be noted that neutralizing the properties of the thesis and antithesis extremes would locate the third member of the dialectical triad, the *synthesis*.

TABLE 2.2: Sample Typology Setting Out Relevant Dimensions and Antithetical Positions

Dimensions	Antonymical Forces	
	Thesis	Antithesis
1. Social	Alienated	Insitutionalized
2. Political	Consensual	Coercive
3. Axiological	Aristotelian	Socratic
4. Economic	Acquistive	Ascetic
5. Cognitive	Rational	Psychogenetic

simply in setting out the multidimensionality of the cultural context, so the precise meaning of the terms can wait [7]. Through the qualitative analytical instruments mentioned earlier, each of these dimensions may be converted to a coded continuum with the specific paired terms representing polar conditions. A specific subject* thus becomes intelligible—at a moment in time—in terms of the specific values assumed on each of the several continua. A structurally integrated "map" of the subject's configurational state its "personality" is then available when we can house the several dimensions within a single analytical framework, using the specific continuum values as base reference points. By way of example, we can develop a highly simplified "map" using only three dimensions (social, economic and axiological), allowing us to remain within the Euclidean space (representations requiring more than three dimension would suggest topological constructs). The map takes the form of a simple iconograph in Figure 2.4, where three different configurational resolutions are pictured—the planes X, Y, and Z.

While the precise definition of the various terms used in the figure must wait until later sections, there are some specific remarks to be made about the iconograph itself. First, the continua have been imposed on the iconograph in such a way that a forced "clustering" takes place. In practice, this means that we try to arrive at an *a priori* expectation about which particular attributes among the several dimensions are likely to be *co-present* in a subject, and order the continua accordingly. In the present case, the expectations are that a Socratic axiology will tend to be accompanied most often by a tendency toward social institutionalization and economic asceticism (or subsistence). Conversely, it is expected that an Aristotelian axiology will usually be found in association with an acquisitive economic posture and con-

*Note that "subject" here could mean anything from an individual to an institution or, at the extreme, the societal system itself.

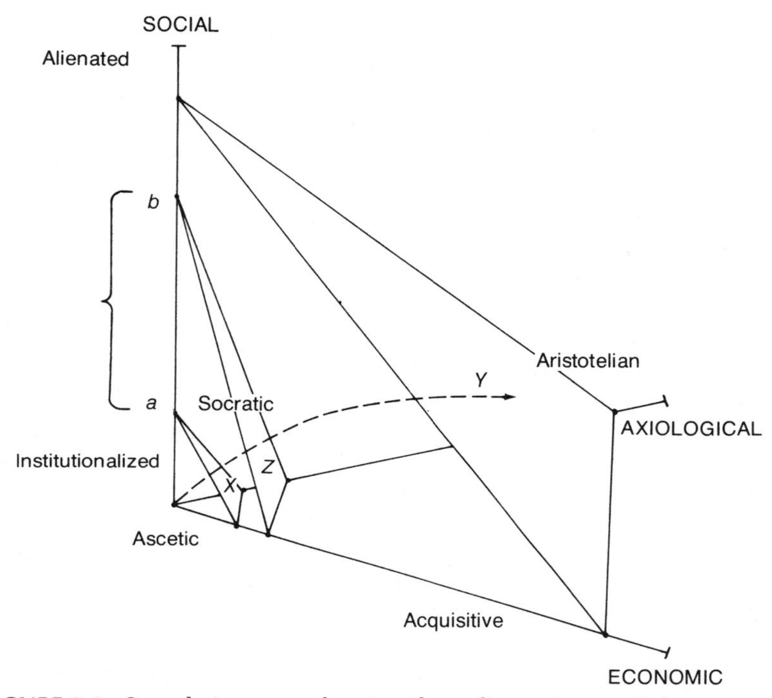

FIGURE 2.4: Sample iconograph using three dimensions and three cases.

sequent social individuation. No attempt is made to force a causality—that is, there is no *a priori* position about one dimensional attribute being the cause of another. Moreover, assertions about accompanying or correlated attributes are probabilistic and plastic rather than axiomatic or fixed. Thus, the clustering is done because the typologies upon which dialectical models are built are perforce metahypotheses. Specifically, if a large number of empirical cases we impose on the iconograph tend to cluster around one or another of the extreme cases identified, or otherwise show a high density for certain attribute associations (subsets), then we have at hand the basis for constructing ideal-type referents which may be expected to exhaust the majority of real-word cases we shall encounter. Thus, by structuring our analytical instruments in a way that reflects normative or theoretical deductions,* as with the expectations about clustering of certain attributes in the above example, we have at hand a mechanism by which the accuracy of

*I might add here that, in a formal dialectical analysis, the generation of the deduced correlations—and the selection of the state variables themselves—would be disciplined by the use of a Delphi process, consensus statistics or some other device that allows us to determine the logical (judgmental) probability of accuracy of these model components.

our theoretical positons may be judged empirically. This, again, is the strategy underlying the development of apodictical models.

Specifically, then, we have two extreme cases and one intermediate case indicated on the above iconograph, associated with the three planes marked X, Y and Z. The X-plane joins points on all three dimensional continua which are quite near the origin, while the Y-plane represents the behavioral state of a subject whose behavioral properties on the three dimensions are the effective *antitheses* of the properties for X. Thus we have isolated two "personality" types. The intermediate case, Z, may be thought to represent an anomaly from the standpoint of our *a priori* expectations, for we would not expect that an individual tending toward an ascetic economic posture and a Socratic axiology would simultaneously exhibit symptoms of social alienation. In short, case Z is an instance of theoretical incongruence, whereas the polarized configurations (X and Y) represent "congruent" cases. Symmetrical configurations, generated by attribute positions on the several dimensions which are roughly equidistant from the origin, generally indicate instances of configurational congruence. Strongly asymmetrical configurations would represent instances where there is an apparent conflict among a subject's predicates or referents, at least when arguing from the standpoint of the theoretical propositions used to generate the continua in the first place. But, again, the extent to which *a priori* propositions are validated—and hence the extent to which meaningful "classes" of personalities or phenomenal ideal-types may be generated—will depend upon the extent to which a clustering around one or several configurations occurs during empirical mapping exercises.

With respect to configurations, the "degree" of difference between behavioral states (e.g., between the psychological inventories of two individuals) may be indicated by the relative distances between the central points on the planes defined for the subjects (with respect to the origin of the iconograph). The "nature" of the differences between two personality sets is reflected in planar disparities—the differences in the morphology of the resultant configurations. Similarly, the disparity between one subject's position (attribute value) on any dimension and some other subject's position for that dimension is largely a matter of reflective measurement. For example, the distance *ba* on the iconograph measures the magnitudinal difference between the social postures adopted by case X and case Z. The degree of incongruity in a behavioral "state" may be measured by noting the differences—with respect to the origin of the iconograph—of the points on the several dimensions which indicate a subject's assigned attributes. Congruent cases will have slight differences, and ideally no differences (because the points describing that individual's position on the various dimensions will all be equidistant from the origin). The locus of greatest difference—that particular dimensional attribute which is most distant from the others with respect to the origin—might represent the personality area in which the therapist might want to begin work. In short, the dialectical epistemology seems to lend itself rather well to a methodology for developing typologies of direct theoretical significance, and using tabular, numerical and iconographic methods in a feedback-based therapeutic process. Hence its managerial implications, implications we shall spend much time exploring in later chapters.

It remains to be said here, still within the realm of configurational analysis, that the techniques we have developed for the highly simplified three-dimensional case could be extended to more realistic studies in n dimensions. What would alter is not the basic analytical exercises involved, nor the fundamental nature of the results sought, but the technology involved in moving from a simple Euclidean three-space to topolgical representations. As the number of dimensions increases, the sophistication of the representational and manipulational tasks increases accordingly (in numerical and mathematical content as well as the logical operations required). But this is a price many of us may willingly pay for the closer approximations to reality that we expect.

DYNAMIC IMPLICATIONS OF THE DIALECTIC

Initially, consider that each of the various numerical or iconographic representations developed for a subject under configurational analysis represent moment-in-time maps of that individual's multidimensional behavioral state. The dynamic component of behavior will then be made available to us in terms of changes which occur—both on the individual dimensions and in aggregate—in a succession of these structural maps produced over some interval of interest. Thus a dynamic dialectical analysis would find us "overlaying" sequentially produced static maps and attempting to codify or functionally describe the trajectories that these changes prescribe. As we now know, surrogate values are available for measuring the magnitude of changes along any individual dimension, and the aggregate change would be expressed in terms of some arc or trajectory described by the configuration moving through n-space (e.g., for the simple three-dimensional case, the trajectory was the path, measured with respect to the origin of the iconograph, of some appropriately defined midpoint on the triangle defined by the values assumed on the three dimensional continua). Thus, through either numerical, geometrical or topological techniques—depending on the number of dimensions—we have an opportunity to capture and summarize behavioral changes simply by reemploying the structural instrument already described, and making sets of sequential comparisons.

Our ultimate ambition here is the same as it was in the structural case. We want to design an instrument which will first of all translate the dialectical epistemology into an usuable set of procedures. This instrument should, secondly, give us the opportunity to isolate any distinct, often-repeated behavioral patterns which can then serve as ideal-type referents (and thereby directly provide dialectical system science with an efficient theoretical base). It is immediately apparent from our reading of the dialectical epistemology that the evaluation of any particular behavioral pattern as functional or dysfunctional—positive or pathological—can only be judged with respect to the subject's milieu. And though we shall have much more to say about this later, I want here to introduce the logic of functional (dynamic, behavioral) congruence, just as I introduced the logic of configurational congruence in the last section.

We can, specifically, conceive of a set of ideal-type behavioral patterns arrayed on a continuum. The particular patterns (the referents) would represent those cases which we expect *a priori* to be most often approximated by real-world behavioral patterns, i.e., they are the "hypothetical" ideal-types whose validity will have to be supported or rejected by empirical exercises conducted under their guidance. Although the individual practitioner may choose any of a variety of dimensions to structure this continuum of behavioral classes, I will choose one of significance for the model-builder—the inherent simplicity or complexity of the pattern as indicted by its amenability to mathematical modeling. At one pole, we would expect to find a behavioral pattern which is easily represented by a usefully simple, deterministic equation (or set of equations). At the other extreme, there might be no way to mathematicize the function, it being too complex and ill structured. Keep in mind that these patterns would all be generated within the context of our sequentially analyzed structural maps, and that these patterns of change are as relevant to the entity in aggregate as to the entity operating on individual dimensions (which are always intelligible in two-space, whereas the aggregate entity operates in n-space). Thus the technology of capturing and codifying the dynamics of the behavioral process will depend on our level of analysis, but the ideal-type referents I shall now define, and the interpetational logics, are level-independent.

The continuum and classes I shall use to initiate the dynamic analysis are again merely for illustrative rather than substantive purposes. The sample continuum is presented in figure 2.5, and I shall

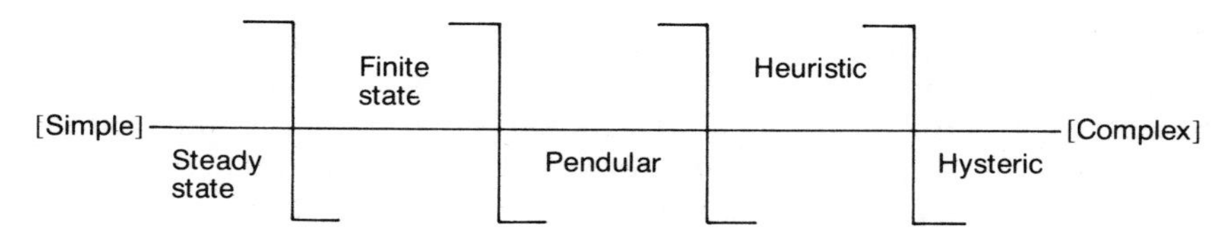

FIGURE 2.5: Sample continuum of behavioral ideal types.

briefly define the various classes in order of their apparent simplicity.

The steady-state modality, as indicated by its position on the continuum, is the simplest behavioral pattern any entity (individual, organization, society, etc.) might be expected to exhibit. Its generic properties have been defined in various contexts by many system authors [8]; we can provide specific apprecation here. Particularly, in terms of behavioral dynamics, a steady-state modality is indicated by gradual convergence on a particular set of behavioral properties that give an indication of remaining stationary in the future—e.g., the differences between successive structural "maps" prepared for some entity are secularly dampening. The finite-state modality, on the other hand, shows a somewhat wider repertoire of behavioral alternatives (i.e., map configurations), but each of the behavioral alternatives is paired invariably with some unique stimulus or event. In short, for each specific starting-state condition (stimulus) there is one and only one behavioral response which emerges. Simply, the finite-state modality is the behaviorist's "null hypothesis", and can be both adequately and efficiently explained by devices such as operant conditoning and its variations.

In the pendular modality, an individual's behavioral states make wide, dramatic and rapid swings back and forth between antonymical conditions, between thesis and antithesis. The pendular process might very well describe certain aspects of "adolescent" behavior, with its flavor of all-or-nothing experimentation. It seems a very appropriate referent for the dynamic behavior of socio-economic or socio-political systems, with shifts in *a priori* referents causing rapid inversions of behavioral postures. Synthetic (syncretic or mixed) states are seldom realized and apparently seldom sought. The net result is a behavioral pattern that is reverberative, but of definite consistency. We shall have much more to say about pendular societal behavior in the next chapter, as it is a critical corollary to the dialectical perspective we are developing here.

The fourth of our *a priori* dynamic referents, the heuristic behavioral modality, emerges as a highly disciplined trial-and-error process with a strong "learning" component (indicating a significant feedback between concept and percept, between ambition and experience, etc.)

There is a constant flow forward toward unprecedented original positions (e.g., the behavioral maps show a constant change toward nonrepetitive structures). But changes are such that it is apparent that the subject has some formal *criterion* for selecting a specific trajectory from among the set of alternatives which presents itself at all points in time. Even if the actual paths selected are indeterminate, the subject apparently has a "method" of choosing which is crudely rationalizable (at least in retrospect). He is equipped with a formal "navigational" system which operates to constrain the search space even in a sea of events or properties entirely unprecedented.

The final modality—the hysteric trajectory—could be conceived of as a product of "undisciplined" trial and error, where the feedback and learning mechanisms are impaired. Even dysfunctional (painful or unproductive) responses tend to be repeated, and no underlying method of procedure or progress can be uncovered. The hysteric subject seems to wake up in a brand new world every morning.

These, then, are illustrations of how an *a priori* continuum of ideal-type behavioral references might be constructed, here employing a dimension of amenability to mathematical modeling. Other dimensions may be employed without altering the fundamental logic that all the referents should be reducible to the same essential terms (e.g., have a common denominator) such that intervals may be defined which bear immediate comparison. It must also be suggested that the continuum against which the behavioral referents are to be juxtaposed—the continuum of ideal-type environments or milieux—might also be set in the same terms as the former, as it is partly the properties of the subject's environment that are used to determine the congruence or incongruence of the behavioral posture it adopts. In our example, then, a continuum of ideal-type environmental referents may be established on the basis of their relative amenability to mathematical modeling, and the particular alternatives that interest us are indicated in figure 2.6.

I have described environmental properties elsewhere in some detail [9], and a detailed understanding is not really important for us here. So we may be brief. In the placid milieu, properties are very simple, are

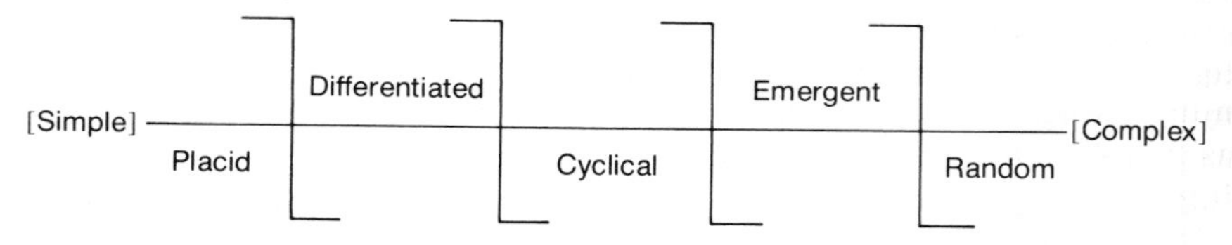

FIGURE 2.6: Array of ideal-type environments on a dimension of analytical tractability.

distributed symmetrically across the field, and change only insignificantly through time. The differentiated milieu is slightly more complex in structure. The properties are variegated and heterogeneous (e.g., a greater diversity of resources, dangers and nutrients). So the distribution algorithm is more complex. It must be modeled using statistical or probabilistic components rather than the deterministic ones appropriate for describing the placid milieu. Properties tend to appear in *clusters*, so that the resident in this milieu must be expected to employ some cognitive instruments in moving across the field if he is to avoid perilous circumstances or locate advantageous niches. There is also the possibility of some limited changes through time in both the properties themselves and their distribution across the milieu.

The "cyclical" environment is interesting primarily in its dynamics. Wide, dramatic alterations are involved which more or less completely alter the properties of the milieu, but these changes are predictable largely because they are periodic and iterative (much as with the various seasonal changes, which essentially invert climatic parameters). These cyclical shifts may be either exogenous or autochthonous in origin, and may be likened roughly to a succession of thesis-antithesis replacements. It should be clear that some cognitive capabilities are required on the part of resident entities, primarily the ability to develop projective inference functions which use leading indicators or some other variables as statistically significant harbingers of change.

The "emergent" milieu is one which is constantly in flux. But the changes that take place lead us to suspect that the environment is becoming something definite rather than simply reverberating or resonating without apparent method or purpose. Neither the structural nor the dynamic properties are fixed, nor are the various states which emerge entirely precedented. At any point in time, the resources of the milieu appear to be ill structured and unamenable to either deterministic or statistical modeling. Yet certain secular properties may be isolated from the superficially non-methodic dynamic functions, and it is these which serve to differentiate the emergent field from the "random" (where changes appear to be entirely without pattern, purpose or precedent). In the strictly random milieu, the past is absolutely no clue to the future, nor can any retrodeductions as to causes and effects be made (as they can with the emergent field). In the case of a random milieu, there is virtually no type of "intelligence" which can be exercised to any advantage. It is, in essence, an environmental *maze*. In the emergent milieu, however, the lack of fixity and the protean quality of the environment suggest a need for alertness, responsiveness, analysis— and a strong dose of resilience and humility. In this milieu, then, the

TABLE 2.3: Evaluated Conjunctions Between Behavioral and Milieu Referents

Modes of Behavior	Milieu Types				
	Placid	Differentiated	Cyclical	Emergent	Random
1. Steady-state	X			Pathological	
				Ineffective	
2. Finite-state		X			
3. Pendular			X		
		Inefficient			
4. Heuristic				X	
5. Hysteric	Pathological				X

cognitive demands on the resident entity are the highest of all our several ideal-type referents. In short, the entity in the emergent milieu must be capable of "strategic" and highly ordered deductive reasoning; merely statistical or associative capabilities are insufficient here. On the other hand, in the random milieu (very rare in the real world), crude trial-and-error behavior is apparently our best and most "rational" recourse.

When the contextual (milieu) factors and the behavioral modalities are both set on continua, we have the basis we need to develop a typology to house and summarize *a priori* assertions about congruent or incongruent dynamics, much as we did in dealing with configurational aspects in the previous section. For the illustrative behavioral and milieu types we have been using, such a typology is given in table 2.3. The main diagonal of the typology represents cases of congruence, where the behavioral modality adopted by the resident entity responds functionally to the properties of the milieu. Although a full defense of this construct is impossible within the scope of this chapter, I have tried to show that the cognitive demands increase for each of the milieu types, until we reach the random case (where no cognitive function promises to be rewarding) [10].

Immediately off the main diagonal we run into two areas which I have marked *inefficient* and *ineffective*. Inefficient pairings of behavioral modality and context would occur when the individual is employing a behavioral modality that is more sophisticated than the milieu really warrants. It is a case of "overkill", or simply a waste of either physical or cognitive energy. The ineffective combinations are more serious, for they impair the fundamental integrity of the entity. Thus, for example, a subject who insists on treating an emergent environment as if it were placid is going to be in for some rude (if not fatal) surprises. At the extremes of the incongruent combinations in the typology we may properly call the situation pathological: the behavioral properties of the entity are so far out of line with the properties of the milieu that successful or even adequate functioning becomes highly improbable. Thus, for example, a subject adopting a hysteric posture in the face of a placid environment must be classified as seriously impaired (i.e., manic); and the adoption of a steady-state behavior in the face of an effectively random environment may be interpreted as self-induced sensory deprivation, also a distinct pathology [11].

Were we now able to evalute a large number of specific real-world behavioral cases, and isolate the milieu properties in which they were resident, we could hope that there would be a clustering of cases along the main diagonal (the vector of congruence) and a dampening of the density functions as the incongruent or pathological domains are approached. We would thus want to emerge from a series of empirical analyses with a strongly peaked "normal" distribution. Were a dialectical system scientist indeed able to devise a set of continua, dimensions and types that would cause such a distribution to be generated (as I have often mentioned, the specific continua, dimensions and classifications I have been using are for illustrative purposes only), then I suggest that this would be tantamount to a *validation of the utility of the proposed methodology* for dynamic analysis (just as a clustering of configurational types would validate the structural method earlier outlined). And, as I hope the reader recognizes, such a validation would prove most auspicious indeed for the future of an integrated social science.

As our concern here is primarily with methodological issues, I shall only briefly point out that our knowledge of the interface between pendular behaviors and cyclical milieux is most limited (and, indeed, we shall be trying in later chapters to improve it). We may speculate here that even a placid milieu can quickly become cyclical when occupied by entities exhibiting pendular behavior. We may also note that many societal contexts tend to be cyclical. And we must, at some point, seek to explain why social, political and even intellectual activities seem so often to reflect this particular pairing rather than more productive alternatives that are generally available. It must be hoped that a strong dialectical presence in psychology will yield the answers, simply because at this juncture it alone would ask the questions. For this is the only behavioral/milieu pairing that is really not adequately treated by

TABLE 2.4: Typology of Conjunctive Referents and Associated Subdisciplines

BEHAVIORAL MODALITY	MILIEU TYPE	ASSOCIATED SUBDISCIPLINE
1. Steady-State	Placid	Cybernetics (homeostatic theory)
2. Finite	Differentiated	Behaviorism (operant conditioning)
3. Pendular	Cyclical	Dialectical psychology
4. Heuristic	Emergent	Cognitive (developmental)
5. Hysteric	Random	Abnormal psychology

an existing subdiscipline, as the typology in table 2.4 suggests.

From this table we see that the dialectical psychologist must take major responsibility for analyzing and explaining conditions at the interface between pendular behaviors and cyclical environmental settings. Hopefully, our work in Chapter 3 will suggest that a comprehension of the substance and dynamics of societal systems depends in large measure on just such an analysis. For we shall see that the development of societal systems is essentially intelligible as a cyclical process, and that pendular behavior is perhaps the dominant modality among a majority of societal populations. In short, the history of society is not really the evolutionary one that many popular theorists propose, but a *pendular* one. And the cause of this pendularity rests, in large measure, with axiomatic and axiological predicates, and with the role played by the societal prophets. But before moving on to strike directly at the problem of pendular behavior and cyclical societal contexts, it is well to take another look at the cognitive context, this time using our dialectical perspective to amplify some of the points we raised in Chapter One.

The Cognitive Corollaries

Because of the work we did earlier, the cognitive implications of the dialectic are already familiar to us. In fact, the basic components are those we defined in some detail: the inductive and deductive engines (the data-driven modality was, clearly, a non-cognitive engine). In the following discussions, then, we are concerned about approaching the inductive and deductive modalities from the dialectical perspective, so that "data" here refers only to direct predicates (i.e., contextual information of empirical, perceptual origin). We shall reach essentially the same conclusions about the inductive and deductive behaviors as we did in the previous chapter, but will here arrive at these conclusions by a somewhat different route. Let us begin with two propositions of distinctly dialectical flavor:

1. The cognitive state of any entity may be represented, at any point in time, by the position its attributes are assigned on three conjunctive continua:
   i. One concerned with the extent to which an individual's behavior is predicated upon a data base (empirical, perceptual) as opposed to a model base.
   ii. One concerned with whether model-based behaviors (to the extent they exist) are inductive or deductive in predication.
   iii. One concerned with the apparent relative influence of the "right" and the "left" brain (where "right brain" indicates allegorical processes and "left brain" algorithmic functioning).
2. If these three dimensions join to produce an instantaneous map of an individual's cognitive state, then the dynamics of cognitive phenomena are accessible to us by a functional modeling of successive changes in the structure of the maps produced over some interval.

These propositions lead, as is to be expected, to another instance where the iconograph may be used as a tentative investigatory heuristic, as shown in figure 2.7. Again, as with our iconograph in figure 2.4, we have a "cluster" effect, such that ideal-type *planes* are formed for three different individuals, with the extreme cases (*X* and *Z*) representing instances of thesis and antithesis, respectively (just as the extreme points on each of the three continua represent conditions of dimensional antithesis and thesis). Again, because of our work in Chapter One (and particularly the summary presented as table 1.1) the following cases are already familiar, and are really just restatements of previous discussions:

CASE X The subject derives most of his inputs from sense data exclusively, without ordering these data into more sophisticated structures such as a symbolic set or a *Gestalt* (where multiple-origin percepts and concepts are structured into an integrated *system* of arguments representing a complex event or stimulus). Given these sense data as input arguments, no cognitive mediation intervenes between the input (stimulus) and response (behavior), the processing being conducted largely at the limbic as opposed to the cortical level. Case *X* would then represent essentially simple behaviors, such as primitive "drives" and the essentially semi-automatic and semi-autonomous functioning of the nervous, respiratory or vascular systems. This is the lowest

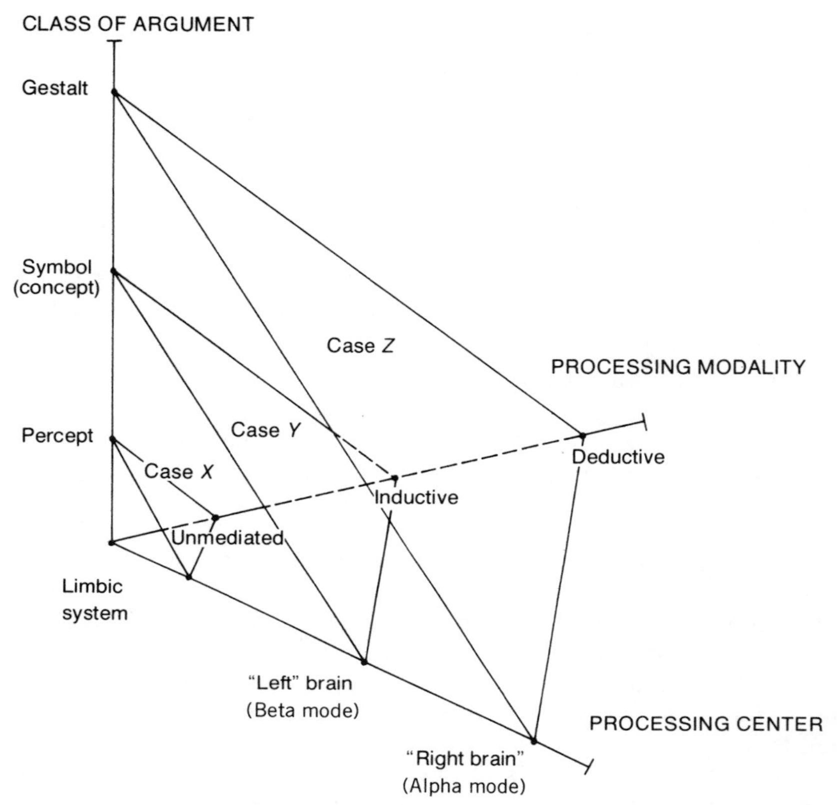

FIGURE 2.7: The cognitive dialectic in iconographic format.

level of behavior available to man, but the highest available to certain low-order animals and seriously impaired psychological casualties.

CASE Y The subject here is capable of translating perceptual inputs into inferences by employing the inherent inductive capabilities, using the amenability of the left brain to certain types of both simple and complex conditioning—experiential-based, empirical learning. Where the form of the input argument is symbolic, the left brain, operating again according to the *a priori* inductive mechanisms, is capable of operating analytically, using the fundamental forms of logic. In this model, language and associative/statistical behaviors are determined. The processing is apparently accomplished linearly, sequentially and algorithmically. In short, here we have the Lockean-Humean-Skinnerian man.

CASE Z The subject here is not totally dependent on the outside world for his inputs; they can be endogenously structured in the form of *Gestalten* which combine experiential, empirical and "original" components into conceptual structures which may have no precedent or source in the natural (external) world. Recent research suggests that the "right" brain is primarily concerned with the processing of these complex units of cognition and also suggests that the processing is neither linear nor sequential. These *Gestalten* are processed by the "right" brain using its apparently deductive (as opposed to inductive) rules. They give rise to behaviors which apparently have no precedent—or, more strictly, which are improbable when viewed solely against the array of exogenous stimuli or inputs available to the subject (his genetic, experiential or empirical informational predicates). At the positive extreme, these essentially deductively driven behaviors are classified as instances of genius, as they add something to the world that was not there before. On the other side, these behaviors become the basis of pathological disorders.

Thus, from the dialectical perspective, we visualize a *range* of behaviors available to man—and a range of cognitive mechanisms—which I think is a considerable improvement over trying to force some single paradigm to do all the work in psychology, that for which it is ill equipped as well as that which it does efficiently. But, as the reader will recall, this is exactly the same conclusion we reached in Chapter One, and essentially the problem which the sycretic paradigm was posed to partly solve.

THE DIALECTICAL DISPUTATIONS

In a broader framework, the dialectical position we have adopted would question popular social-science presumptions in three areas, all of them interrelated:

1. The tendency to view all systems as instances of cybernetic behavior.

2. The tendency to equate the quality of a system with its consistency and homogeneity.
3. The use of a fundamental unit of analysis predicated on "shared values".

These three tendencies are related, and all have their origin in the preferences of the inductive-instrumental basis of empiricist social and behavioral science. As exhibited by the strict behaviorist, for example, there is a tendency to view all societal phenomena as purely adaptive and reactive, so that society is the slave of its environment or the victim of circumstances. That is, the possibility of *a priori* predicates determining system or individual behavior is *a priori* excluded. Now, to the extent that all systems are assumed adaptive, the fundamental mechanism of behavior is presumably cybernetic. The system's activities are thus determined by the type of mechanisms we described in Chapter 1. The rationality of a system, or its quality (past or present), thus becomes intimately tied up with its ability to maintain an in-phase relationship with the emergent properties of its environment. In short, a system is "good" to the extent that its properties reflect the properties of the environment. The system is denied *a priori* the opportunity to affect the environment. Causality is thus a one-way street: all the determinants of system structure and behavior are exogenous (just as all the determinants of human nature were presumed exogenous by the behaviorist school). We have already disputed this point of view, and elsewhere I have suggested in detail that there are practical limitations on the cybernetic process which virtually prevent it from being the origin of most real-world systems; the cybernetic process simply cannot accommodate the complexity of the milieux in which most non-primitive systems reside [12]. Thus, again, the dialectical perspective insists that until actual investigation proves otherwise, we allow systems (and individuals) the opportunity to determine their own environment—to structure, in part, their own milieu. There is nothing to gain from presuming one-way causality except theoretical neatness, and this theoretical neatness is often gained at the expense of scientific integrity.

Now the preference for presuming a cybernetic basis for all societal systems leads directly to the second proposition of traditional social science: the tendency to equate system quality with consistency and homogeneity. Consistency here refers to the behavioral (dynamic) dimension of systems. It means stability of system structure through time. There is thus an inbred distrust or devaluation of system change, or at least a derogation of the utility or therapeutic quality of change. Alvin

Gouldner, in a very impressive and erudite study of sociological history and methodology, has traced this preference for a stable system back to the post-Revolutionary French origins of western sociology [13]. The early French social scientists were intimately acquainted with the chaos and disruption brought about by changes due to the Revolution of 1789, and became intensely interested in the role of tradition, stability and the *status quo* as system determinants. At the same time, they tended to view change as dysfunctional. The "ideal" system, from their point of view, thus became one which had the ability to maintain itself in a given structure despite external perturbations--one capable of exercising homeostasis and maintaining equilibrium. Now, homeostasis and equilibrium are distinctly cybernetic referents, so we have a strong axiological link with the formulators of behaviorist psychology and modern sociology, not to mention classical economics.

Once the preference for a stable system was established, two very interesting prescriptions were evolved which have much current currency. First, there is the assumption that the perils to system integrity—the sources of change, challenge, etc.—all exist *outside* the system [14]. Hence the common perception of systems as homeostatic, attempting to maintain internal (endogenous) consistency in the face of continual external pressures. Thus, little attention has been paid to the internal sources of disruption. An amazing situation has resulted: the basic defintiion of a social system is built around the concept of a group of individuals united by what is commonly referred to as a set of "shared values". The presumption here is obvious: a system which is homogeneous (whose members are all similar in terms of their "cultural" base) has a better chance of resisting change and challenge than a system whose members have diverse cultural referents. Thus, among both ancient and modern sociologists, there is a too frequent tendency to equate societal quality and probability of survival with homogeneity of culture. And, as suggested, this preference is so deep-seated that the basic unit of sociological analysis is often defined as a system that indeed has such homogeneity. In short, by defining a social system in terms of shared values, we automatically exclude endogenous perils and *a priori* lend credence to the proto-sociological assumption of the value of consistency and the exogenization of perils.

Now, just as our dialectical position was reluctant to accept the presumption that the origin of all causality lies outside the system, so must we be reluctant to accept the assumption that the origin of all challenge, peril and change lies outside the system. The dialectical perspective demands that we define a new unit of analysis, one not making any automatic assumptions about homogeneity. For homogeneity, like any other system property with which we shall be concerned, is properly a *variable* and not a defining attribute. To try to define a unit of analysis with the essential property already built in is a great disservice to science and society [15]. Therefore, for the remainder of this book, we shall consider a societal system to be defined *as a collection of individuals or objects whose expected value of interaction exceeds some critical threshold* (where expected value is a product of probability and intensity of contacts). This basis for defining systems allows us to treat those entities which behave like systems, but have diverse cultural bases—which do not encompass a membership all predicated on the same or a similar set of *a priori* concepts or percepts. More importantly—and certainly more realistically—the potential for conflict and dissolution may now exist *within* a system. And the potential for resolving conflict is no longer restricted to vapid and often infeasible pleas for creating homogenous units [16]. Indeed, as we shall later try to show, one of the critical keys to the improvement of the quality of systems lies not in their reduction to units where all components are similar (e.g., clusteration), but in the development of methods to exploit and amplify diversity—in the development of "syncretic" societal systems. At any rate, we shall have ample opportunity to elaborate on the implications of homogeneity and heterogeneity in later chapters.

In summary, then, the three propositions supported by the dialectical platforms are these: (1) causality of system structure and behavior must be presumed to be Janus-faced (both external and internal) until experimentation or observation indicates otherwise; (2) the potential for change cannot be presumed to rest solely outside systems, nor can homeostasis (adaptivity) be presumed to be the *only* rational form of societal strategy; and (3) the basic unit of analysis for system studies, and particularly for inquiries into complex systems must be defined in non-normative terms—specifically, we have elected to employ a unit predicated on probability of contact rather than on the more popular concept of shared values.

As a final note on the dialectic perspective, let me state that it should serve as the link between our comprehension of individual behavior and our emergent concept of collective or societal behavior. Much of the preceeding discusson in this chapter was meant to forge such a link, but we may now be a great deal more explicit. For, to a great extent, this present section will mark the point where we cease to talk about the individual as an entity unto himself, and begin to talk about him as significant only within some societal context.

The key for us will be one of the simplest of all possible variants of the generic dialectic perspective: the relationship between aspirations and achievements. Intially, *aspirations* may be thought of as representing some set of properties which serve as referent targets for an individual's behavior. A referent, in our framework, may be virtually any form of ambition. It may, for example, be an ideal-type personality toward which the individual strives (e.g., I want to be a calm, disciplined person; I want to be a spontaneous, warm person). It may involve a material ambition which may be expressed quantitatively (e.g., I want to be number one among the world's figure skaters; I want to be a millionaire by the time I am thirty; I want to have a biceps that measures eighteen inches; I want to be president of my company or dean of my faculty). Or it may involve the postulation of some "state" which is expected to hold some sort of fulfillment (e.g., I want to be Mary's boyfriend. I want to be secure; I want to loved, or admired, or respected, or feared, or mysterious—whatever).

Thus, the range of aspirations that may be defined by an individual is great. But we must generally confine ourselves to instances where two properties are present: First, aspiration proper is a product of cognitive mediation, or deliberate, conscious definition. Second, it must define a property, condition or state which will demand some exertion if it is to be achieved. In this sense, then, aspirations do *not* define needs or urges which develop in the physiological or id region (e.g., data-driven factors), nor can they define tautological ambitions (e.g., I want to be myself) or states which cannot be defined precisely enough for us to know when we have attained them (e.g., I want to be "self-actualized"). Thus, in addition to the above, a true aspiration must be definable in concrete terms such that we may know when it has been attained; or at least it must be possible to specify, in terms of reasonably unambiguous properties, the condition of the situation toward which we are aiming. Many of the popular psychological cliches would be eliminated by these criteria: happiness, self-actualization, being oneself, having good feelings about other people. Either they lack sufficient precision (mainly because they specify not conditions, but compound states whose components are ambiguous), or they do not involve directed investment of energy or intellect.

In the simplest terms of the dialectic, aspirations thus set targets, and achievements are the real or perceived positions we occupy (at some specific point in time) with respect to the aspirations. In its rudimentary form, the dialectic thus poses human existence as a confrontation between aspirations and achievements, and when this rudimentary postulate is expanded, it may give us some conceptual insights into complex behavior. Particularly, the initial question of signficance for us is this: what are the origins of the set of aspirations an individual might hold? Within our analytical framework, there are three possible answers: (1) They may be intrinsic, having been defined *ab intra* by the individual himself. (2) They may have emerged from the environment, and hence be *realistic* or empirical in origin. (3) They may have been conditioned by the cultural context, and hence be a product of *aprioristic* engination. Now, were we interested in abnormal or Freudian psychology, for example, then we would confine our concerns to *ab intra* aspirations, and describe all behavior as due to drives, urges, compulsions, etc; as being spawned somewhere in the numinous nexus of body and soul. Or, were we behaviorists, we would look to the tangible properties of the environment for all aspirations, confining our concept of "exogenous" determinants to those which may be directly measured. But as proper students of societal phenomena, we might be primarily concerned with those aspirations of *a priori* engination. And indeed, as has been suggested from the first, our pivotal proposition in this book is that aspirational behavior is very much more significant in scope and depth than is generally supposed. In short, for our purposes, *the direct and pressing link between the behavior of individuals in collectivities is through the mechanism of culturally conditioned aspirations*. The four remaining chapters of this book attempt to define this link in the most explicit way possible, and to provide both conceptual and empirical defenses for the proposition itself. Here, however, we may proceed to investigate the implications of this proposition in its generic or theoretical significance, as a prelude to the work which follows.

In terms of their operational significance, the cultural predicates serve to bound *a priori* the range of *legitimate* aspirations an individual may define for himself. In more extreme cases, the individual's aspirations and his *normative* behavior are completely prespecified (as with the exigetical behavior modality.) In other cases, only abstract "principles" are provided. Although in the next chapter we shall be more specific about this, the general situation that we look for is illustrated in Figure 2.8. In simple set terms, the above figure describes the following situation. Initially, W may be thought of as representing the universe of aspirations available to an individual within a societal setting—that is, all possible aspirations which might emerge intrinsically, empirically and through cultural determination. Now, L represents that proper subset of behaviors which are *explicitly legitimate*, given the cultural predicates of that societal system. The subset X, on the other hand, represents aspirations that are *specifically proscribed*—that the *a priori* predicates condemn. Finally, C is the complementary subset, $W - (X \cup$

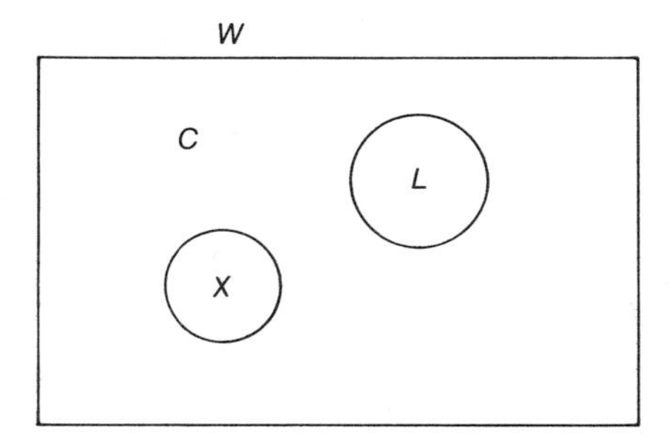

FIGURE 2.8: The cultural constraint function.

L). These aspirations are neither specifically proscribed nor prescribed, and thus may not even be defined within the terms of the particular cultural engine in operation.

Now, to a great extent, cultures can be rendered distinguishable from one another—or analytically unique and identifiable—by noting the specific contents of the L and X sets. That is, a culture is to be distinguished in terms of the aspirations (or more generally, behaviors) which it proscribes and/or prescribes. And the specific sets of proscriptions and/or prescriptions now become the basis for a psychology of collectivities or societal systems. Thus, in the chapters which follow, we will be spending a great deal of time trying to define an array of different cultural predicates by seeing what aspirations and what behaviors they condemn. In operational terms, then, *culture becomes significant only to the extent that it makes certain activities more probable than others*, and cultures may be characterized in terms of the probability distributions that are imposed on the universe of activities in which an individual might engage [17]. Prescribed behaviors thus become probable, given the cultural engine, and proscribed behaviors become improbable.

Now some societal systems have evolved cultural predicates that specify legitimate aspirations that can be achieved by virtually everyone, independent of any objective performance criteria. Indeed, as we shall later see, this is the essential logic of (and the essential benefit offered by) cultures predicated on deterministic religion and sentimentality (i.e., stressing affectivity and spontaneity in behavior rather than purposivity), and extentialistic systems. On the other hand, there are other forms of society or culture that are characterized by open-ended aspirations that can never be completely exhausted, e.g., the systems rationalized by the Protestant ethic in its prototypical form, or those societal systems lacking any axiological or ethical constraints (secularized or instrumental societal systems). In the latter systems, anxiety may be very high, but real levels of achievement are also high. So, in rudimentary terms, the dialectical relationship between aspirations and achievements would tend to pose health and happiness—or complacency and material sophistication, etc.,—as competitive ends. And, for reasons which we shall try to make very clear in the coming chapter, a synthesis between complacency as thesis and performance as antithesis is an improbable societal state. Thus, the societal dialectic is most often an incomplete one: we find societal systems pursuing either thesis or antithesis, on virtually all dimensions, while instances of synthesis remain scarce indeed.

In summary, then, culture becomes operationally significant insofar as it serves to restrict the aspirations or behaviors that individuals pursue. The mechanism by which culture operates is the *a priori* specification of legitimate aspirations (prescriptions) and illegitimate aspirations (proscriptions), or perhaps some inferentially related mixture. One culture is distinguished from another in terms of the specific L and X sets it defines, and also in terms of the mechanisms it employs to confine the population to the legitimated aspirations (or to legitimated behaviors). The empirical differences between societal systems are thus in part due to differences in the cultural predicates they labor under, and these differences become specific in terms of the activities individuals normally undertake, and those they tend to avoid.

Notes

1. The iconograph, as used here, is simply a diagram for a continuum of indiscrete vectors, and hence is itself a vehicle for dialectical comprehension. For more on dialectical appreciation, see Nicholas Georgescu-Roeyen's arguments in *The Entropy Law and the Economic Process* (Cambridge: Harvard University Press, 1971, pp. 45–59). The iconograph does not obey mathematical laws or have pretensions to mathematical rigor. Rather, it is strictly an analytical artifice.
2. For a note on the need for conceptual integration that takes a somewhat different tack, see Francisco Varela's "Not One, Not Two" (*Coevolution Quarterly*, no. 11, September 1976, pp. 62–67).
3. See Paul R. Erlich, "Coevolution and the Biology of Communities" (*Coevolution Quarterly*, no. 1, March 1974, pp. 40–45).
4. See Alvin Goulder's book *The Coming Crisis in Western Sociology* (New York: Basic Books, 1970).
5. For exposition of a cogent yet radical morphogenic model, see Anthony Wilden's *System and Structure: Essays in Communication and Exchange* (London: Tavistock, 1972, pp 351–394). Also see pp. 7–66 of Walter Buckley's excellent *Sociology and Modern Systems Theory* (Englewood Cliffs: Prentice-Hall, 1967).
6. Mathematically inclined readers will find an excellent if somewhat technical analysis of "fuzzy set" theory in L. A. Zadeh's "Outline of a New Approach to the Analysis of Complex Systems and Decision Processes" (*IEEE Transactions on Systems, Man and Cybernetics* SMC-3, No. 1, January 1973).
7. A similar view of culture in terms of antagonistic forces is developed and projected into the institutional realm by William Irwin Thompson. See both *At the Edge of History: Speculations on the Transformation of Culture* (New York: Harper and Row, 1971) and *Passages about Earth: an Exploration of the New Planetary Culture* (New York: Harper and Row, 1974, pp. 30–55).
8. See, for example, the appropriate section in Chapter 2 of my *Systems: Analysis, Administration and Architecture* (New York: Van Nostrand Reinhold, 1975).
9. For more on this, see Chapter 4 of my *Systems* [8] or the seminal paper by Emery and E. Trist: "The Causal Texture of Environment" (*Human Relations* 18, 1965).
10. The task of random choice in a random environment is discussed in terms of a cultural practice, often bemeaned as mere superstition, by Omar Moore in "Divination: A New Perspective" (*American Anthropologist* 59, 1965).
11. For a fine discussion of the fit between observer and environment, see D. G. Kovolovsky's *An Ecological and Evolutionary Ethic* (Englewood Cliffs: Prentice-Hall, 1974).

12. For an analysis of the cybernetic paradigm, see my article, "Some System Theoretic Limits on the Cybernetic Paradigm" (*Behavioral Science* 20, no. 3, May 1975).
13. See Gouldner [4].
14. For an excellent discussion of this theoretical bias, see Walter Buckley's *Sociology and Modern System Theory* (Englewood Cliffs: Prentice-Hall, 1967).
15. The political-ideological repercussions of this bias are discussed by Wilden ([5], pp. 217–229). Also see Herbst's treatise, *Alternatives to Alientation* (Mennen Asten, 1976), Chapter 1, footnote 13.
16. For a complaint on homogeneity as it intrudes into anthropological exercises, see Anthony F. C. Wallace's *Culture and Personality* (New York: Random House, 1970, pp. 27–38).
17. This argument seems to gain some support from the comparative population psychologies that are of interest to some anthropologists. See, for example, Robert A. Levine's *Culture, Behavior and Personality* (Chicago: Aldine, 1973).

PART II
MODELING

As the subjects we study become more and more complex, a priori comprehension becomes more and more essential. That is, in attempting to study societal phenomena, we must seek to equip ourselves with some deductive or theoretical scheme that will lend a tentative order. In practice, such schemes would be developed under the rubic of qualitative analysis. Eventually these deductive "front ends" would be linked to quantitative constructs, and these in turn tested using the instruments of experimental science. This is a very general portrait of the problem of modeling societal systems; the two chapters constituting this part of our work will deal with this activity in some detail. Particularly, Chapter Three attempts to show how the fundamental structures of societal systems are essentially responses to the behavioral predicates of the individuals involved. When collectivized, these behavioral predicates give rise to what is popularly called an ideology; more technically, we are dealing with proper cultures. But not all societal systems are cultural, per se. Therefore, we also have to try to define the bases by which non-ideological systems become organized and maintain themselves. Finally, we have to suggest just how our a priori *(deductive) model components might be subjected to reality testing, and the requirement here is that we develop a quantitative linkage of the type we have been urging. Once this is done, we are led, in Chapter Four, to the most obvious instrument of societal-system analysis: the ideal-type. The purpose here is to try to generate an array of ideal-type societal referents that, as fully as possible, exhausts the real-world societal systems that are apparent to us from empirical (historical) study. These ideal-types would thus serve as the basic theoretical vehicles for societal-system science, and to promote scientific efficiency. Moreover, the ideal-type constructs add a dynamic component to the structural models developed in the fourth chapter. And then we have the other modeling problem, to link these essentially qualitative ideal-types with quantitative implications. This we do by establishing a behavioral matrix that sees societal transformation in terms of simple Markov processes, and thus allows the technician to take his proper role in the study of societal phenomena. In general, then, the substantive modeling exercises that we illustrate in these two chapters are responsive to the methodological criteria developed in Part I, and may thus serve to lend a bit more weight to those essentially abstract propositions.*

3
The Determinants of Societal Structure

In this chapter we shall be concerned essentially with the origins and implications of societal structure, and the task of "mapping" or modeling it. From the dialectical perspective gained in our last chapter we know that societal systems may be driven by three classes of predicates (behavioral engines): (1) realistic, (2) intrinsic; and (3) aprioristic. What most distinguishes proper societal systems from other organized phenomena we might study—e.g., formal organizations or functional institutions—is their essentially *a priori* predication. Realistic systems, as we shall later see, are somewhat trivial, being merely direct responses to environmental properties; they hold little challenge or interest for the societal-system analyst, and indeed contain only an insignificant proportion of the world's population. Intrinsically driven systems, similarly, are extremely rare and, being almost impossible to model, are also of little interest to the student of societal systems. Thus our primary concern with those societal systems driven by *a priori* predicates reflects two realizations: First, such societal systems contain the vast majority of the world's population; secondly, they represent the essential analytical and modeling challenge for the societal-system analyst. And as we begin work on what might be called the ideological or cultural dimension of societal systems, we shall find that there are plenty of complications and qualifications to concern us, but some opportunity for realizing order and predictability as well.

THE CASE FOR QUALITATIVE DISCIPLINE

As was implied in our Preface, the scientist setting out to model societal systems or societal phenomena in general must have some comprehension of the rules and substance of qualitative analysis. It would, of course, take another large volume to treat this subject adequately. Here, however, we can devote a few pages to some of the more rudimentary propositions. But even at the level of rudiments, our skill in the qualitative analysis area lags far behind our quantitative sophistication. From the beginning, however, it must be confessed that the social and behavioral sciences (and perhaps this holds true for the so-called "hard" sciences as well) have been fundamentally and demonstrably successful only in solving the simpler problems with which the world is faced. For, again, the strictures of empiricism, behaviorism, cybernetics and analytic methodology virtually obviate attack on significantly complex issues. As a result, social, economic, political and behavioral problems have been left largely to poets or popular social commentators. For empirical science does not recognize normative or *a priori* predicates of behavior and therefore cannot deal with them. It is thus that we have a much finer appreciation of the behavior of the atom than of man; we comprehend, with greater precision, the properties of deep space than we do the properties of societal systems. Modern science, like modern man, has become more attentive to technology than to teleology, and more concerned with instrumentalism than integrity. In the simplest sense, science has taken the world apart, and done so thorough a job at this that no one can any longer see it in its real perspective. What we have is a collection of bits and pieces—manageable within the quasi-mathematical laboratory environment which modern science prefers—but no effective comprehension of those instances of organized complexity that are the more significant but also the more elusive aspects of this world in which we live.

For the complex societal system is something different than the mere amalgamation of its social, economic and political parts. Additivity, by and large, is a property to be found only in artifacts or other deliberately "engineered" products. This "something different" must be found in the nexus of sociological, economic and political appreciations, but also beyond it. And when we go beyond the simple nexus, we move into the domain of proper societal-system analysis.

The key methodological vehicle in societal-system analysis is thus very different than the empiricist-analytic engines which drive normal social science. In fact, it demands a return to the long-disregarded *deductive* arm of science. For again, as a subject becomes more complex, it is less likely to be captured by the artifices of inductive science: by reductionism, empirical observation, analytic attack and the process of hypothesis-free experimentation (or by experimentation which derives its hypotheses from observations). Not only is it less accessible empirically, but it is less amenable to experimental manipulation, and less likely to permit measurement (either directly or through surrogation). Therefore, as a phenomenon becomes increasingly complex, the value of a *priori comprehension of the whole* increases. It is thus that the deductive method becomes the *prerequisite* for the use of inductive-empirical technology. For the deductive process allows us to specify how we might put back together what subsequent analytic analysis takes apart. Thus, the recommended procedure for societal-system analysis is to develop a tentative deductive structure which will be used to guide and discipline inductive-empirical analysis. The deductive method thus complements the inductive; it does not replace it.

Of the deductive instruments available to us, the *hypothetico-deductive* method emerges as most attractive [1]. It may be used to develop tentative heuristics, which are of both substantive and methodological significance. In the substantive dimension, the heuristics seek to constrain the search space over which we must wander, employing various *a priori* restraints (derived from theoretical, logical, normative or perhaps even intuitive sources). The heuristic front end thus artificially suppresses some of the complexity of the system being studied, and outlines certain investigatory trajectories which have the highest *a priori* probability of proving fruitful. In its methodological aspect, the heuristic also sets the rules by which a favorable trajectory may be distinguished from an unproductive (or diseconomic) one. We follow the recommended trajectories by employing the instruments of inductive science—observation, measurement, manipulation—and use the results of the inductive-empirical processes to modify our *a priori* assumptions housed in the heuristic (or hypothetico-deductive constructs). Thus, the method of societal systems analysis becomes comprehensible in terms of *action-research* or "learning" paradigms [2]. And the essential difference between the hypothetico-deductive method and normal science thus exists primarily at the point of initiation. While normal science, following Bacon's lead, prefers hypotheses that are elicited from facts (e.g., observations), the hypothetico-deductive exercise is initiated by hypotheses of *a priori* origin. For complex phenomena, the hypothetico-deductive front end thus seeks to answer the essential questions: what to observe, what to measure, and how to manipulate.

For those not familiar with the arcane arguments of modern epistemology, the obvious question might arise: How does one determine what to observe, what to measure and how to manipulate (or what to manipulate) in the absence of such *a priori* constraints? Well, one doesn't. It is merely that some scientists choose not to recognize *a priori* constraints. And there is also what might be called the phenomenological bias: the advancement of hypothesis-free research and analysis. Under this thesis, the scientist runs across his subjects by accident, or simply chooses to wake up in a new world every morning. But for most scientists, the question of what to study and how to study it—and what kind of results to generate—is pre-answered by the "school" to which he belongs. Therefore, one might be a scientist all one's life, and never answer any of the questions that the hypothetico-deductive process poses. This, essentially, is the point made by Kuhn in his classic study of the role of paradigms in scientific enterprise [3].

But if the hypothetico-deductive process is so obviously innocent, and so apparently promising as a vehicle for complex system analysis, why is it so seldom employed? There are basically two reasons, both quite straightforward. First, scientists have only recently begun to realize that the world is a more complex place than Newton, Laplace, Bacon and other prophets—even Einstein—proposed. If we extend the implications of these great scholars, we see them searching for fundamental relationships of a symmetrical and hierarchical order—a single set of "algorithms" that determine all events. The general idea was that the world and all that's in it were *inherently deterministic*, and that stochasticity and unpredictability would untimately be made to disappear as knowledge approached a theoretical maximum. And if the world is inherently deterministic, then the appropriate methodology (at least for a patient science) is trial and error, observation, measurement and manipulation, with science being the respository for generalizations all induced from specifics. In such a world, apriorisms (and deductions in general) are simply gratuitous, and therefore of no interest. Now, we cannot here fruitfully question the presumptions of Newton, Laplace, etc.; being teleological, they cannot be either validated or invalidated in the temporal world. But many scientists have begun to suspect that if the day of maximal knowledge is coming at all, it is much further away than was once supposed. Therefore, the *efficiency* of normal inductive science is being increasingly called into question, if not its fundamental rectitude.

The second reason why the hypothetico-deductive method has been largely ignored is somewhat simpler, and perhaps even more compelling. The tendency has been to condemn *all* deductive science because of certain excesses associated with a restricted (and inferior) form of deductive enterprise: rhetorical science. This was the process employed when science was a handmaiden to philosophy, which was in turn the servant of sacerdotalism. The deductive-rhetorical process really sought to provide answers without experimentation, answers which might have arisen from intuition, revelation, casual speculation or any of a number of other undisciplined deductive processes. The bases of all rhetorical schemes are unvalidated first premises offered as indisputable facts, coupled with an attempt to induce affection for the scheme through elliptical or parabolic logic. But the hypothetico-deductive method, first of all, demands that all its constructs be apodictical. That is, they must be phrased and designed in a way that permits their eventual validation or invalidation through empirical means. In short, as with normal science, fact is the ultimate arbiter of scientific utility. The second aspect of the hypothetico-deductive method is that its constructs are ruled at all points by logical (judgemental) probabilities, and thus subject to the same type of confidence analysis as are the constructs of inductive-empirical science. Thus, hypothetico-deductive efforts are distinguished from deductive-rhetorical ones by the discipline that attaches itself to the former.

It should be clear even from this very brief discussion that the hypothetico-deductive logic is a key requirement for the development of adequate interdisciplinary models (which are, in turn, the prerequisites for an integrated, non-parochial social science). Interdisciplinary interfaces—in the context of operating societal phenomena—are too protean and rapidly changing to respond well to hypothesis-free investigation or to be "induced" from isolated experiments. What we need, then, is the general rules of organization to which social, economic and political phenomena respond—the algorithms by which the societal system becomes an elaboration rather than a product of its parts. And in this infant stage of interdisciplinary science, the interdisciplinary interfaces must initially be forged of tentative heuristics, and only much later provided with detail and authority by empirical science. In short, the problems of the world will not wait the time required to let inductive-empirical science work its way upward to the cross-disciplinary connections.

We have, of course, a history of interdisciplinary adventures. For the most part, sadly, they have been dramatically unsuccessful. The primary reason for this is that representatives of a particular discipline have tended to predominate, with the variables and models of the other disciplines being entered as exogenous factors, or "givens". In a proper interdisciplinary exercise, variables from all relevant disciplines have an *a priori* probability of becoming determinants (i.e., no prior assertions are made about variables from one disciplne being the "cause" or engine of variables from other disciplines), and the attempt is made to make as many factors as possible *endogenous*. That is, the number of factors that are determined externally to the model is minimized, so that the full repertoire of possible interrelationships becomes accessible to the researcher or analyst [4].

Such models will tend to be more complicated, at least in their structure, than traditional constructs which place heavy emphasis on manipulating only a few variables at a time, leaving the majority external. But, by the same token, true interdisciplinary models have a higher probability of representing realities, for they are capable of reflecting more of the critical interrelationships and complexities that the real world itself houses.

Such models, of course, could quickly become unmanageable, incorporating too many variables and trying to encompass too many interconnections. Indeed, it is partially for this reason, and not totally because of a deliberate neglect of interconnections, that we have so few interdisciplinary constructs. For if one has only the traditional model-building and analytical techniques at his disposal, even the most sincere interdisciplinary ambition swiftly falls afoul of practical constraints. However, there are some techniques available to us that assist in the abbreviation of complex realities, which means that the complexity of a model need not fully reflect the complexity of the subject we are trying to describe or predict. Through the use of such techniques—notably the general system instruments of abstraction, ideal-type formulation and isomorphic analysis—we can treat complex subjects without having to develop models which are incomprehensible and unwieldy. Thus, as we shall see, it is possible for the social scientist to abridge the correlation between structure of models and structure of reality, and to do this without paying the traditional price of oversimplification or lack of explanatory-predictive authority.

The art of abstraction serves us in two different ways. First, it allows us to consider the process of allegorization: the attempt to develop predictive-descriptive models which are cast in terms qualitatively different than those of the real-world subject itself. Although this technique has a long and honored tradition in the social sciences, its heights have been reached in physics and engineering. There, for many centuries, reality has been transformed into mathematical substance, with the mathematical formulations serving as a shorthand. The result is that complicated subjects may be effectively captured using constructs that themselves have both fewer variables and less complex interrelationships than they would if we attempted a one-to-one modeling process, where every aspect of reality was inserted into the model without any morphological transformation. Thus, abstraction is a variation on the surrogation process, where we try always to find terms and units of analysis that have an adequate probability of reflecting reality, without duplicating it [5].

The second advantage of abstraction is that, in many cases, it permits us to find points of fundamental similarity among subjects which are obscured when we work at the superficial or basic level. These points of similarity, when made clear through the process of abstraction, are known as *isomorphisms*. To the extent that abstractions lead to the recognition of isomorphisms, the cause of scientific efficiency is served, for more than one subject now becomes explicable in a single analytical framework [6]. Naturally, the extent to which such efficiency can be realized depends on the extent to which real isomorphisms are present, and on the extent to which our abstraction processes are properly disciplined. There is always the danger that we shall work at a level of abstraction that is too high, so that our conclusions are too general or obscure to be of any practical use. Within this section on modeling, we shall be guarding against that danger, but it is a difficult one to avoid entirely. As Anatol Rapoport has suggested: ". . . seek simplicity [but] distrust it" [7].

The process of formulating and manipulating ideal-types has a significant history, yet there is seldom the recognition that ideal-types carry distinct theoretical implications, and that they must therefore be bound by the laws of strict deductive inference. Rather, they often tend to be employed as isolated discriptive constructs. To be employed properly, however, they must be developed as, and treated as, adjuncts to an emerging theoretical system. This involves, in practice, the generation of ideal-types that are Janus-faced. One face searches the theoretical-deductive superstructure of science for points of articulation and speculative reinforcement; the other face is attentive to the real world and to the necessity for empirical validation. Thus, the proper ideal-type is simultaneously of rhetorical and empirical significance. We shall be as explicit as possible about the extent to which the ideal-types we shall be building in later sections meet these criteria.

In the case of the search for relevant isomorphisms, we have the direct and most pertinent of our modeling ambitions: the attempt to show that, when raised to a proper level of abstraction and when cast into proper ideal-types, many social-science subjects become intelligible in the same generic terms. That is, the basic differences between the social, political, economic and behavioral subjects of the social sciences tend to be softened when we operate at a higher level of abstraction and conduct a vigorous but temperate search for isomorphisms. For some relationships become clear which in the absence of the heuristic "masks" we shall be wearing would remain beneath our level of comprehension. Finally, to the extent that isomorphic relations can be found among social, political and economic components of complex systems, the cause of an integrated social science is made more plausible and accessible.

To try now to lend these abstract arguments some life and substance for the reader, we can begin to attack the problem of societal modeling *per se*, beginning with the attempt to define the concept of culture.

THE CULTURAL CONSTRAINTS

Once we allow *a priori* predicates to play a role in societal systems, we can use the dialectical perspective to forge a link between individual and collective phenomena, as follows shown in figure 3.1. The set of axiomatic and axiological predicates to which any societal system responds may be thought of as constituting its *cultural envelope*. But, like individuals, collectivities of any form or magnitude may also have a constant conflict between realistic and aprioristic determinants, and therefore between the adaptive and exegetical behavioral modalities. Again, realistic-adaptive mechanisms imply that the contents of the direct (empirical) predicate should determine the response, yielding the collective (or societal) counterpart of the operant-conditioning process. The exegetical modality produces a culture-sensitive re-

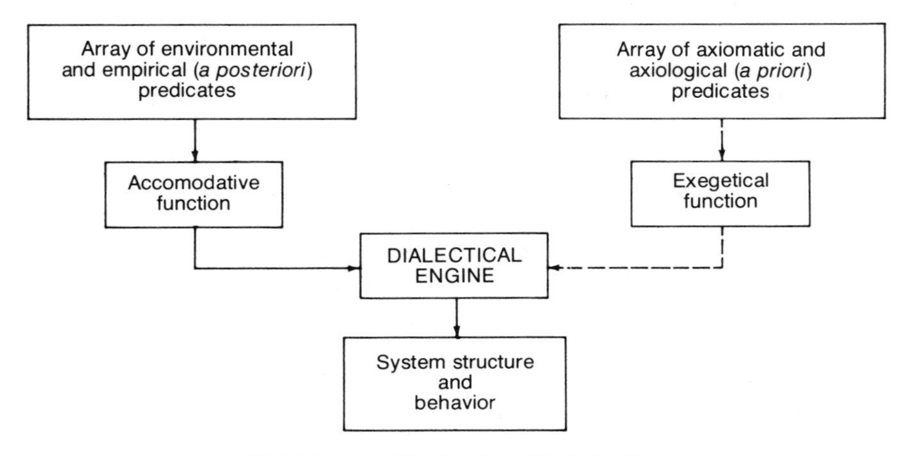

FIGURE 3.1: The "cultural" dialectic.

sponse. In the general sense, this means that some systems may have their structure and behavior determined not simply by their environment but by empirically transparent factors in the form of values or affectively oriented predispositions, or by deductive or theoretical apriorisms.

The existence of cultural mediators means that the two systems may very well develop entirely different structural and behavioral properties in response to the same set of environmental properties. Thus, cultural predicates may be thought of as driving the collective equivalent of the *assimilative* function which we earlier allowed individuals. The assimilative function, we recall, allowed individuals faced with the same set of direct predicates—the same stimulus—to interpret the stimulus in different ways, depending on the contents of their experiential or "cultural" bases. Here we shall merely be allowing collectivities (i.e., societal systems) the same privilege. The cultural dimension may thus be thought of as the conceptual "filter", and thus a potential *cause* of societal structure and behavior, and not always merely an effect. In this regard, consider the following set of propositions:

1. Every societal system is simultaneously of cognitive, social, economic and political significance.
2. The actual properties of these several dimensions are responsive to two sets of forces:
   i. The set of environmental and empirical predicates that constitute the tangible aspects of the milieu of the system.
   ii. The *a priori* (axiological and axiomatic) predicates that constitute the cultural envelope to which constituents are responsive.
3. From the dialectical perspective, there is a constant conflict between the empirical and *a priori* predicates for determination of the system's properties. Two polarized ideal-types may be postulated:
   i. An *accommodative* system gives constant predominance to the empirical predicates, and thus becomes primarily explicable as an adaptive system (one whose structure and behavior are highly correlated with the historical and emergent properties of the milieu).
   ii. An *acculturated* system is one whose properties are not well correlated with the empirical properties of the milieu.
4. A majority of systems we encounter in the past and present will tend to be at least partially acculturated. This is because a primary engine for system evolution and development is the *prophetic process*: the development of normative referents through some sort of deductive exercise.

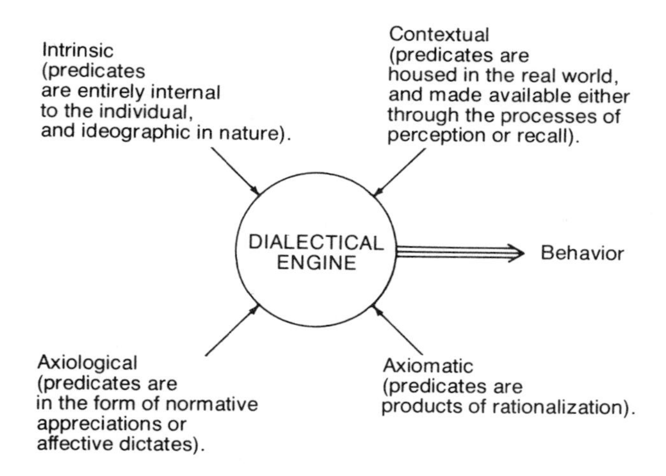

FIGURE 3.2: The several predicate sets.

5. Among systems which are manifestly acculturated, there will tend to be a strong clusteration, such that a few generic (ideal-type) system referents may be defined, which a majority of real-world systems tend to approximate. This reflects a lack of fundamental variety among societal systems, given a certain level of abstraction.
6. This lack of variety is due to faults of the prophetic process. Particularly, prophets have tended to produce only polarized system referents rather than more sophisticated, synthetic (syncretic) alternatives.
7. Finally, the lack of variety of real-world systems tends to be aggravated by the fact that the majority of system activities (decisions) are determined using only the simplest cognitive modalities: association and exegesis.

Our defense of this set of arguments will take us the remainder of this chapter. But before we proceed, let us summarize what we already know. Initially, we have suggested that there are basically four types of predicate sets to which disaggregated societal phenomena may respond, as in figure 3.2.

Now as our concern in this section is with well-ordered collectivities, we may temporarily abandon our interest in the intrinsic bases of behavior. For they would tend to produce essentially atomistic amalgams (i.e., systems of only temporal or transitory significance, or very ill-bounded, poorly articulated collectivities). In fact, instances of real-world societal systems tolerating significant injections of intrinsic behaviors are very rare. So our interest here must focus on the primary bases: the contextual, axiological and axiomatic.

Recall from the last chapter that culture exercises a constraining influence on individual behavior. It does this by limiting the set of aspirations that individuals may legitimately hold, thus making certain behaviors more probable than others. In substance, then, the cultural dimension of any societal system consists of a more or less bounded set of proscriptions and/or prescriptons. To the extent that some set of individuals are susceptible to these *a priori* dictates, we have the basis for a proper *collectivity*. The implication is that the behavior of the members of a collectivity will be highly correlated, so that we have a group of individuals who respond to any given predicate with significantly similar responses.

But not all proscriptions and prescriptions are of cultural significance. Some may be empirical in origin. This means that they are derived and formalized on the basis of collective experience with the real world. For example, certain collective behaviors may have emerged as responses to iterative survival conditions (e.g., agricultural or hunting technologies), or specific behaviors may be proscribed because they hold peril of some kind which earlier generations discovered (e.g. incest). The point is, then, that some proscriptions and prescriptions may owe their origin to acquired empirical information, or to transgenerational accommodation to the exigencies of the environment (driven, perhaps, by some variation of operant conditioning) [8]. Now, for our purposes here, such proscriptions and prescriptions will not be considered as elements of the cultural base. Rather, they will be grouped under the heading of collective *lore*, and as such will be left primarily to the anthropologists and empirical sociologists.

Thus, those proscriptions and prescriptions that are deemed proper components of the cultural base will all have an aprioristic rather than adaptive origin, and will be products of one of two types of essentially deductive processes: (1) axiological speculation, which develops constructs of normative or affective significance; or (2) axiomatic reasoning, which develops constructs of logical or rationalistic significance. The processes involved here are already familiar and are resummarized in figure 3.3. We can see that we are back on ground with which we are essentially familiar. The process by which axiological predicates become translated into behavior is the exegetical modality which we discussed in Chapter One. The axiological base contains a set of values or ethical/moral referents, which suggest how people *should* behave in a given situation. Thus, the direct predicate (the circumstance) invites a

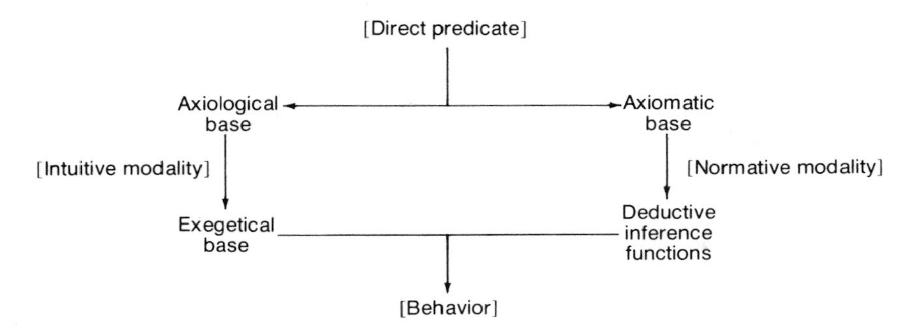

FIGURE 3.3: The axiomatic/axiological process.

set of secondary predicates from the axiological base, and the real-world behavior then becomes in part a product of *a priori* engination. The axiological base differs from the axiomatic in this way: Whereas the components of the axiological base are products of moralistic, or ethical or value-dependent imperatives, the components of the axiomatic base are purported products of directed logic. For example, the Ten Commandments are axiological predicates, whereas the assertion that the economic welfare of a nation is best achieved through the free interplay of market forces (i.e., an extension of the Smithian "invisible hand") is an axiomatic predicate.

Now, the problem for us here is to investigate the origin of *a priori* predicates. What we have proposed thus far is that they do have some effect on behavior. But we have thus far treated only (and only in a somewhat abstract way) the case where the axiological and axiomatic predicates are already available to the individual. We have not discussed the process by which axiological and axiomatic bases gain defintion and moment. Hence our concern with the *prophet*. For it is he who creates the various aprioristic predicates that the majority of us internalize and execute.

First, however, we want to say something about the concept of the prophet as an individual of rather special attributes, and try to get a somewhat disciplined view of the prophetic process.

AXIOLOGICAL PROPHETS AND MORAL ORDER

In the most direct sense, the prophet sets the upper limits of man's comprehension, and thus bounds the range of societal forms that the real world houses. The importance of the prophet is often understated. For the currently popular view of the origin of societal structure—and hence of social, economic and political sectors within societal systems—is that it somehow arises as a set of induced accommodations to the vagaries of the natural environments of the groups. It is our contention, however, that the prophets create the masks and spectacles through which realities are perceived by most of the rest of us. If the spontaneous theory were correct, so that every societal form was basically a response to unique environmental conditions, then there would be much greater heterogeneity among societal systems than there actually is. But the fact is this: when we trace societal systems back into nakedness and night, we see just a very few fundamental forms, all of which will be defined in detail in later chapters. Moreover, the process of societal dynamics—the trajectory of societal change—appears to be largely *pendular* rather than evolutionary. This means that truly unique societal forms are seldom seen. Rather, the societal stage is usually occupied by one or another variant of those generic societal forms which have been with us practically since the dawn of history. In short, most of the societal concepts we think are new and original are really just old schemes going in and out of style, so that societal systems have not really *evolved* through the ages, but simply reintroduced themselves.

The cause of this pendularity—the failure to evolve new societal forms and the limited number of forms available—is found in the nature of the prophetic process itself. As we now begin to explore some of the key properties of this process, there are three arguments that are crucial. First, the assertion that, at heart, *the prophetic process is reactive rather than creative.* Secondly, that *the actionality of any prophetic position* (the probability of its being implemented) *is positively related to its essential simplicity.* And, finally, that *the simplicity of any prophetic position acts against its persistence* in any pure form. To show something of the prophet's position in the normative political subsystem, consider the following [9]:

1. The attributes of the societal system evolve from the people, largely through their articulation of a set of laws. These laws, in the last analysis, are merely the codification of *what the people want.*
2. To execute the law, an individual is elevated to the position of *king,* or its local equivalent. His role, essentially, is to see that the will of the people, as embodied in the structure of law, is exercised. Now, to the extent that there is a diversion of interest from the principle—to the extent that the king is tempted to act on his own rather than the people's behalf—there emerges a discontinuity between principle (the spirit of the law) and practice. The qualitative difference be-

tween principle and practice may be thought of as *what the people will tolerate*.

3. Now we come to the two roles of the prophet. Sitting outside the formal social, economic and political envelope, he is presumed to be the objective guardian whose function is first of all to minimize the difference between principle and practice. But, secondly, his is the default or residual responsibility of assessing the *quality* of the law —which means, of course, that he is the critic of what the people have decided is right or valuable, or what they have come to tolerate [10].

Against the background of this highly simplified system, the effects of any flaws in the prophetic process (and especially the considerations we just mentioned) become immediately apparent. Let's first consider reactivity. This is associated with any prophetic platform to the extent that it basically represents a challenge to some existing societal scheme. This is true of the platforms of virtually all axiological prophets: Socrates, Jesus, Shankara, Mohammed, Mao, et al. For all, there was first the deploratory process—the projection of a dismal future if existing societal practices and properties were allowed to persist. These prophets were not, however, mere forecasters. The variables employed were uniquely of their own composition, often highly idosyncratic in formulation. Their extrapolative scenarios were not so much statistical as rhetorical in engination. And in all cases there was the postulation of an alternative future, whose realization depended on a rejection of the status quo and its replacement with an effectively polarized system of values.

Where there was instrumentalism, the prophet brought the transcendent aim. Where there was resignation among the people, the prophet proposed responsibility. Where there was anxiety and titillation, the prophet offered solace through sacrifice and merit through self-denial; where there was sacrifice and constraint, the prophet preached the virtues of the sensual and praised spontaneity. In short, if there existed any critical attribute, the prophet proposed its complement or inverse. In all cases, the prophets chastised the people, their leaders or their priests. For, to the axiological prophet, the faults of society were always faults of character, complex interplays of self-delusion, weakness of will, cowardice and disinvolvement. The prophet's target was thus always the masses, and his ammunition always moral, rhetorical and normative.

The moral aspect of axiological prophecy emerges from its defense on other than logical or empirically demonstrable grounds. Indeed, most axiological prophecies stress acceptance on *faith* rather than on validation. The axiological prophecy rests ultimately on the assertion that certain forms of behavior are *right* without too much effort expended in defining the basis for this attribution. In some cases, the concept of what is right is suggested to be of divine origin, through revelations made to the prophet. In more sophisticated societies, the concept is defended with respect to some cosmological or teleological interpretation of the Creator's will. In other cases, axiological systems will merely be quasi-deductive structures predicated on ideographic first premises (e.g., the demo-Christian platform built from the assumption that "all men are created equal").

The *rhetorical* component of axiological prophecy stems from the fact that the prophet's purpose is not so much to understand behavior as to change it. Thus, the vehicle of axiological prophecy is always persuasion, restricted primarily to the affective or emotional domain (whereas, as we shall see, axiomatic prophecy attempts to make its appeal through logical convictions). Indeed, the key ingredient for the successful axiological prophet is the facility for excitation. In some cases this involves the ability to create similes and metaphors which reduce complex issues to a set of clearcut, polarized alternatives. (Political cartoonists are also masters of this art.) In other cases, prophetic utterances are poetic, allegorical or even parabolic, as with the striking if somewhat ambiguous parables of Jesus or the aphorisms of Mao.

Finally, the *normative* component makes the prophetic platform practicable, giving it its substantive significance. The normative aspect eventuates in a set of specific behavioral signals, either proscriptive or prescriptive. The normative dictates may demand that people cease doing what they have been doing (as in most of the Ten Commandments) or that they begin to do what they have not done before (as in Mao's exhortations). In terms of socio-psychological significance, we thus have the following:

1. If the normative dictates are prescriptive, they generally find their appeal in legitimating something that people have wanted to do, but that carried a historical shame or guilt. They thus serve to give vent to aspirations or impulses which were repressed under the previous regime. In either case, the prescriptions gain their appeal by lowering behavioral dissonance (either by reducing inherent anxiety or by releasing repressed energies).
2. If the normative dictates are proscriptive, they may serve to curtail the anxiety associated with permissive systems. They do this by se-

verely restricting the legitimate behavioral repertoire available to individuals (by constraining the feasible search and solution space, as it were). Thus, anxiety is reduced because the instances requiring a decision become fewer and fewer. Also, such dictates can serve the function of alleviating apprehension, largely by providing some sort of order where license once prevailed. A proscriptive axiological platform may often be the basis for lending an artificial certainty to existence, but may also be the basis for propagating or constraining the behavioral properties of the system itself.

The components for the most rudimentary appreciation of societal dynamics are now at hand (albeit in a somewhat abstracted form). Prescriptive prophecies will tend to acquire appeal after a period of repressive, deterministic order (usually maintained through the offices of some proscriptive axiology). Now, these prescriptive platforms generally leave the individual enormous behavioral latitude, which makes strong and constant demands on his discrimination. Little is forbidden, much allowed. Life becomes not merely a matter of realizing certain available opportunities, but of having to constantly forego alternatives (because of constraints on time, energy, etc.). Additionally, there is a distinct impairment of the ideology and dogma. Realism replaces rationalism, license replaces restraint, and existence takes on distinctly stochastic and relativistic overtones. Beyond a certain point, license (accompanied as it is by uncertainty) become disutile. Institutions that promise to give purpose or permanence then become more attractive, despite their curtailment of behavioral freedom. At some point, then, the cry goes out for certainty, for order—and the answer comes back in the form of the proscriptive prophecy. Once this is activated, license is replaced with constraints, doubt with enveloping dogma. The future once again can be seen to be a function of the past, and order is seemingly restored. But beyond some point, the repressions that always accompany purpose, permanency and certainty begin to grate more and more. Faith in God, faith in country, faith in the future and the other common artifices of socio-psychological security and certainty now come to be condemned as naive. The institutions once so highly valued for their aura of permanancy and placidity now become viewed as agents of exploitation and constraint. At this point, the prescriptive prophet re-emerges, usually with a resounding cry for liberation, license and latitude. And the people are now ready to listen, with gradually deepening attention. Thus the prescriptive axiology replaces the proscriptive. The proscriptive will be back, for license and latitude eventually mean anxiety and uncertainty. Here, then, is the axiological origin of socio-behavioral pendularity. Of course, no new prophetic platform succeeds without a struggle from those who hold either status, power, or wealth under the existing system. But the dynamics of pendularity suggests that the shift *will* be made [11].

In no case, however, can socio-behavioral dynamics be interpreted simply as a reification of the Hegelian or Marxian concept of a historical dialectic. The essential reason is this: there is no convergence on a synthesis. Rather, the thesis (the culture) gives way to the antithesis (the counterculture), and so on. There has been no historical tendency for a truly synthetic modality to develop and maintain itself; there has been no society in which there was a persistent syncretism between prescriptive and proscriptive properties. Thus, as we shall try to show more fully in later sections, neither the Hegelian dialectic nor the Toynbean evolutionary model is as relevant a historical "mask" as the pendular engine. The reason, again, rests with the prophetic process. As was suggested, axiological prophets gained their significance largely through *reactive* constructs, and the simplest form of reaction is the logical inverse. Therefore, essentially the same attributes of a societal system are maintained through time, with the attributes merely given antonymical "values" by successive prophecies. This process of logical inversion is not a creative or evolutionary one, for no fundamentally new elements are added to the societal mix. Thus, the history of civilization may be comprehensible largely in terms of parametric changes imposed on a more or less constant set of structural properties [12].

There is a somewhat indirect defense of this assertion which we may offer here. The probability that any prophetic process will actually cause a societal change is directly related to its simplicity. Of course, their moral predications or premises may themselves be very abstruse or erudite. But the normative dictates or demands they make—the actual prescriptions or proscriptions—must generally be unambiguous, unqualified and easily comprehended if they are to be practicable. Any equivocations would merely confound the rhetorical process and diminish the probability of a prophetic platform gaining currency and popular affection. For this reason, the exegetical function has been very important, and plays a dominant part in the axiological drama. For the ruminations and conceptual constructs of the great prophets have often been quite arcane in their original formulation. Indeed, as we earlier suggested, many of the prophetic platforms were originally presented in parabolic or allegorical form. Now, were this abstruseness to be maintained, such platforms would have little probability of being understood or accepted by the mass of population. So, in most cases, a *sacerdotal* system of some kind has sprung up to mediate between the prophet and the people. It is the sacerdotalists who transform the abstractions into specific behavioral dictates or constraints. The analytical engine of sacerdotalism is, of course, exegesis. And it is, therefore, through the exegetical engine that prophetic constructs become popularized and operational. It is thus, also, that many of the prophets would hardly recognize the proscriptive or prescriptive systems that have been sponsored in their name. It is thus that Marx, about to die, proclaimed *"Je ne suis pas Marxiste."* At any rate, we shall have much more to say about exegetical sacerdotalism and societal proscription in later sections.

Moving on, we spoke earlier about the inadvisability of predefining societal systems as instances of shared values. From the perspective we now have, it appears that at any time, any viable societal system must contain at least three sets of individuals, who are responding to different axiological predicates: (1) those values that constitute the predominant axiological base (the predicates for the culture *per se*); (2) those antagonistic (inverse) values representing the axiological base of the emergent counterculture; and (3) a set of transitory or "syncretic" values, which are ephemeral and usually unstable mixtures of the dominant polar positions. There is a critical qualification, however. It is not specific individuals that cause the shift from one societal form to its inverse; individuals may stay loyal to the particular axiological envelope they were raised in. It is, rather, a collective engine at work—sometimes generational, sometimes socio-economic or ethnic, etc. Thus, the pendular process is not a simple extrapolation of individual into societal behavior. Pendularity (except in rare, exceptional cases) has no real relevance for the isolated human being, but is a distinctly systemic phenomenon.

Of course, not every generational, ethnic or socio-economic group will play a pendular role. In any society, there are intervals where stability reigns, and where some cultural thesis may remain in force for several or perhaps many lifetimes. In these instances, there does indeed seem to be a shared value base to the society, and the normal categories of sociology are pertinent. But, as a rule, the probability of persistence of any axiological platform will be positively related to the recency of its ascendence. Simply, as a particular prophetic platform ages, the probability of its inverse emerging increases under the pendular engine*. Now, there is the possibility that the period of the pendular cycle—the duration of the process which finds one axiological platform being completely displaced by its inverse or complement—is not a constant. Much is made these days of the opinion that the world is

*See Section 6.2.

becoming more complex, and that as a result change is proceeding at an every-increasing rate. If this is true, it means that the period of the pendular cycle is shortening, so that any given prophetic platform enjoys a successively shorter expected duration. But there is another aspect to the assertion that change is proceeding at a more rapid rate, and this may involve merely a local phenomenon. We may expect that the rate at which the antithetical axiology exhausts the thesis—the rate at which the counter-culture displaces the culture—will accelerate with time. Thus, we may advance the concept of a societal "learning curve", describing the tendency for societal transitions to both broaden and intensify in the manner of any concatenative process. As we shall shortly see, there is some evidence to suggest that much of the developed world is involved in a gradual deflection from an era of relative discipline and dogma-driven order to an age that will be relativistic, permissive and protean. If this is indeed the case, then we are somewhere on the accelerating portion of the learning curve, and the rate of change will indeed be increasing. Once the substitution is made, however, we may expect the rate of societal change to diminish again, and the cycle will then appear to have lengthened.

There is one final point that should be made before we begin to discuss axiomatic prophecy. Because the axiological prophecies are usually just inverses or complements of each other, the shift from one form to another cannot be expected to increase the net, long-run satisfaction available to the members of any society.* In real terms, then, the marginal benefits associated with a counter-culture are potentially no greater than those associated with the culture against which it has set itself. In the short run, however, there is a distinct incremental benefit associated with the transition of axiological forms—the high marginal utility of the substantive changes which occur. If the shift is from a proscriptive to a prescriptive axiology, the utility of the licenses is perceived to be very high, and only begins to decline in the longer run, as they become more common. If, on the other hand, the shift is from a prescriptive to a proscriptive platform, the initial evidences of order and certainty are valued very highly. In either case, the diminishing returns do not set in until some time has passed (and, as we shall later see, the intervals involved may be very different for different system sectors).

At any rate, in terms of our general model of the pendular process, the shift between axiologies will appear to be rather like trading off one set of benefits against another of more or less equal magnitude. Indeed,

*We shall make much of this point in discussing strategies for societal management in Chapter Six.

as I have already implied, a shift from a proscriptive to a prescriptive axiology, or the reverse, merely results in the displacement of one set of problems by another. In a later section of this chapter, we shall see this argument take a more precise form. For the moment, however, we can suggest that the real utility of pendular swings does not warrant the enthusiasm with which some welcome them, and the vehemence with which others resist them. But because axiological platforms are usually unamenable to any objective proof or disproof, they tend to be defended with great vigor, and just as vigorously declaimed by their adversaries. It is an abiding if somewhat deplorable trait of human nature that we are always more urgent in the defense of what we doubt than what we really know. And this trait leads us to the axiomatic prophet.

AXIOMATIC PROPHECY AND RATIONALIZATION

When we enter the domain of the axiomatic prophet, we leave moral argumentation for what appears to be an attempt at logical engination. The axiomatic prophet is an architect of more or less formal deductive systems. Such prophets include Marx, Marshall, Maslow, Weber, Keynes and Freud—builders of all grand theories in the formalistic tradition.

There are many, often quite subtle differences between the axiological and the axiomatic prophet. Perhaps the most significant is this: the axiological prophet, as we just saw, gains his social moment by generating schemes that eventuate in precisely formulated *codes* of behavior. Originally, however, he may have produced only vague parabolic or aphoristic utterances. Now, as suggested, if the axiological prophet produces works that are highly arcane or turgid, then the reduction of the abstractions to specific rules and principles of behavior is accomplished by their apostles, usually through the process of exegesis. At any rate, the end result of axiological prophecy is a set of specific behavioral dictates that need not be (and, indeed, seldom are) arrayed in any formal order.

The products of axiomatic prophecy, however, are usually sets of theoretically significant *principles* of behavior, deduced from sets of ordered premises. For example, the dialectical materialism of Marx, the human-relations postulates of Maslow, the neo-classical economics of Marshall, the economic thesis of Keynes, and the democratic-liberatarian political philosophy of Locke and his apologists are all theoretical structures which gain empirical significance only to the extent that specific behaviors may be deduced from the *a priori* premises of the deductive system. To clarify this point, consider the table 3.1.

Here I have drawn components from the Marx and Maslow platforms just for illustrative purposes, and have used a bit of philosophical

TABLE 3.1: Components of Axiomatic Systems

Components	Illustrative Examples: Marxian System	Illustrative Examples: Maslowian System
1. First premise	The natural state of society exhibits a symmetry in the distribution of economic prerogatives.	The drive for self-actualization is the highest and constant objective of human beings.
2. Axioms	The Hegelian dialectic is an accurate representation of the engine of societal development.	Industry is an authoritarian jungle.
		The happy worker is a productive worker.
	The victory of the proletariat is assured by the logic of the dialectic.	
3. Internal deductions	Labor and capital are arrayed in an inevitable conflict.	The happy worker is one who has control over his own productive activities.
	The outcome of the conflict will result in the imposition of a permanent and pervasive socialism.	Self-actualization demands self-regulation.
4. A priori prescriptions	Eliminate distinctions between agriculture and industry, between mental and manual labor, and between town and country.	Equalize power between labor and management
		Proceed with innovations such as job enrichment, job enlargement and other aspects of industrial democracy.
	Centralize economic planning.	
	Destroy capitalist predilections and institutions.	Eliminate all aspects of coercive control.

license in "parsing" their schemes. At any rate, within the framework of axiomatic prophecy, a deductive system begins with one or more first assumptions. Axiomatic systems in the prophetic domain are always characterized by a first assumption of distinctly substantive implications. Thus Marx presupposed (perhaps wrongly as we shall point out later) that the prototypical societal system exhibited a natural symmetry in the distribution of economic and socio-political prerogatives. Maslow's pivotal assumption was that there was an impulse toward self-actualization, common to all men, which was constantly in exercise.

The point is that these first premises are distinctly *aprioristic*: they are largely beyond either proof or disproof, at least in the forms offered by their authors. This fact means that, right from the start, the deductive systems of the axiomatic prophets are effectively *non-apodictical*. It serves to distinguish prophetic axiomatic platforms from formal axiomatic platforms, such as those of pure mathematics, or of artifacts such as chess or music. Formal axiomatic platforms have no need of first premises, as they really have no pretensions to either representing or influencing reality. They begin, so to speak at the second step: with the specification of axioms *per se*. But for the axiomatic prophet, the specification of axioms is a second phase of deductive system-building, even though it may often be quite independent of the first step. Indeed, in most of the more popular axiomatic platforms there is an *elliptical* relationship between the set of first premises and the set of axioms. In the particular cases presented here, this elliptical quality is evident. The Marxian axiom that the Hegelian model represents reality really has no necessary connection with the arguments of the prior premise. In the same way, the assertion that industry is an authoritarian jungle is not a necessary conclusion from the Maslowian first premise. Prophetic axioms thus tend to take the form of bald truisms, whose validity is not in question and whose proof is either not offered at all, or is dependent upon the semantics of the axiom itself—tautological. In the simplest terms, then, deductive constructs developed by most axiomatic prophets tend to have a first premise that becomes a *cause* of the remainder of the structure, and this cause is often itself axiological in predication. Following this, when we try to break out the axiomatic bases for the prophetic system, we find axioms that should really be stated as hypotheses (with the first premises then properly becoming metahypotheses). But prophets are not applauded for their uncertainty; therefore, any hypothetical qualifications surrounding popular axiomatic systems is quickly erased and veneered with a most aggressive certainty.

Moving on, in the case of formal axiomatic systems, we move from the specification of axioms to the level of *laws* (e.g., idempotency, complementarity, as defined in mathematics). These tend to specify certain unambiguous functional constraints: like the rules of chess, they bound the "moves" permissible under the axiomatic system. Every move that one might contemplate in mathematics, music, chess, etc., thus may be evaluated in terms of its permissibility given the set of laws *qua* functional axioms. To establish the validity or invalidity of an action, we thus have recourse to the formal logic of *syllogisms* (or to any other formal logic we might impose, such as Boolean algebra or the predicate calculus). The validity of an action may thus be determined unequivocally with respect to the laws of the system. The axioms and laws thus serve as a concatentative and interconnected set of premises from which a binary conclusion may be drawn: an action is either permissible, or not. Different logic may impose different qualifications, but actions are evaluated according to deterministic criteria. Thus, in the confines of a formalistic axiomatic system, the creation of new forms or structures or implications is always a disciplined process, disciplined specifically by the supra-laws of deductive logic itself.

But in axiomatic prophecies (prophetic logics), we often find the proper syllogism replaced with an *enthymeme.* An enthymematic construct is one which generates conclusions, but where one or more real premises are left unspecified. For example, let us consider a sequence of internal deductions drawn from the Maslowian system (and I believe this a fairly good representation of what Maslow had in mind):

1. The happy worker is a productive worker.
2. The happy worker enjoys freedom from exogenous controls.
3. The productive worker is therefore a self-regulated worker.

The enthymematic quality of this set of statements transpires when we try to equate happiness with self-regulation. As some social theorists have pointed out, it is entirely possible that freeing workers from exogenous controls may frustrate them and actually increase anxiety. The happy worker may in fact be one laboring under explicit constraints rather than one having broad license. Thus, it is *logically* possible that for some individuals, self-actualization implies the existence of constraints and the freedom from having to make decisions or regulate one's own affairs. It is therefore possible that efforts at introducing industrial democracy may actually be counter-productive beyond a certain point. As a very wise scholar once pointed out, the implicit equation of happiness with lack of regulation may reflect the industrial psychologist's own proclivities, and not necessarily those of the subjects he studies [13]. In summary, then, the Maslowian scheme excludes a logical possiblity, and therefore produces an enthymematic rather than a properly syllogistic deductive system. As a result, the entire Maslowian scheme becomes fundamentally hypothetical in nature, and those who pursue its implications without regard to the possibility of error in premises merely court dysfunction or surprise.

There is also an enthymematic quality to the Marxian scheme, but of a somewhat different order. Note that its ultimate implication is that the cause of the dialectic can best be served by maintaining and even exacerbating the conflict between capital and labor. But for those enjoying a relatively objective perspective, the cause of socialism may also be served by setting up a nonconstant-sum game between capital and labor, such that both may benefit together. The "victory of the proletariat" is left largely undefined in Marxian analysis (we don't know exactly what it entails). But we do know that the victory of the proletariat cannot be *exclusively* associated with a zero-sum game, for there is a logical possibility that a nonconstant-sum game would also arrive at the postulated "socialistic solution." Thus the Marxian system also becomes probabilistic, despite its deterministic pretensions. And, just as with the Maslowian scheme, those who attempt to draw policy inferences from its substance do so at the risk of ultimate embarrassment and dysfunction.

Many grand societal prophecies thus tend to generate *non sequiturs* when implemented. A social scientist may obey the axiomatic dictates set out, impose the prescriptions implied, and obtain an effect that is not at all contemplated in the basis of the structure. That is, the logical flaws in the constructs tend to make them poor predictors of real-world responses to actions imposed under their precepts. And the reason for this, ultimately, is that most axiomatic prophecies have components that are insufficient as *descriptors* of socio-behavioral realities.

We can see even from this abbreviated discussion that there are ample opportunities for irrationality to enter into the structures of the axiomatic prophets. At the risk of being somewhat redundant, I list here some of the more common symptoms of axiomatic insufficiency:

1. The tendency toward ellipticism, such that the relationships between components of the conceptual structure are either incompletely specified or insufficiently qualified.
2. The tendency to preface a deductive structure with an unvalidated first premise that casts the axiomatic prophetic platform into a non-apodictical mold. This means that the structure's substance must be accepted more on faith than on evidence, much as with axiological constructs.
3. The tendency to employ terms which are ill-defined, or defined only idiosyncratically (such as the Maslowian use of the term "self-actualization"). The employment of ill-defined variables means, in effect, that the structures are difficult to attack on objective grounds,

but always possible to defend by merely reading in different implications.

4. The tendency to extend beyond their limits models or assertions that hold logical (or empirical) validity only within some range—that is, the tendency to generate unbounded formulas or theses. For example, there is really no attempt anywhere in the Maslowian system to suggest what might be the magnitude (or quantitative aspects) of the relationship between self-regulation and productivity. (Beyond a certain point, would not freedom from exogenous controls necessarily lead to a loss of productivity?) In general, there is the almost constant failure to take propositions to their logical conclusion, and specify ranges or limits of relevance. As a result, there is a distinctly non-operational quality to most axiomatic prophetic platforms.
5. Finally, there is the tendency we mentioned earlier: the failure to provide deductive inferences which exhaust the logical possibilities. This is indicated by concentration on a single conclusion from a set of premises, where there really exists some array of alternative conclusions which could be drawn. This gives the axiomatic prophetic system a distinctly evangelistic or adversarial quality.

It is important to understand that the faults just listed are perils of any deductive exercise having pretensions to explain real-world phenomena. It is also important to see that most axiomatic prophets have had such pretensions. They are the progenitors of modern social and behavioral science, either directly or by accident of history. This is true essentially of Plato, Aristotle, Augustine, Maslow, Marx, Lenin, Mao, Locke, Marshall, Keynes, Freud, Weber, Hegel, Skinner, Smith, and a regiment of less notable theorists. And it is through these prophets—and their apostles and successors—that the bounds of societal consciousness have been fixed, at least in large part. For it is they who have given us the conceptual masks through which we process reality, and have thereby restructured the "real world" in an evidentiary sense. Just as the moralistic assertions of the axiological prophets set the limits of spiritual and affective behavior, so the efforts of the axiomatic prophets have set the parameters of the socio-economic, socio-political and politico-economic forms which man has evolved. Local, environmental conditions (e.g., contextual predicates) have often led to variations on the themes of the prophets, but the basic melodies are of ancient familiarity.

It is interesting to note that most of the axiomatic prophets sought to defend their societal schemes with reference to the prevailing theories of the natural world. For example, the capitalist economic theory of Adam Smith had intellectual roots in eighteenth-century biology; the social Darwinists drew their inspiration from nineteenth-century evolutionary theory. To a certain extent, then, the work of the axiomatic prophets may be explained as an attempt to inject properties of nature into the societies of men, on the assumption that what was operative in the natural domain was the ultimately functional referent for the artifices of civilization.

However, there was always a less than complete comprehension of the natural processes they sought to define and emulate, and the concept of the natural world changed through time. Thus, the social Darwinist's biological assumptions are not valid within today's evolutionary theory, nor is Adam Smith's biology the same as mid-twentieth-century biology. Yet the appeal of the axiomatic societal schemes seems to remain long after the validity of the naturalistic premises has disappeared. The explanation of this phenomenon is perhaps revealing. Biology proceeds as a more or less rigorous, nomothetic science, where the acceptance or rejection of a theoretical component rests on effectively objective grounds. But the socio-behavioral sciences—the inheritors, apologists and executors of the grand theoretical schemes of the axiomatic prophets—cannot pretend to the same epistemological tradition. We are, therefore, quicker to accept—and less willing to reject—any construct which gives us personal satisfaction or which holds normative appeal. And, because most of the axiomatic platforms suffer from the several logical or constructual defects mentioned above, they tend to be effectively indefensible on strictly objective grounds. Therefore, they tend to become objects of affective significance both to socio-behavioral scientists and the population at large. Thus, the errors that so often enter into the axiomatic prophetic process tend to blur the distinction between axiological and axiomatic constructs.

It will be seen in the next chapter that there are substantive differences in the societal structures determined by axiological as opposed to axiomatic platforms. As a rule, however, where an axiomatic prophecy holds sway, the social and spiritual sectors of a society are dependent for their significance on (or subordinate to) the economic and political sectors. Where an axiological modality reigns, economic and political aspects of society are subordinate to, or corollaries of, the social and spiritual ordering (as in ascriptive systems) [14].

The normal societal system may thus have a predominating axiological platform *and* a predominating axiomatic base, complementing each other in that they pertain to basically different sectors. An anomalous situation may emerge when there are two axiological or two axiomatic platforms, arrayed in competition. And as we have already mentioned, the conflict between platforms is likely to be severe, largely because no prophetic platform can escape the depth of affection and defense which attend basically *a prioristic* positions. Thus, despite the axiomatic positions' pretensions to logical discipline, conflict is not likely to be resolved objectively, for the platforms are often fundamentally irrational. Therefore, axiomatic prophetic conflicts are most often going to take the same form as axiological ones, rife with recrimination, condemnation and the failure to separate the individual from the belief. But we shall have more to say about this later.

A competitive relationship between two prophetic platforms is likely not to last long (though it may be intense), and be due mainly to problems of phasing. In what we might call the *normal* situation, the axiological and axiomatic platforms will tend to be adequately correlated. That is, the socio-spiritual predicates will not be entirely inconsistent with the politico-economic predicates. They will tend either to support each other, or to largely ignore each other. But there are times when interference might take place. For example, let us suppose that there is in force a proscriptive axiology being challenged by a prescriptive axiology, and that the axiomatic platform dominating the political and economic sectors is congruent with the proscriptive platform. Thus, the behavioral dictates of the emergent prescriptive axiology are likely to be inimical to the interests of the prevailing axiomatic platform. For example, the counter-cultural movement in the United States (characterized by a disavowal of social differentiation, consumption, industrial discipline, etc.) is inimical to the interests of the prevailing axiomatic base (industrial capitalism). The dominant axiological platform is still a variation on the Protestant ethic, which distinctly supports capitalist, industrialist ambitions. Now, were the counter-cultural axiology to dislodge the Protestant ethic—and were the capitalist axiomatic base not to yield to, say, a socialistic one—then we could have an *anomalous cultural context:* one where the axiological and axiomatic predicates work against each other, so that the system as a whole becomes something *less* than the sum of its parts [15]. Thus an anomalous cultural context is counter-synergetic, in that the socio-behavioral and politico-economic sectors are operating at cross-purposes. The usual situation would be, of course, that a shift on the axiological dimension would initiate a sympathetic shift on the axiomatic, or the reverse. But the shifts may not be accomplished at the same rate. Therefore, there is always a potential for localized out-of-phase conditions between the socio-behavioral and politico-economic sectors. Any comprehension of the dynamics of socio-behavioral systems must involve an appreciation of this possibility, and we shall have an opportunity to discuss it in greater detail later on.

At any rate, there is now a chance for us to become a bit more precise about the cultural context *per se*. For the concepts of axiological and axiomatic predicates may be dealt with at a different level of analysis than we have employed thus far.

THE DIMENSIONS OF CULTURE

Again, the cultural envelope within which a system labors will comprise a set of axiological and/or axiomatic predicates. As we shall later see, some societal systems are subject only to axiological predicates, some only to axiomatic predicates, and others to both at once. In general, however, we may suggest that when we know the cultural predicates to which a system's population is subject, some portion of that system's behavior is *explained*. That is, when we have access to the axiological and/or axiomatic apriorisms, the *cause* of certain activities is analytically accessible to us.

For our immediate purposes, the behaviors predicated on intrinsic and realistic bases may be thought of as those events *unexplained* by axiological or axiomatic predicates. In the statistical sense, the sum of the unexplained events constitutes a variance—a sort of residual error term. This idea is sketched in figure 3.4.

The probability distributions are arrayed over a set of events that rep-

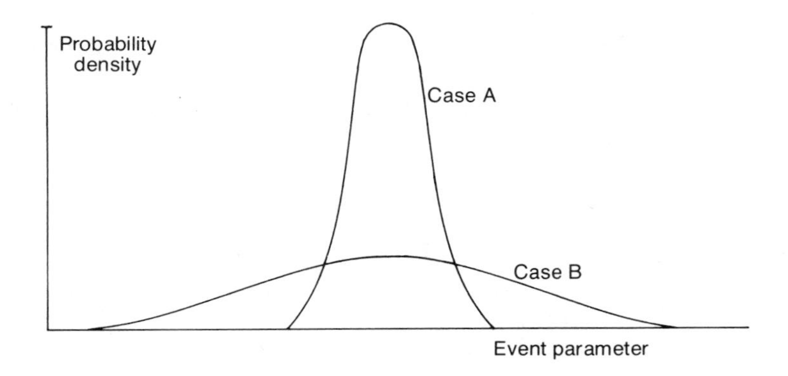

FIGURE 3.4: The variance-based cases.

resent alternative responses to some predicate which has arisen (e.g., the responses are conditional with respect to the predicate). The implication is that events (responses) are similar if they are close together on the horizontal axis. Case A shows the situation where there was a strong clustering of responses, such that a considerable majority of the system population all responded in much the same way. In Case B, the responses of the population were distributed over a much wider range, involving both a greater number of alternatives exercised, and a greater dispersion. Now, as a general rule, a system with a powerful cultural engine will tend to show tight clustering for a wide range of predicate response sets. That is, where the aprioristic constraints are operative, we expect to find a significant proportion of the population responding in much the same way to a wide variety of different predicates. On the other hand, where the system is an essentially realistic one—or where there is a strong intrinsic component—we will expect greater variation in responses to any particular predicate. Let's amplify these propositions.

What we are specifically concerned about in this section is what we shall call the *coefficient of cultural control* [16]. There are at least three variables of which the coefficient of control is composed:

Extent The proportion of all circumstances in which *a priori* predicates are potentially operative. In specific terms, this means the extent to which all circumstances requiring some sort of system response (or response by the population of the system) have a relevant axiological or axiomatic predicate. Thus, cultural bases will differ in the degree to which they try to determine the behavior of the population, some being more limited than others in their implications. In some systems, virtually all possible contingencies will be covered by some sort of *a priori* dictate; in others, *a priori* predicates will rule only a small proportion of activities.

Intensity The proportion of the system population susceptible to *a priori* predicates. A system where a significant proportion of the population is affected by *a priori* predicates is said to have strongly cultural determination. A system where only scattered pockets of the population are so affected is one with weak cultural determination.

Potency The extent to which *a priori* predicates—for susceptible individuals and appropriate circumstances—provide the only or the major basis for behavior. This variable measures the magnitude of determination. In cases where the *a priori* predicates are simply idle references (e.g., where they receive largely just lip-service or abstract or rhetorical deference), the cultural base may be considered a weak engine of behavior. On the other hand, where the *a priori* predicates exhaust the bases of behavior, they are highly potent.

Now, in moving from the individual as the unit of analysis to the system *per se*, we have another concern: the extent to which the *a priori* or cultural base is heterogeneous or homogeneous. Recall from an earlier discussion that the concept of a system for our purposes here is defined in probabilistic terms (as instances where the probability of contact between individuals exceeds some specific threshold). All we are interested in is the extent to which a certain group of individuals interact (or the probability of their interaction), and we need make no assumptions about whether or not they have shared values, or other substantive presumptions. Thus, again, it is possible for us to have a unit of analysis where the cultural base is diverse (or heterogeneous), whereas many studies in the tradition of Western sociology enter shared values as a virtual prerequisite for a socio-behavioral system. So, once we allow the cultural base to be a system variable rather than a premise, we have another attribute with which we need be concerned:

Diversity — The extent to which the cultural base of a system is partitioned rather than homogeneous. Specifically, we are here concerned with the extent to which a single set of axiomatic and axiological predicates rules, rather than the system's housing two or more distinct cultural engines.

Now, as we earlier suggested, it is one of the characteristics of *complex* systems that there will be some degree of cultural diversity, such that different segments of the population (or at least those susceptible to *a priori* determinations) will respond to different axiological-axiomatic bases. And whenever we have a heterogeneous (partitioned) cultural base, we must consider other variables:

Distribution — The algorithm by which the cultural base is partitioned. There are basically two modalities which need concern us:

Clustered, where the several subcultures follow neat demographic lines, partitioning the population into well-bounded and essentially exclusive segments.

Stratified, where there are no neat, *a priori* predictable grounds along which the several subcultures are defined, so that cultural predicates are distributed throughout the population in a complex way.

Macro-Relationship The nature of the interrelationship between the different cultural bases (or subcultures, subsystems, partitions):

Complementary, where a symbiotic or supportive relationship exists.

Competitive, where the partitions are real or potential adversaries.

These, then, are the six major dimensions of culture. Specification on all these dimensions is required if we are to assess the structure of a system with respect to its cultural properties. (We would, of course, have to add specifications on the realistic and intrinsic dimensions were we to attempt a complete inventory of a complex societal system).

One can see immediately that the coefficient of cultural control will generally be greater—that is, the system's behavior will be easier both to manipulate and to predict—if certain properties predominate on the cultural dimensions: those which act to reduce the variance of behavior within the system. In table 3.2 we have isolated this set of properties, and paired them with those that will act to decrease the coefficient of control (increase the variance).

The overall implication of the table is that the properties that appear in the left column all support the cause of behavioral control and prediction—and, to the extent that we have the contents of the cultural base at hand, analytical and managerial simplicity as well. On the other hand, the properties in the right-hand column are those that serve to introduce variance into the system's behavioral repertoire, making prediction and control much more difficult. Let's explore these assertions in a bit more detail, elaborating on the table in the process.

Initially, we can see that behavioral predictability or tractability is improved by any mechanism that determines that one from among any set of alternatives will be *the only probable response*. In this regard, consider figure 3.5. Obviously, a system that could be modeled in terms

TABLE 3.2: An Elaboration of the Culture Dimensions

	Polar Properties	
Dimensions	Low Variance	High Variance
1. Extent	**Exhaustive:** mystical or ideological systems (e.g., theocracies)	**Limited:** anarchic, adolescent or instrumental systems (e.g., technocracies)
2. Intensity	**Significant:** ritualized or ascriptive systems (e.g., primitive subsistence systems)	**Attenuated:** materialistic or secularized systems (e.g., hedonistic cultures)
3. Potency	**Strong:** moralistic, ethical systems (institutionalized cultures)	**Weak:** Utilitarian systems (relativistic cultures)
4. Diversity	**Homogeneous:** acculturated systems (e.g., ethnic cultures)	**Heterogeneous: ad hoc** or differentiated systems (e.g., immigrant-based systems)
5. Distribution	**Clustered:** segmented systems (e.g., federations)	**Stratified:** reticular systems (e.g., agglomerations)
6. Macro-relationships	**Complementary:** symbiont systems (prototypical ecosystems)	**Competitive:** tribal or legislated systems (zoned cultures)

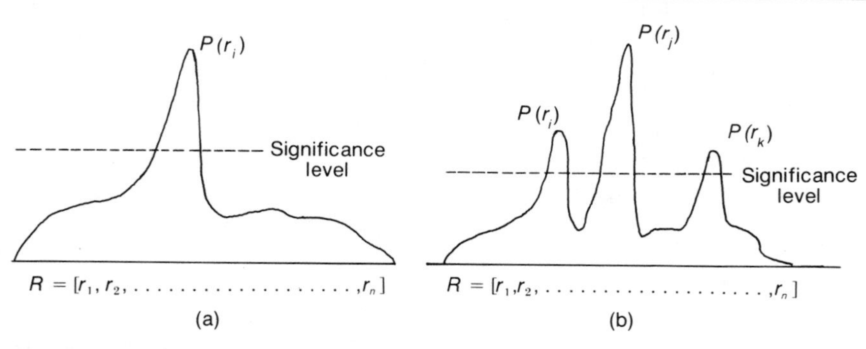

FIGURE 3.5: Stimulus/response. (a) The deterministic case, where one and only one response has any significant probability of occurring in the face of any given predicate. (b) The stochastic case, where, for any given predicate, more than one response becomes significantly probable.

of figure 3.5(a) is simpler than one that would have to be described by figure 3.5(b). In the former, a vast majority of the population all met a stimulus or condition with essentially the same response. In the latter, there were several distinct responses which gained predominance, indicating that the system may be essentially non-homogeneous, or that the cultural control coefficient may be weak. In this instance, then, the figures may be thought of as representing not probability distributions, but frequency distributions. That is, we imagine that we have collected empirical data on the responses of the population to some real-world, historical stimulus or event, and from these data constructed the distributions.

Now, when we treat the figures as proper probability rather than frequency distributions, we get a somewhat different interpretation. These figures then become *generic* (e.g., ideal-type) referents which could be analyzed as follows: The system described by Figure 3.5(a) would be expected to exhibit a strongly clustered response to any stimulus or event that might arise in the environment. That is, it is a *finite-state* societal system, such that there is expected to be only a single response for every unique stimulus or event demanding action. The stochastic distribution, on the other hand, suggests that there may be several responses—each having some significant probability of occurrence—with respect to any event or stimulus demanding collective action. The critical assertion, then, is this: Properties in the left-hand column of Table 3.2 will tend to give rise to a case such as that described in (a), whereas properties in the right-hand column of the table will tend to lead to (b). In short, systems with a strong or significant cultural base will tend to be better behaved—and be more easily modeled—than systems where the coefficient of cultural control is weak or insignificant. Again, to the extent that a large proportion of the system population is suceptible to *a priori* discipline, then we may expect that the behavior of the individuals will be highly correlated, i.e., the responses will be similar. The degree of similarity will be enhanced to the extent that the *a priori* predicates completely determine behavior, leaving little or no latitude for intrinsic or realistic engination. Now, our ability to predict what response will occur is increased to the extent that the system is one in which only one cultural base predominates. However, should the cultural base be a partitioned one—such that there are several different *apriori* engines at work—then the cause of system simplicity is served to the extent that the various subcultures are in a complementary (e.g., symbiont), rather than competitive, relationship. Given the former, it is possible to make logical inferences about aggregate behaviors, for complementary relationships tend to be relatively stable. Thus, for example, if a stimulus arises and we know the *expected* response of some subculture A, then the conditions of complementarity will often allow us to make predictions about the response of some subculture B (or any subset of the various partitions arrayed in the complementary set).* Thus, in terms of analytical tractability, complementation may allow us to use *regression* logic to predict responses even when a cultural base is diverse. On the other hand, the existence of a competitive relationship among subcultures or partitions may introduce conflicts; for a competitive or adversary role implies a *strategic* interrelationship. In such a case, the behavior of the competitive subcultures may be entirely opportunistic, and hence highly stochastic. And stochastic structure not only makes a system more difficult to describe or analyze; it also makes it more difficult to manage.

The logic behind this assertion is as follows: Initially, when a system is ruled by intrinsic or realistic predicates, there is a greater opportunity for diversity of response to any event or stimulus that arises. The intrinsic behaviors are by definition unprecedented, idosyncratic and therefore largely beyond *a priori* assessment. In fact, intrinsic phenomena—driven by opportunistic or heuristic algorithms—generate a third system case, as diagrammed in figure 3.6. In this case, the set R constitutes the array of responses which we are able to predefine, none of which is significantly probable as the true behavior. Therefore, the significant probability goes to some response X, which is *a priori* undefined, and merely an unstructured entity in the set of residuals (which represents an aggregate error term).

Again, the properties on the right-hand side of table 3.2 are all those which argue for intrinsic or realistic responses, or which serve to weaken the effectiveness or determinacy of the cultural base. Thus, first, when only a very limited number of circumstances or stimuli are cov-

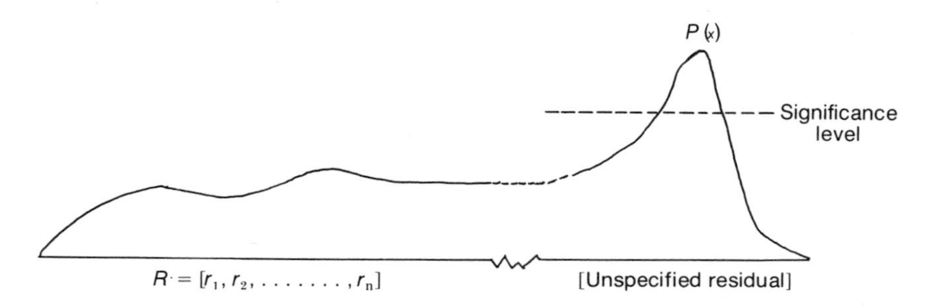

Figure 3.6: Stimulus/response: The indeterminate case, where there is no response which can be assigned any significant probability of occurence in the face of any given predicate, implying that some non-predefinable response will be the true behavior.

*For more on this, see the Appendix to this chapter.

ered by *a priori* predicates, the necessity for turning to either the intrinsic or realistic bases is increased. By the same token, when the *a priori* susceptibilities of the population are highly attenuated (so that only a small proportion of the population is amenable to rationalistic predicates), or where the potency of the *a priori* predicates is weak (that is, they are there, but used only for rhetorical or abstract reference rather than as real determinants of behavior), the recourse to intrinsic or realistic predicates is increased. And we have already suggested that cultural diversity, competitive macro-relationships and incongruent micro-relationships serve to dilute the determinacy of any axiological or axiomatic bases which may be operative.

To conclude this chapter we now introduce some simple set symbology, and see if we can grasp in a somewhat more precise way the structural implications of cultural predicates.

The Cultural Set Concepts

In the case of the cultural predicates of societal behavior and structure, simple set concepts may serve us well. For they allow us to draw a graphic portrait of the operations of the cultural engines. To begin, then, consider the simple diagram in figure 3.7. Very simply, the universal set U contains the specification for all activities undertaken within the confines of some societal system. The subset A are those behaviors explicitly proscribed by the dominant axiology. The complement of the proscribed set is A^c, the subset of all behaviors not explicitly proscribed, i.e., the set of all *a priori* uncondemned activities.

Now let us specify another collection of sets in figure 3.8. This time, however, we shall be concerned with the population of the organization or system, not with the behaviors. In the simplest terms, the set X is the part of the population obeying the proscriptions established under

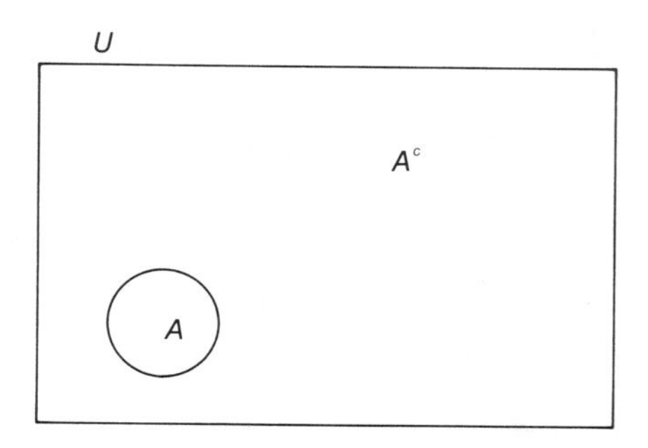

FIGURE 3.7: The behavior sets: proscriptive case. U is the universal set of all behaviors, $A \subset U$ is the subset of proscribed behaviors, and $A^c \subset U$ is the subset of non-proscribed behaviors.

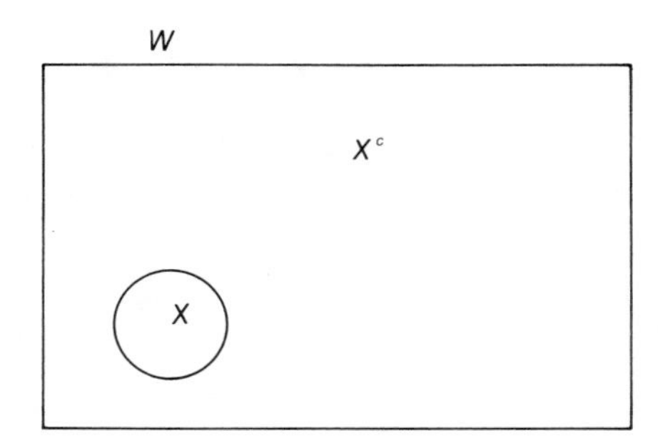

FIGURE 3.8: The population sets: proscriptive case. W is the universal population set for the system (e.g., all individuals), $X \subset W$ is the subset of all individuals susceptible to the particular prescriptive axiology, and $X^c \subset W$ is the subset of non-susceptible individuals.

the axiology A, with the population subset X not committed to the proscriptions. This does not mean that all members of X^c consistently violate the proscriptions. It simply means that they cannot be presumed to be constrained *a priori* from their exercise. For example, suppose one of the proscribed behaviors is the eating of meat on Friday. The presumption is that all members of X will obey this constraint, but it does not imply that no members of X^c will eat fish or cheese, etc.

Now, in practice, there are several different categories of behaviors which have to be considered. So we can add another component to our simple set concepts, this time in the form of a domain O. We shall consider O to be a partition of the domain of behaviors—i.e., the set U—so that each behavior, belongs to one of its members, as follows:

$O = [f, i, n, m, t, s]$,
where
f = proscribed behaviors (forbidden),
i = imperative behaviors (prescribed),
n = normative behaviors (implied),
m = non-normative (contradictory) behaviors,
t = tolerable behaviors,
s = sanctionable (*a posteriori*) behaviors.

Now, every societal or organizational system may be expected to have a value-assignment function, G which will act to associate some element of O with every member of U, the universal set of empirical behaviors. What we are interested in, then, is the *mapping* which takes place under the value-assignment function. In short, we are interested in the subset of the Cartesian product of the sets O and U determined by G:

$$G \subset \{u \in U; o \in 0\}.$$

To a certain extent, societal differentiation may become comprehensible in terms of differences in the function G, in addition to any differences in the set U of activities undertaken within the system. Thus, two organizations or societal systems may differ either because their basic activities are different—e.g., $U_m \neq U_n$—or because their evaluation of the activities contained in $U_m \cap U_n$ are different (that is, because $G_m \neq G_n$). In the latter case, the set of activities performed by system m is the same as that performed by system n, but the values attached to certain of these activities are different.

To a certain extent, this is the implication of the transfer between a proscriptive and a prescriptive axiology. In the most extreme case, as was described earlier, there is a complete lack of correspondence between the values. That is, the value-structures of the proscriptive and prescriptive axiologies are effective contraries. By this we mean that some activity u carrying a value f under the proscriptive function G_m carries a value i under the prescriptive function G_n, and vice versa. More simply, $G_m(u, o) = f$ if and only if $G_n(u, o) = i$. If, for example, there is a proscription against helping one's neighbor (stemming from the dictate that "God helps those who help themselves"), then the prescriptive transformation would turn this activity into an imperative (e.g., God loves those who help their neighbors). Again, then, in the simplest case, prescriptive axiologies merely reverse the value placed on activities falling into the set A of the proscriptive system. Thus, we have figure 3.9. Now, this is the *simplest* prescriptive case, because the sets A and B are identical. That is, they contain the same behaviors. But the value associated with the elements of the set B is the class i, [$G(b; o) = i$ for all $b_i \in B$], whereas that associated with those of the set A in Figure 3.7 was f.

The simple prescriptive case is the fullest expression of the pendularity associated with the prophetic process. But there are other cases where the transformation is incomplete, or where the proscribed and prescribed activities are not identical. In this instance, the interpretation of the above figure would be that $B \subset A$, or B is a proper subset of A, and not identical to it (i.e., there are some elements in A not in B). In some cases, we might find the association between the set of proscribed and prescribed behaviors quite weak, so that $[A \cap B] \rightarrow 0$. This expression means that some measure of the size of the intersection of sets A and B approaches zero, so that the two sets are effectively disjoint or independent. That is, very few activities proscribed by the prior proscriptive axiology have been translated into imperatives by the emergent prescriptive axiology. As a general rule, then, we can assess the *degree of pendularity* by considering the size of the intersection between the two activity sets, $A \cap B = v$, with a high value for v indicating that the prescriptive axiology is highly reactive and thus suggesting that the societal or organizational system is undergoing strong cultural pendularity. In this regard, consider figure 3.10.

To a certain extent, the potential for conflict within a society is linked to the value of v, the extent to which the same behaviors are given opposite values by different segments of the system or organizational population. Where the two axiologies are respectively condemning and demanding *different* activities, there is some room for accommodation. But there is also a time factor to consider. As a general rule, cultural change proceeds slowly, that is, the pendular process follows something like a learning curve, as in figure 3.11. What the curve suggests is this: Initially (in stage I) the prescriptive challenge proceeds very slowly, so that dv/dt (the derivative) is very small. In stage II, the rate at which the prescriptive axiology is exhausting the proscriptive increases: the change in v is accelerating. Eventually, during stage III, we get the situation where $A = B = A \cap B$, so that the prescriptive engine has completely taken over. Finally, in stage IV, the pendular process proceeds to a reversal, the prescriptions gradually being transformed into proscriptions again. This reversal of the process is due to a factor we mentioned earlier: the tendency for societal or cultural change to proceed too far in any given direction. Thus, we reach stage IV when the permissiveness associated with the prescriptive axiology begins to introduce disorder or uncertainty. The result is an increase in the perceived value of the order associated with the proscriptive system, and a return to discipline. We shall explore the implications of these various stages in a later chapter; here it is only the learning curve itself which concerns us.

A process of cultural change may stop at any point on the curve. Where v is small, the change process was a weak or attenuated one. If it reaches a point where v is higher, the process is more complete. Moreover, the morphology of the learning curve for a particular context is important. For example, if the rate at which the prescriptive set exhausts the proscriptive (or vice versa) is very high, we may talk about a cultural *revolution*. Where the rate is very low we speak of cultural *evolution*. As a general rule, the potential for conflict or severe disorientation depends on the rapidity of the changes, or on the average value of dv/dt through the interval of change. When we later speak of the *cultural derivative*, it will be this of which we are speaking.

Now let us consider another aspect of our set constructs. Recall that

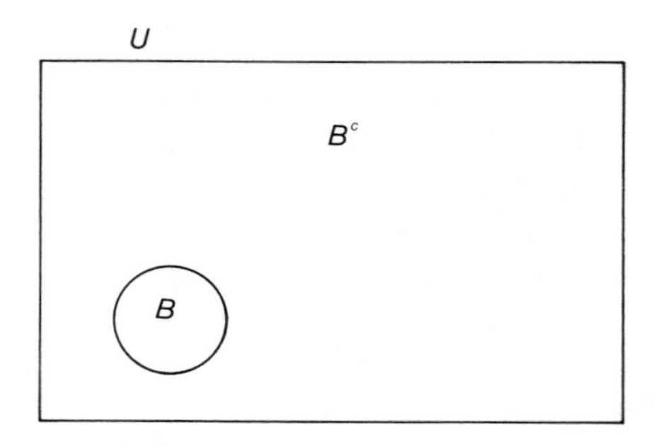

FIGURE 3.9: The behavioral set: the simple prescriptive case. U is the universe of all activities, $B \subseteq A \subset U$ is the subset of prescribed behaviors, and $B^c \subseteq A^c \subset U$ is the subset of non-prescribed behaviors.

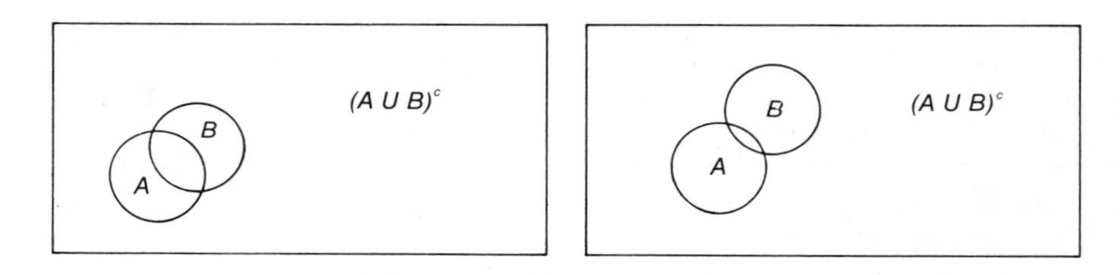

FIGURE 3.10: (a) The strongly reactive case (v high). (b) The weakly reactive case (v low).

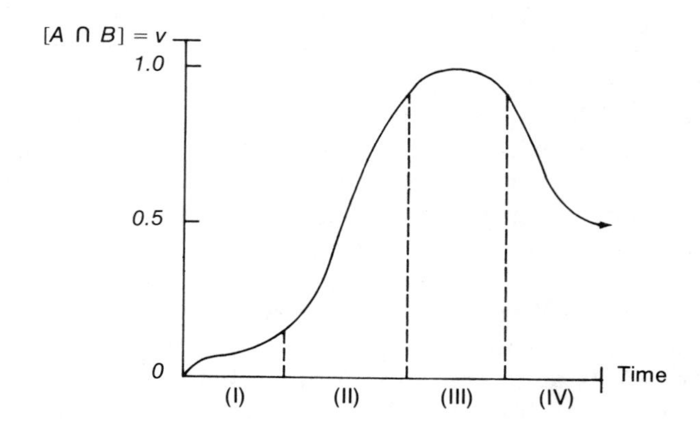

FIGURE 3.11: The cultural learning curve.

in Figures 3.7 and 3.9 there was a *complement*—the range of behaviors which are not included in either a proscriptive or a prescriptive subset. Yet the cultural value function will be operating on these as well: on the set A^c or B^c. But of the various values available to be assigned, f and i are preempted, for these refer to proscribed or prescribed behaviors, respectively. Thus, the complementary behaviors—those lying outside the range of axiological specification—must be assigned one of the four remaining values in the set O: They may be normative (n), non-normative (m), tolerable (t) or sanctionable (s). We shall now explore the implications of these value assignments in a bit more detail.

Initially, the value n, referring to *normative* behaviors, reflects the condition where a behavior is neither explicitly proscribed nor explicitly prescribed, but is rather a direct result of some prior predicate, and consonant with the "spirit" rather than the letter of some aprioristic base. Specifically, normative behaviors are *expected*, given the cultural predicates. In the axiological domain, normative behaviors would be products of exegesis directed on the proscriptive or prescriptive platform—behaviors that would have been proscribed or prescribed except for the fact that they are natural corollaries and hence treated implicitly. Normative behaviors have more meaning when we consider the axiomatic base of culture, for they are the activities that are consistent with the principles resident in the axiomatic base. For example, given an essentially axiomatic platform such as capitalist theory, Marxian socialism or anarchic socio-economics, there will be a large range of behaviors which may be defined as direct products of deductive inference (or logical exegesis) from the principles explicated in the axiomatic prophecy. These behaviors are not explicated (i.e., they are neither proscribed nor prescribed), but they are *implied*. Now, those activities which carry a negative implication—e.g., behaviors clearly inconsistent with the axiomatic principles—would constitute the set of non-normative behaviors.

Tolerable behaviors, on the other hand, are not necessarily derived from any aprioristic base (axiologial or axiomatic), but are deemed inoffensive. That is, they are not proscribed, prescribed or implied, and thus escape *a priori* evaluation altogether. Sanctionable behaviors, on the other hand, are those which are neither proscribed, prescribed nor implied, but which—when they emerge—are condemned. In every proscriptive system—and in axiomatic systems as well—the *a priori* constraints do not anticipate the complete range of dysfunctional or contradictory behaviors which ingenious individuals can define. In a proscriptive system, all behaviors which are not explicitly condemned are usually considered permissible. In an axiomatic system, if the normative behaviors must be inferred from the prophetic base, so must dysfunctional or undesired behaviors (and to the extent that the principles of avoidance are explicated, axiomatic systems take on a proscriptive quality, usually by restricting the means that may be legitimately employed to obtain an end). But neither the proscriptive nor the axiomatic systems can tolerate all unexpected behaviors that emerge. So the right is reserved to condemn—on an *ad hoc, a posteriori* basis—those innovations that slipped through the proscriptive lacunae, but that would have been condemned had they been preconceived by the proscriptive or axiomatic agents (e.g., the priesthood or the legislators). As such, sanctionable behaviors have a very limited life. They may be tried once or twice, but their exercise usually means that they will be entered eventually into the list of proscribed behaviors (the set A) or into the non-normative set M of behaviors associated with any axiomatic system.

In summary, then, the set of universal behaviors (activities) within any system may be partitioned as follows:

$$U = A \cup B \cup N \cup M \cup T \cup S$$

where

$$G(a) = f \quad \text{for} \quad a \epsilon A,$$
$$G(b) = i \quad \text{for} \quad b \epsilon B,$$

etc.

We shall return shortly to the concept of the various "values" assigned behaviors within the sphere of culture, but first let us make some entirely natural extentions to our basic set concepts. Initially, there may be more than one axiological set operating within the

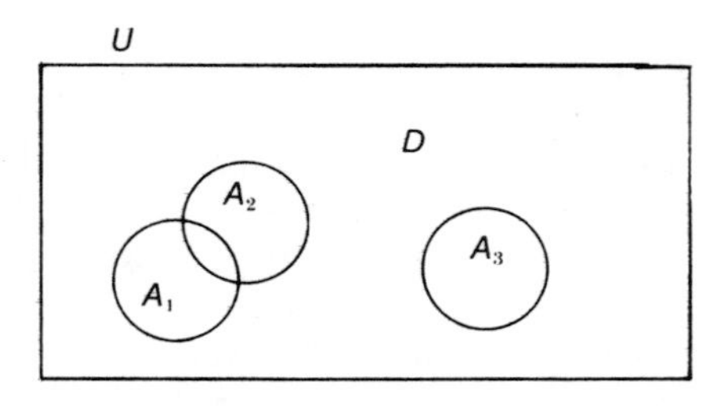

FIGURE 3.12: Multiple proscriptive axiologies. U is the universal behavior (activity) set, $A_i \subset U$ are subsets of proscribed activities, and $D = U - (\forall A_i)$ is the set of behaviors whose value $\neq$ f.

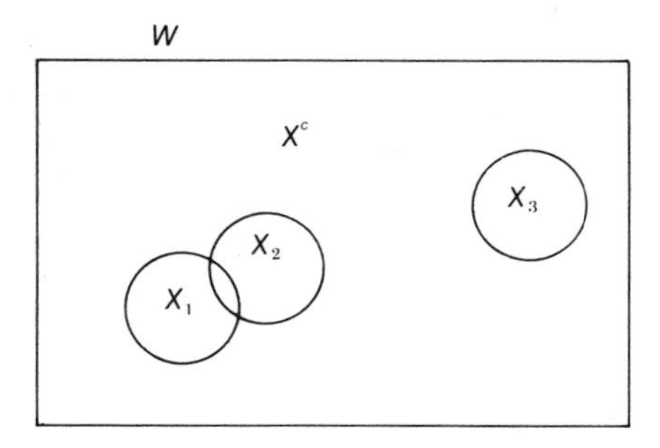

FIGURE 3.13: The population set for multiple proscriptive axiologies. W is the univeral population set (all individuals in the system); X_1, X_2, and $X_3 \subset W$ are the population subsets susceptible to one or another of the proscriptive axiologies, and $X^c \subset W$ is the subset of all individuals in the population not susceptible to any axiological proscriptions.

confines of any system, and these axiologies may be either proscriptive or prescriptive. Where we are concerned only with proscriptive axiologies, we might have a situation like that in figure 3.12. In this case, A_1, A_2 and A_3 are all identifiable proscriptive axiologies, each rejecting some subset of the universe of behaviors, A_1 and A_2 share some proscriptions, so that $A_1 \cap A_2 \neq \emptyset$. The proscriptive axiology A_3, however, shares no behaviors with the other two axiologies, and is therefore disjoint: $(A_1, A_2) \cap A_3 = \emptyset$). The subset D is the range of behaviors not proscribed by any axiology. Each of the axiologies would partition D in a unique way. For example, the complement of A_1 would be $D \cup A^2 \cup A_3 - (A_1 \cap A_2)$, etc. Where multiple proscriptive axiologies were in evidence within a system, there would also be a partitioning of the population as in figure 3.13.

Essentially the same set situation would emerge, of course, were we considering prescriptive axiologies or axiomatic engines, or some combination. Though there are some very interesting conflicts which can occur when we have proscriptive, prescriptive and axiomatic predicates all operating in the same context, we shall defer a treatment of these until late in the next chapter. Here, however, we may explore some of the direct implications of our set constructs with respect to the dimensions of culture, especially with respect to the summary given in table 3.2.

Recall that the first dimension of culture was *extent*: The proportion of all behaviors susceptible to cultural (*a priori*) determination. In terms of our simple set symbology, this would be measured as the ratio $[A \cup B \cup N \cup M \cup S] / [U]$. Where the value of this ratio is high, we suggest that the system or organization is highly acculturated, i.e., a majority of activities are subject to *a priori* prohibition, approval, etc. Where the ratio is low, this would indicate that the system is essentially *realistic* (i.e., places great emphasis on response to *a posteriori* or environmental predicates) or one which allows idiosyncratic (intrinsic) behaviors great latitude. In short, the measure of the weakness of cultural engination is the ratio of T/U where T is the residual factor and represents the index of *system tolerance*.

Similarly, we may get a rough idea of the degree of *repression* inherent in a system by taking the ratio $[A \cup M \cup S]/[U]$, this being the size of the set of all proscription (axiological and axiomatic, explicit and potential) relative to the universal behavioral set. So also, the degree of cultural *directivity* may be measured by $[B \cup N]/[U]$, which indicates the extent to which behaviors are prescribed by *a priori* predicates. Continuing, the index of a system's algorithmic quality—the efficiency of its cultural mechanisms—may be reflected by $[S]/[U]$, though we must say more about this index in a later section. Another interesting ratio is $[N]/[A \cup B]$ which expresses the relative strength of axiological versus axiomatic predicates. This ratio we shall call the *cultural balance*. Finally, the ratio $[A]/[B]$, indicates the extent to which either proscriptive or prescriptive predicates dominate. When $[A] \gg [B]$, we say that the culture is a *passive* one, more concerned with what may not be done than with what should be done. When $[B] \gg [A]$, the cultural predicates tend to *activate* the population. In the next chapter we shall have more to say about these ratios as well.

The second cultural dimension of interest to us was *intensity*, defined as the proportion of the population susceptible to cultural determinations (either axiomatic or axiological). In this case, intensity would be measured as $(\Sigma_{r=1}^{n}[X_r])/W$. This is the number of all individuals susceptible to any type of cultural determinations, relative to the population as a whole. When this ratio is high, then we again say that the system is acculturated. When it is low, a significant majority of the system population is being driven by realistic or intrinsic behavioral engines. Now we may be a bit more explicit. Let's see r take on any of three values:

1 = proscriptive axiology,
2 = prescriptive axiology,
3 = axiomatic engine.

Now let us define the basic population unit as $X_{r,j,}$, where j indicates some specific platform within the subset r (the jth member of the rth cultural subset). Thus, the intensity associated with any specific cultural engine would be $[X_{r,j}]/[W]$. Then the intensity of any particular modality—axiological proscription or prescription, axiomatic engination—would be given by $\Sigma_{j=1}^{m}[X_{r,j}]$. With these ratios at hand, we can make comparisons between the intensity of any cultural engine and any other, or between the various cultural modalities—e.g., the ratio $\Sigma_{j=1}^{n}[X_{1,j}] / \Sigma_{j=1}^{n} [X_{2,j}]$ would yield a value for the intensity of proscriptive relative to prescriptive axiologies. On the other hand, the ratio $\Sigma_{r=1}^{n} [X_{r,1}] / \Sigma_{r=1} [X_{r,2}]$ would reflect the comparative intensity of two different proscriptive axiologies within some system.

We may approach the third dimension of culture—*potency*, or the strength of the cultural predicates—on two levels. At the aggregate level, potency may be represented as a product set relating all behaviors susceptible to cultural determination with all individuals susceptible to axiomatic or axiological predicates: $(A \cup B \cup N) \times U / U \times W$. But there is another level on which we must operate: that of individual *utility*. Each of the behaviors (or activities) within a system may be thought to carry some *index of satisfaction* for each member of the population. This index of satisfaction is what economists call utility. Let us now expand our notation and denote each individual within a system by $x_{r,j}$. The index r, as before, refers to the general cultural modality to which that individual is susceptible (i.e., a value of 1 indicates a proscriptive axiology, 2 a prescriptive axiology, and 3 an axiomatic platform). The index j refers to the specific axiology or axiomatic platform to which the individual is responsive, within set r. The index x identifies the individual *per se*: he is the xth individual connected with a particular cultural engine within the specific cultural modality. Now, each individual will have a utility function that determines the satisfaction index he attaches to each activity within the system, or at least those of interest to him:

$$h\,(r, j, x \in X_{r,j}; u_i \in U)$$

However, the utility that any individual places on any cultural predicate is in part conditioned by the axiological or axiomatic position to which he responds. But each activity or behavior may also be thought of as exercising an appeal in its own right, (as having an independent (unmediated) utility. Now, the potency of any cultural predicate (proscription, prescription or principle) can really only be tested indirectly through a surrogation analysis.

A brief example will help, although we can by no means do more than merely approach the issue here. Suppose an individual who operates a drug company is offered an opportunity to make $Y by selling a quantity of drugs to an unlicensed buyer. Suppose, further, that were he to sell the same quantity through legitimate channels, he would receive only $V ($V<Y$). Let us further suppose that no conflict enters his calculus in the form of any expectation of being caught and punished. Now, as we have suggested, the regulation of commercial affairs is usually dominated by an axiomatic engine of some kind (e.g., the Judeo-Christian ethic or the tenets of capitalism translated into legalistic restraints on the means by which commercial ends can be obtained). If the individual refuses to carry through with the transaction, the net difference $Y-V$ in dollar terms is a lower bound on a surrogate measure for the potency of the axiological dictate under which the individual labors.

What is being suggested here—and will later be made explicit—is that the potency of any cultural predicate has an empirically transparent force in terms of its effect on the psychological welfare of the individual. The implication of the example is that there was perceived to be a psychological penalty involved in the transaction were it carried through, and that this penalty—if translated to dollar terms—would have been at least equal to and perhaps greater than the material advantage to be gained (i.e. the difference between $Y and $V). For the strict realist, material gains are all that count, as he is not susceptible to any *a priori* constraints or dictates. Thus, whenever we have an individual presumed to be susceptible to some cultural determinant, we must look for the tradeoff between moral (or psychological) and material utilities, the former generally being measurable only as a reflection of the material opportunity cost the individual is willing to suffer. In our little example, then, the utility functions of the individual eventually worked out in such a way that the perceived advantages in taking the illegal (or immoral) profit disappeared, that is, $h(u_r) \leq h(u_k)$, where u_r is the material satisfaction and u_k the perceived moral satisfaction.

There is one further point to be made here: we can see that the utility function is really a more detailed version of the general value function we introduced earlier: $G(u_r \in U, o_r \in O)$. The value function placed each activity or behavior in a generic *category* (e.g., proscribed, normative). The utility functions $h(x_{r,j,k}; u_r \in U)$ now seek to assign a *magnitude* of satisfaction to the behavior or activity. Now, the issue is this: to what extent is the utility (satisfaction index) assigned by the individual independent of the categorization undertaken by the cultural engine to which the individual is responsive? Let's work this through a bit. Take, for example, the act (or behavior) of intoxication. Most axiomatic positions that have any current popularity condemn intoxication on "logical" grounds, e.g., as an activity which diminishes the individual's health, sociability, etc. Virtually all proscriptive axiologies condemn it out of hand as "immoral". But it cannot be denied that intoxication has a strong perceived satisfaction for many individuals. Let us suggest then that though the cultural value function has assigned this activity to the class f $[g(u_r o_r) = f]$, the individual's utility function assigns it a positive utility: $h(u_r) > 0$. Now in a Freudian world, these two calculations are independent. The individual (realistic) evaluation of the satisfaction to be obtained from intoxication exists independent of its cultural devaluation. Now, it was precisely the conflict between these two forms of evaluation which concerned Freud and most neo-Freudians, who described it as the competition between the "drives" of the individual and the sanctions or constraints of society. The individual is like the proverbial donkey between two bales of hay. If he elects to vent his urge and get drunk, he is condemned to repent at leisure through the mechanism of *guilt* (if not through a hangover). If, on the other hand, he stays his impulse and refrains, he is condemned to suffer the *frustrations* of a repressed drive.

Now, in utility terms, the exercise or rejection of intoxication is determined by the perceived magnitude of the expectation of guilt through the contravening of a cultural dictate, in comparison with the magnitude of the frustration. But because frustration is real and tangible, and because expectations of guilt are always probabilistic and amorphous, there may be a weighting which makes restraint improbable. Moreover, certain societal mechanisms have been evolved to alleviate guilt even should it arise (e.g., most of the prescriptive axiologies perform this service, as do certain institutional mechanisms such as Catholic confession and the sharing of one's guilt in Quaker meetings). I shall not make much of it here, but all indications are that the expectations of guilt do not really operate to constrain behavior in the way that Freud and many modern psychologists and psychiatrists suggest. Individuals, by and large, are not prone to make complicated extrapolations or review the second-order consequences of activities. Therefore (and I offer this more as a speculative hypothesis than as a firm proposition) it seems likely that cultural predicates, operating through the value function, serve to *discount, a priori*, the perceived utility of culturally condemned or proscribed behaviors. That is, the fact that an activity belongs to the class of forbidden behaviors ($u_i \epsilon A \subset U$) acts to dilute the utility that would be associated with that behavior by the individual not susceptible to cultural determinants. In short, for any $u \epsilon A$, the perceived utility of the culturally susceptible individual will be lower than that of the realistic or idiosyncratic individual. Therefore, to some extent at least, cultural proscriptions gain their moment through the *a priori* dampening of the individual's utility functions, and not solely through the more commonly recognized process of the individual's forecasting the fact that guilt will follow a contravention. When men are at a sales conference in a big city, and the possibility of a long night of comradely drinking holds itself out, some may say: "Oh, it isn't worth it". Indeed, this is precisely what the man means: given the *a priori* reduction of the utility of such activity caused by the operation of the cultural proscriptions, it will be "worth it" only if the situation promises exceptionally great benefits. Thus, once again we have the potency of the cultural predicate accessible to us in terms of the material opportunity cost foregone: it is equal to or greater than the utility the individual places on the activity from which he refrains. But, as mentioned, we shall wait to explore this issue further in another chapter.

The fourth dimension of culture—the *diversity* of the cultural base—depends on two factors. The first, and most obvious, is the extent to which different axiological or axiomatic platforms are operating within the confines of the system. Secondly, we are concerned about the extent to which these various platforms have similar value functions. If they tend to value the same behaviors in the same way, then diversity is dampened despite the existence of multiple cultural engines. On the other hand, if one cultural engine tends to proscribe behaviors prescribed by another, then diversity is amplified. If, however, there exist both a proscriptive and prescriptive axiology within a system, and there is no overlap, then the cultural structure is a simpler one. As a general rule, then, what we are looking for in any system—on the diversity dimension—is the result of the specific value functions for any operative cultural engines. That is, each cultural engine—axiological or axiomatic—will have a specific $G_r(u_r; o_r)$, which assigns each behavior to one or another of our six categories (*f*, *i*, etc.). Now, to the extent that the results of the evaluation function are isomorphic—that is, the extent to which $G_1 = G_2 = G_3 = \cdots = G_n$—we have cultural homogeneity. To the extent that the evaluations differ, we have cultural heterogeneity. Naturally, where there is only one cultural engine, we have only one G and therefore have the highest level of homogeneity.

The fifth of our dimensions concerns the *distribution* of cultural predicates across the population. Recall from an earlier discussion that we identified the population susceptible to any specific cultural engine as $X_{r,j,}$ where r referred to the modality (proscriptive, prescriptive or axiomatic) and j referred to the specific axiology or axiomatic platform. Now, where the distribution of cultural predicates is *clustered*, we get a situation where the cultural predicates are distributed according to some *natural* demographic factor: race, religion, age, ethnic or geographic basis, etc. That is, with any demographic cluster d_r there will be associated some population subset $X_{r,j}$ (or, in cases of less complete correspondence, merely an X_r). In set terms, then, clusteration is indicated by the condition $D_i = X_{ij}$: the demographic and cultural subpopulations are the same. Now, where the cultural base is *stratified*, the cultural predicates are distributed throughout the population on other than demographic grounds, and therefore the analytical complexity of the cultural modeling task will increase.

Cultural clustering exists, for example, where we have both a proscriptive and a prescriptive axiology in operation, but the proscriptive one appeals predominantly to those above the age of forty, while the prescriptive one appeals mainly to those between the ages of eighteen and forty. In another situation, we may have two proscriptive axiologies in operation, say Puritanism and Quakerism. Now, where the *potency* of these cultural engines is very high, they may lead to a demographic clusteration, e.g., Quakers and Puritans tend to cluster together in well-bounded communities. In the former example (the proscriptive and prescriptive axiologies clustered with respect to age) the cultural engine responded to a natural division of the population. In the latter, the cultural engine *caused* the demographic clusteration. In either case, we get the demographic-cultural correlations we expect.

Where a cultural engine is weak—i.e., where it is largely just an ancillary aspect of existence—or where the cultural engine has no demographic correlation (e.g., the Marxian ideology operating within affluent nations, without real regard for socio-economic, geographical or racial bases), then we have a stratified case, which is often very difficult to model with any precision. In short, as we have often suggested, cultural engines do not always lead to (demographically) viable partitionings of systems, a point we shall return to in the next chapter. Here we must turn to the last of our cultural dimensions: the nature of the macro-relations within the cultural context.

To a certain extent, we have already treated some macro-relations—those between subcultures within some systems—in earlier discussions (particularly those associated with Figures 3.12 and 3.13). Very briefly, whenever there are two or more axiological engines operating within a culture, the probability of competition (and eventually of conflict) increases to the extent that they have a significant intersection: $[A_1 \cap A_2] \gg \epsilon$. Where the axiologies are both of the same modality—both proscriptive or both prescriptive—competition arises from a perversity of human nature. We tend to be most aggressive against those who only partially share our values or interests than against those who completely deny them or have no interest in them. Thus, the promise of competition between (say) two Christians (a Protestant and a Catholic) is generally greater than between (say) a Protestant and a Buddhist. Where we have a proscriptive and prescriptive axiology both operating within the same domain, conflict will again be increased to the extent that the one is trying to proscribe some of the same behaviors prescribed by the other. To the extent that the intersection is negligible $[A \cap B] \ll \epsilon$, there is some opportunity for complementation or at least mutual tolerance.

There is a rationale behind these assertions though necessarily a hypothetical one. It is housed within what we may call the *second-order difference* model of conflict. It suggests this: whenever two groups of individuals come into contact, the possibility of a conflictful relationship is enhanced to the extent that first-order differences are not offset by second-order differences. Let's illustrate this with a couple of examples. Suppose we have two groups of individuals whose first-order differences are economic: one group is wealthy, the other is poor. To the extent that there are offsetting second-order differences—say the members of the rich group are all fat or all old and the members of the second group are all svelte and young—the probability of conflict is reduced below that which would occur were both groups old and fat or both groups svelte and young. The mechanism by which conflict is abated (although this does not necessarily mean that a cooperative or complementary relationship will result) is that the members of the second (poorer) group are able to offset their material disadvantage in two ways: (1) they may have the satisfaction of knowing that they are better looking; or, (2) they may suppose that when they get older, they too will be rich.

Similarly, suppose again that we have two groups, one rich and another poor (the first-order difference). But let's say that the poor group are all religious people and the rich group all atheists. The second group is unlikely to come into conflict with the first because the second group may rationalize the first-order difference in the following way: "Those people are getting rich only at the peril of their mortal souls". Conflict is avoided because the second group sees a compensatory calculus at work, perhaps deriving from the biblical dictate that it is harder for a camel to pass through the eye of a needle than for a rich man to get to heaven. In the coming chapter, we shall make much of this concept of the compensatory calculus, for it is at the heart of the appeal offered by virtually all proscriptive axiologies.

The assertion that conflict is more likely between members of two Christian sects than between either and a Buddhist emerges directly from the difference model. The Protestant and the Buddhist have religion as the first-order difference; so do the Protestant and the Catholic. But in terms of second-order differences (e.g., race, geographical area, ethnic background), the Buddhist is clearly more distinguished from either of the Christians than they are from each other. That is, the Protestant may suggest that the reason that the Buddhist is not a Protestant is that he came from a different land, that his skin is yellow or that his race is unenlightened. On the other hand, the chances are that the Protestant and Catholic share many second-order characteristics: both may be white, both may come from the same district, etc. In this case, then, there is no evident rationalization why the Catholic should not be a Protestant and vice versa. Here, then, the potential for conflict is very high.

This second-order difference model finds its direct corollary in the societal context through the operation of the value functions [the $G(u; o)$'s by which cultural engines assign *a priori* values to certain behavior or activity subsets]. Contrary to the obvious idea, our model would suggest that the more similar the evaluations performed on activities by two different subcultures, the greater the probability of conflict. The rationale derives directly from our above discussion, with this generic addition: the more closely the value systems of two subcultures approximate each other, the less ostensible reason there is for the existence of two subcultures. In short, the more tenuous the distinctions between any two cultural systems, the more likely there is to be an aggressive and competitive relationship between the two.

In summary, then, a useful way of analyzing cultural conflict (or collective conflict in general) is to consider the fact of differences between subcultures or subsystems as constituting the first-order difference—one group calling itself Catholic, another Protestant; one Marxist, another capitalist. The second-order difference then is measured by the difference in the value functions, the respective $G(u; o)$'s. To the extent that these evaluational functions are similar (which implies our previous condition that they have a significant intersection of behaviors in which they are interested), the probability of competition and conflict is increased. To the extent that the value functions are different, the rationale for the existence of two separate subcultures gains material weight and irrational conflict may be avoided. Again, it must be suggested that the second-order difference model is merely a hypothetical construct, only introduced here because we have such a paucity of models designed to deal with cultural conflict. We shall therefore have to defend the model later in this book, when we discuss cultural dynamics in the coming chapter.

Appendix to Chapter Three: Ideologics

by Stephen E. Seadler

The root of "culture" is *cult(us)*, whose meanings include tilling and worship, and which is derived from the past participle of *colere*, to cultivate, worship, dwell. Not surprisingly we find that the word "cultivate" has the same root as "culture", signifying both worship and tilling — which is understandable in view of early man's close association of magic and religion with working the land to grow crops.

This leads us to the English word "cult," and we find dictionaries defining it in a sociological context as a group having a sacred ideology and a set of rites with sacred symbols. Being aware of the ancient origins of this religious emphasis, and correcting for dictionaries' simple addition of ideology, rites and symbols, we update the meaning for contemporary purposes by defining "cult" as *a group having an ideology and congruent sets of behaviors and customs together with unifying symbols*. The English stem "cult" now properly constitutes the basis for the construct "culture", which in turn constitutes the basis for the construct "society".

Turning next to "predicates", we have only to note that it is derived from the root *praedicāt(us)*, past participle of *praedicāre*, to declare publicly, assert.

Pulling all of this together, we find that *a society organizes and operates on the basis of public assertions of an ideology and its congruent sets of behaviors and customs together with unifying symbols*.

By employing the concept of congruence we defer the question of causality, that is, of whether ideologies determine behaviors and customs, or vice versa; or whether they are interactively interdependent, and if so, which is predominant under what circumstances. Instead, etymologically and historically, we leave matters in the irrefutable statement that *ideologies are at the core of and integral with the cultural predicates of societal behavior and structure*.

This prelude not only sets the conceptual stage for what follows, but seeks also to treat the cultural bias in the United States among academicians as well as the public against explicit ideological considerations. This bias has resulted in ideological neglect whose dysfunctional severity has been at least proportional to ideological salience. In fact, ideological salience itself is generally unrecognized, or where recognized ignored, or where not ignored pronounced dead or otherwise dismissed.

Let us, therefore, engage in some much-needed ideological considerations, and begin by returning to the preludial conceptual stage, and noting that the profound and pervasive primacy of ideology calls for a special, new science to deal with it on both the theoretical and operational levels. However, the word "ideology" is confusingly ambiguous in that it represents both process and substance — ideology is the study of ideology. Consequently, the term *ideologics* (ī′ dē•ō•lō′•gics) is used to denote this new science of the nature, structure and dynamics of ideologies, and of their roles in individual and societal phenomena. Ideologics thus joins the family of behavioral sciences, along with economics, history, psychology and sociology. In fact, ideologics is so fundamental to all of them* that it can be looked upon as the *basic* behavioral science.

Ideologic methodology falls into three broad categories, of which the first two are theoretical-analytical. Category 1 comprises adaptations of extant methodologies, such as multivariate statistical procedures. Category 2 comprises new methodologies, and presently includes the emerging field of *ideologic topology*. Category 3 comprises applications and operations. The remainder of this Appendix will be devoted to brief presentations in each of these categories.

Cluster and Factor-Analytic Applications

A good methodology with which to begin Category 1, in that it serves to introduce fundamental concepts, is ideologic cluster analysis, an application of one member of the class of multivariate statistical procedures for the behavioral sciences. The objectives of this technique differ from those of other multivariate techniques, such as multiple regression and correlation analysis, in that it delineates similarities between observations of sampled individuals, and thereby mathematically defines population sets in ideological terms.

*With the exception of extreme schools of behaviorism, wherein all "mentalism" is dismissed, cognition excised, sapience denied, and *homo sapiens* reduced to mere *Hominidae*—a homunculus of purely stimulus-response mechanisms, to be conditioned and reconditioned according to the dictates of behavioral elites. However, this is not science, but the transformation of laboratory data into ideology—authoritarian ideology at that—and should be treated by ideologics along with other nineteenth- and twentieth-century authoritarian ideologies—all of which are based on core sets of psuedo-scientific transformations, extrapolations and distortions.

"Individuals" can be entities such as companies, industries, unions, media, people, cities, and nations. The observations (measurements) of these individuals on a number of characteristics provide a score for each individual on each characteristic, which scores together provide a "profile" of him (it).

The measurements are obtained from a battery of m tests, or a single test of m distinct parts, each of which contains a set of measurements representing a particular characteristic, or dimension. A "test" can be any feasible instrument or method, such as a survey questionnaire (oral or written), or content analysis of speeches, articles and broadcasts. With such data one can proceed to a number of cluster-analytic objectives, of which ideologics selects the ability to determine—without advance assignment of observations to groups—what clusters (groups) of observations (individuals) are similar with respect to their ideological profiles.

The procedure sets up a hypothesis as to an individual's group membership in the form of a likelihood function $P(H_j|X_t) = P_{ij}, j = 1,2,\ldots, g, i = 1,2,\ldots, N$, which reads: The probability of the hypothesis H_j that individual i is a member of group j, given that his score vector X_i, is P_{ij}. Centours (centile contours) are computed which describe a multivariate normal distribution of the m test scores of the i individuals in terms of hyperellipsoids, each of which represents a centour, i.e., the locus of points of a particular frequency, and is derived from χ^2.

This is the same χ^2 distribution as is usually so designated. In this application it is called the "classification", and is a function $X_i'D^{-1}x_i$ of the distribution's dispersion (variance-covariance) matrix D, and an m-element vector of deviation scores, $X_i = [X_{1i}-X_1, X_{2i}-X_2, \ldots, X_{mi}-X_m]$, and its transpose X'_i. For instance, the last element is the deviation of individual i's score on test m from the mean of all scores on test m. Each test j in the battery of m tests assesses a different dimension, or characteristic, and is represented by one of the m orthogonal coordinate axes in the m-dimensional test space. The larger the value of χ^2, the larger the size of the hyperellipsoid corresponding to it—that is, the larger the proportion of the group's membership that falls within it. Therefore, the smaller an individual's χ^2 for a particular ideological group, the closer he is to its centroid, and the more certain he is to be a member of it.

D is an estimate of the dispersion of the population from which the sample was drawn, and within that, D_j is an estimate of the dispersion of the groups $j = 1,2,\ldots,g$ within the sample. An individual in the sample has an *a priori* probability p_j of belonging (in terms of test scores) to group j, determined from estimated real-world relative frequencies of membership in groups.

Pulling the foregoing together, with the aid of Bayes's theorem, we obtain an expression of the form $P_{ij} = P(H_j|X_i) = f(p_k, |D_k|, 8^2{}_k)$, $k = 1,2,\ldots,j,\ldots,g$, $i = 1,2,\ldots, N$, for the probability of membership of individual i in group j.

With this procedure it becomes feasible mathematically:

1. To define an ideology by defining membership in it in terms of real-world measurements of various characteristics of individuals, which can be people or groups of people.
2. To define and determine the significant (that is, well-populated) ideological dimensions and the degree of significance of each.
3. To define and measure (a) the distance between ideologies, (b) an individual's position within an ideology, and (c) an individual's distance from: (i) the center of his own ideological group, (ii) the center of another ideological group, and (iii) another individual in his own or another ideological group.

An advantage of cluster-analytic ideologics for some purposes is that the same m-test battery or m-part test is applied to all individuals in the sample; that is, no prior ordering or assignment of the m is made among the individuals in the sample. However, delineation of the m characteristics is conceived and designed in advance, on the basis of expert knowledge of the ideologies involved. In some cases, though, it is desirable to determine both the number and the composition of dimensions after having conducted exploratory research. Again, mathematico-analytic ideologic means are available.

In this approach expert knowledge is employed in the advance design of the detailed measures, but dimensionality is subsequently defined, this time by one of a number of multivariate techniques generically known as *factor analysis*. Here the primary objective is, in the ideologic adaptation, to determine the fewest, simplest and most meaningful ideological dimensions latent in the set of measures. The resultant dimensions can then be employed in a wide range of ideologic applications. Therefore, let us now turn to this methodology.

We begin with a sample of N "individuals" i (same range of entities as before) that have been measured on n observed variables X by means of some test or battery of tests. A "variable" can be a test or any item or combination of items on a test, such as a question or group of questions on some aspect of ideological belief, or an element of ideological content in speeches, articles and broadcast programs, with values obtained by any of many available rating or scaling techniques.

The value of the variable X_j for individual i is represented by X_{ji}, and treated by the usual statistical techniques to obtain its standardized value z_{ji}. The set z_{ji} for all i ($i = 1,2,\ldots,N$) is the variable z_j in standard form. The intercorrelations among all the variables are then computed, and the resultant correlation matrix R becomes the primary datum, the foundation for analysis of the ideological domain of concern.

The method of *principal components* within this genre of analysis is based on the simple model consisting of n linear equations

$$z_j = a_{j1}F_1 + a_{j2}F_2 + \ldots + a_{jn}F_n = \Sigma a_{jp}F_p, \qquad j,p +1,2,\ldots,n.$$

These express each of the n observed variables in terms of n uncorrelated components, or factors, $F_1,F_2,\ldots,F_n$, and their factor coefficients, or "loadings", a_{jp}. Geometrically, this means that the standardized values for the N individuals on variable j locate a point for that variable in the N-dimensional sample space, that a line from the origin to that point is the variable's vector representation, and that the cosine of the angle between that vector and the vector for any other variable is the correlation between the two variables. The objective is to define the factors F_p and compute their loadings from the primary datum R expressed by this model.

The N-dimensional sample is thus reduced to an n-dimensional test space in which the test scores (variable values) form a hyperellipsoid. The procedures compute the principal reference axes, or components, that uniquely define the factors F_p that the variables are measuring in common. These factors will, upon content-based expert interpretation, constitute the ideological "constructs", or dimensions, we are seeking, and will then be given appropriate labels.

This definition is accomplished by computing that axis of the hyperellipsoid along which the projection of variable-points produces maximum variance, then constructing a second axis orthogonal to the first along which the remaining variance is maximized, then constructing a third axis along which, etc., and so on until new axes produce only insignificant amounts of variance. Not only will the principal axes (components, factors, dimensions) of the ideological domain of concern have been located thereby, but they will have been reduced to the smallest number required to define the domain.

Expert content knowledge then reenters the procedure to interpret the meaning of the resultant factors F_p on the basis of their loadings a_{jp} in relation to the nature of the variables z_j. In the event that the first set of derived principal-component axes is not considered substantively satisfactory in terms of ideological interpretation, they can be rotated repeatedly until a substantively satisfactory set is located. A variety of rotational schemes have been devised for this, differing in objectives, criteria and properties.

Whether we focus primarily on defining ideological groups, as with cluster analysis, or on defining ideological dimensions, as with factor analysis, we become involved with sets of ideological elements. These elements can be analyzed and dealt with from another, quite different perspective.

Ideologic Topology

Although derived from general topology, ideologic topology (ideotopology) bears a fundamental difference: Rather than developing as a pure (mathematical) science, ideotopology develops as an applied science, and, furthermore, as a mission-oriented applied science. Ultimately a rigorous formulation may evolve, and if so, the progression will be similar to that of other branches of science—for instance, probability theory evolving from games of chance, and the related trio of number theory, set theory and topology evolving from everyday affairs involving ten fingers (*digits* in Latin) and the arrangement of objects in places (*tópos* in Greek).

An essential difference in the progression of ideotopology is that it does not start with raw experience and then creatively abstract and mathematize, but, rather, starts with already highly developed, rigorous mathematics and then adapts the mathematics so as to achieve a predetermined operational goal: to develop real-world ways of dealing with ideologies and their implicit and explicit manifestations. The motivation for choosing this route is the critically urgent need for operational ideologics—the province of the third category of ideologic methodologies. That need, in turn, leads to the need for ideotopology, to which we now turn.

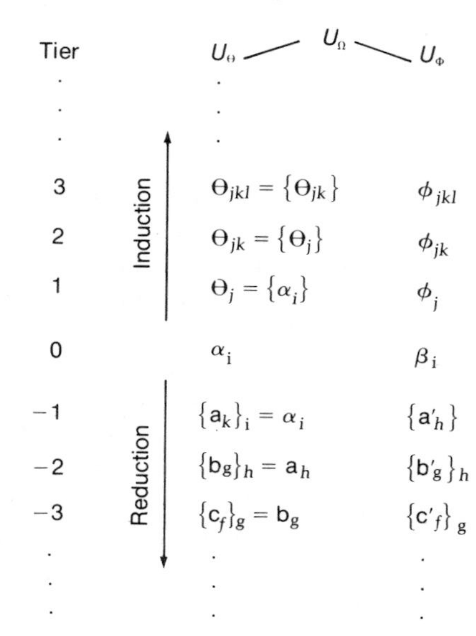

FIGURE 3.14: Specific subscripts designate specifiic sets in a lower tier.

We begin by defining the universe of discourse U_Ω in terms of a hierarchy of sets and elements, as illustrated in figure 3.14. At any level of the hierarchy a class of sets is denoted by Θ with primed subscripts, e.g., $\Theta_{j'}, j' \in \{j'\}$, and a family (collection) of classes by C_{k}', Θ_{j}'. This provides for substantative specificity while maintaining notational economy. Reduction of the set elements into successively lower components is also provided for by an "archeoarchy"* of more primitive elements, represented by lower-case Latin letters. For instance, at tier -1 elements are represented by a_h's, a particular subset of the set of which constitutes the element α_i at tier 0.

Some sets in U_a will have properties that distinquish them as *knowledge sets*, and those are denoted by Θ. Other sets In U_Ω will have properties that distinguish them as *ideological sets*, and these are denoted by ϕ's, with the same induction-reduction schema as for Θ's. Classes and families are similarly denoted by Φ and $C\Phi$, respectively, the latter with primed subscripts.

The choice of Θ and ϕ às set symbols is significant. Knowledge is based on theory, and traces back conceptually to the Greek *theoría*, whose first three letters are those of *theta*, θ. Ideology is a branch of philosophy, which traces back to the Greek *philosophia*, whose first three letters are those of *phi*, ϕ. That there are distinctions between θ-sets and ϕ-sets is fundamental to ideotopology, and explication of the distinctions is one of its objectives.

The choice of α to represent the basic elements in set building and of Ω as the subscript for the universe of discourse is also significant, and derives from the expression "alpha and omega", signifying "the beginning and the end".

The collection of all functions f operating in U_Ω is represented by $\mathcal{J}(\Omega)$. Some f's operate so as to generate Θ's, and others operate on Θ's; both kinds are therefore called *knowledge functions*, the collection of which is denoted by $\mathcal{J}(\Theta)$. Other functions in U_Ω operate so as to generate ϕ's, and yet others operate on ϕ's; they are called *ideological functions*, the collection of which is denoted by $\mathcal{J}(\Phi)$. That is, $\mathcal{J}(\Phi)$ includes functions of the type $f : \Theta_{j1} \rightarrow \Theta_{j2}$ and $f: \phi_{j1} \leftarrow \phi_{j2}$. These collections also include, respectively, functions that generate Θ's and ϕ's from α's, i.e., of the type $f : \{\alpha_i\} \rightarrow \Theta_j$ and $f : \{\alpha_i\} \rightarrow \phi_j$. In addition, $\mathcal{J}(\Phi)$ includes functions of the type $f : \{\beta_i\} \rightarrow \phi$ and $f : \{\{\alpha_i\}\{\beta_i\}\} \rightarrow \phi$.

A class $\mathcal{T}$ of subsets of a set Θ defines a *topology* on Θ iff (= if and only if) $\mathcal{T}$ satisfies certain axioms, and the members of $\mathcal{T}$ together with Θ define a *knowledge topological space* (or, simply, *knowledge space*), $(\Theta, \mathcal{T})$. Similarly, a class $\mathcal{T}$ of subsets of a set ϕ defines an *ideological topological space*, (or, simply, *ideological space*), $(\phi, \mathcal{T})$.

The formation of topological spaces provides formal coherence to a

*Coined from the stem *archeo-* "primitive".

body of knowledge and to an ideology, and makes available (accessible) to ideologics the formalisms, principles and procedures of general topology for adaptation and conversion to operational ideologics.

Among the useful procedures thus made available are, for instance, the search for and analyses involving *functional continuities*, *homeomorphisms*, *bases* and *space connectedness*, among many others.

Functional continuity determines whether a mapping from a topology in one set to a topology in another set also works in reverse, i.e., is reciprocal. Homeomorphisms distinguish from among many functions possible between two spaces, each in a different set, those functions that preserve some aspect of the structure of the two spaces. Bases are analogous to the common understanding of the basis for saying such-and-such.

A class B of subsets of a topology T on a set X forms a basis for T iff: (1) X is the union of (includes all elements appearing in all of) the sets in B, and (2) the intersection (overlap) of any two sets is the union of members of B. Conceptually, therefore, we may say that a base B comprises a pervasive foundation for the knowledge space (Θ,T) or the ideological space (ϕ,T).

Connectedness is a property of a space provided that it is not disconnected. An ideological space (ϕ,T), for instance, is disconnected if ϕ is the union of two disjoint sets in T, that is, two sets with no intersection.

Another procedure holding promise for future development is *metrization*, which means finding a *metric*, or distance function d, on the ordered pairs of the elements of a set X. For instance, the topology T induced on a set ϕ by a metric d would be ϕ's metric topology, and ϕ together with its metric topology T would be its *ideometric space* (ϕ,d).

The potentials of ideotopology are clearly immense, in itself and in conjunction with category-1 methodologies.

For instance, it makes possible delineation of and operations with sets of cultural predicates, the extraction of their proscriptive and prescriptive axiologies, and, in conjunction with cluster-analytic ideologics, the delineation of their associated population sets, as sketched in the preceding section, thereby enhancing further work on the cultural-conflict model therein hypothesized.

Another area of aid to research will be in flagging and explicating *ideotopological fallacies*. One type might be considered an analogue of the well-known *ecological fallacy*, whereby major errors are made in nonexperimental research when one takes findings at one level of analysis and makes inferences as to what is or would be the case at another level of analysis. This type of fallacy is a consequence of the mathematical phenomenon that in the absence of complete randomization the correlation coefficient increases rapidly with the size of data groupings. However, this solely (merely) concerns the units of measurement, e.g., whether the data are in terms of people, counties or states; in any case we are still measuring the same variables (income, education, attitudes, values, etc.). There are other types of fallacies, profound ones, in the realms of substance and intellectual processing—beyond mere logic, beyond fallaciousness in the usual sense—that would be detectable by ideotopology. Some of this occurs in professional research, but by far the most egregious and lethal occurrences are in U_Ω, the universe of ideological discourse.

Applications and Operations

Which brings us to the third category of ideologic methodologies: applications and operations. Within this category, two classes are presently evident: (1) research in the behavioral sciences, and (2) real-world operations. Within the second class three types are evident: (1) cooperative proceedings, (2) adversary proceedings, and (3) negotiations.

The feasibility of category-3 methodolgies is greatly enhanced by modern technology. Large computers can store U_Ω and perform ideotopological procedures. In addition, they offer computer-based teleconferencing (CBT), an asynchronous remote-participation mode of conferring, with computer-resource access. Participants key their theses into terminals or consoles at their mutually remote locations at their convenience; their arguments are stored in a central computer and retrieved later by other participants via their terminals or consoles at their convenience, all participants having parallel access to the central or satellite computer data banks, computational programs and models.

The CBT technique inherently imposes discipline on the proceedings, and also filters out interpersonal and rhetorical "noise". This discipline and filtering in conjunction with ideotopology approaches achievement of the philosophers' quest since ancient times for discourse based on "pure reason".

Let it be clear, however, that the repetoire of Category-3 ideologics is not restricted to mathematics and "pure reason". It includes, as appropriate, rhetorical, educational, programmatic, political and informational dimensions, albeit based, again, as appropriate, on principles, procedures and findings of Category-1 and -2 ideologics.

It is well worth a further moment to clarify this point for mathematical readers and to allay anxieties on the part of non-mathematical administrators of our national and international destiny. In this writer's view, the concepts of "pure reason" and "value-free science" are illusions, and are not encouraged by mathematicians or physicists who comprehend the highly creative nature of all knowledge, and hence of the impact on science of human values, aesthetics, sensibilities and metaphysical inclinations. As I have recounted in an earlier essay,

> In his Herbert Spencer Lecture at Oxford in 1933, Einstein expressed his conviction that the basic concepts and laws of physical science are fictitious, free inventions of the human mind, and that any attempt to derive them from experience is doomed to failure.[17]

Even regarding the nature of set theory, which underlies general topology, we find that

> . . .the notions of a class . . . are fundamental in logic and mathematics . . . The use of good judgment in determining when a statement form is acceptable in defining a class seems to be unavoidable.[18]

Conversely, however, human judgment is a primitive creature, hopelessly inadequate for the needs of contemporary civilization, without the aid of mathematics. Ideologics could prove to be a major enhancement to the human intellect.

For the foreseeable future, the greatest value of ideologics will be in resolution of and defense against threats to the peace. In this perspective, the most urgent national and international need is for the development of "ideological defense systems", that is, systems designed to defend the peace by dealing with the ideological foundations of conflict. Such development would constitute a major response to the oft-quoted passage in the UNESCO Charter: "Since wars begin in the minds of men, it is in the minds of men that the defenses of peace must be constructed". Such development would also enhance the viability of socio-behavioral systems.

Notes

1. I have described this process in Chapter Six of my *Systems: Analysis, Administration and Architecture* (New York: Van Nostrand Reinhold, 1975) and in the last chapter of my *Administrative Decision Making: Extending the Bounds of Rationality* (New York: Van Nostrand Reinhold, 1977).
2. For a treatment of the precepts of action-research, see the articles by Eric Trist in *Experimenting with Organizational Life: The Action Research Approach* (edited by Alfred Clark, New York: Plenum Press, 1976).
3. Thomas Kuhn, *The Structure of Scientific Revolutions* (Chicago: University of Chicago Press, 1962).
4. Herbst has noted that insistence on consistent unidimensional understanding of human behavior underlies the totalitarian behavioral logic of hierarchical organizations. Though he refers to understanding individual attributes (e.g., good and evil), this may well extend to logics for understanding emergent system characteristics (e.g., economic, sociological), where insistence on dealing with only one or another aspect of the subject does real violence to our comprehension of realities. For more on this, see his *Alternatives to Hierarchies* (Netherlands: Mennen Asten, 1976).
5. The possibility that formal quantitative methods may edge quite closely to rhetorical processes is explored by Diesing in his *Patterns of Discovery in the Social Sciences* (Chicago: Aldine-Atherton, 1971 pp. 29–124).
6. For a classic interpretation of the role of isomorphisms, see Ludwig von Bertalanffy, *General System Theory* (New York: George Braziller, 1968, pp. 80–86).
7. Quoted from A. Rapoport's "The Search For Simplicity", in *The Relevance of General System Theory* (edited by Laszlo; New York: George Braziller, 1972, pp. 13–20).
8. For an elegant study of a case where realistic (empirically predicated) functions become imbedded in an apparently aprioristic base, see Roy Rappoport's *Pigs for the Ancestors: Ritual in the Ecology of a New Guinea People* (New Haven: Yale University Press, 1974).
9. See Chapter 1 of James Parkes's *Whose Land? A History of the Peoples of Palestine* (New York: Taplinger Publishing Co., 1971).
10. To some extent, Max Weber's *charismatic* leader—as a source of concepts which later become reified in terms of cultural codes—is a useful adjunct to this discussion. See the Henderson and Parsons's translation of his

Theory of Social and Economic Organization (New York: Oxford University Press, 1947).

11. For an analysis of the dialectical appreciation of history, see Chapter 6 of Louis O. Mink's *Mind, History and Dialectic* (Bloomington: Indiana University Press, 1969).
12. The concept of the various levels of analysis of systems (of which the parametric is one) is given in Section 5.1 of my *Systems: Analysis, Administration and Architecture* (New York: Van Nostrand Reinhold, 1975).
13. A remarkably precise and ingenious look at the precepts of the human-relations school is given by George Strauss in "Some Notes on Power Equalization" (in *The Social Science of Organizations*, edited by Leavitt; Englewood Cliffs: Prentice-Hall, 1963).
14. For an excellent discussion of the implications of rationalistic as opposed to instrumental bases for economic structure and behavior, see Walter Weisskopf's *Alienation and Economics* (New York: E. P. Dutton, 1971).
15. For an analysis of the importance of subcultures in conflict generation and resolution, see Stephen Hallmark's "Subcultures as a Focus of Analysis" in *Protagonists of Change: Subcultures in Development and Revolution* (edited by Said; Englewood Cliffs: Prentice-Hall, 1971, pp. 10–22).
16. For a specifically sociological interpretation of the concept of control, see the recent reissue Edward Ross's *Social Control: a Survey of the Foundations of Order* (Cleveland: Case-Western Press, 1969)
17. Seadler, Stephen E. "Ragnar Granit and the Sense of Ideas," *The American-Scandinavian Review* 44: No. 4, New York: The American-Scandinavian Foundation, December 1956. This essay reviewed the electrophysiological research of Granit (Director of the Nobel Institute for Neurophysiology) on the processes of sensory perception from the perspectives of epistemology, axiology and political philosophy.
18. Graves, Lawrence M. *The Theory of Function of Real Variâbles*. New York: McGraw-Hill, 1946, p. 7.

4
Ideal-Types and Societal Dynamics

In this chapter, we shall be trying to develop an array of ideal-type systems. The ambition is to construct a set of abstract referents against which the population of real-world societal phenomena may be compared. The ideal-type referents thus purport to be of both descriptive and predictive significance. They become the major substantive components in a theory of societal dynamics.

The obvious purpose of ideal-types is to serve as a mechanism for producing generalizations about very complex phenomena. Their efficiency is determined by the degree to which a larger number of specific phenomena become essentially comprehensible in terms of a smaller number of abstract constructs. Their utility, as with any scientific model, must be assessed with respect to the rightness of the predictions they generate. Therefore, ideal-types may generally be thought of as meta-hypotheses, and must be susceptible to validation or invalidation as would any hypothesis of normal science.

But we also want to demonstrate the operational aspects of the model-building technology we first introduced in Chapter One. Thus, our ideal-types will become equivalent to state variables, be lent a dynamic component and finally be transformed into quantitative terms (and placed in a Markov-type framework). In this way, we shall demonstrate a link between qualitative and quantitative analysis, and also show how *a priori* (deductive) constructs may be made apodictical: accessible to empirical validation by the procedures of normal experimental science.

A SYNOPTIC IDEAL-TYPE: THE MODALITIES OF ORGANIZATIONAL BEHAVIOR

In this chapter we shall be concerned with two genre of ideal-type constructs. First is what we shall refer to as the *synoptic* ideal-type. This, basically, serves merely as a vehicle for summarizing and establishing the gross relations between the heretofore isolated components of a theoretical or empirical base. It does not amplify or pretend to create new knowledge; it merely serves to impose some sort of structure on existing inferences or data. As we are operating primarily in the deductive domain—as befits our interest in qualitative analysis—our synoptic ideal-type will serve as a repository for the various deductive inferences we have previously generated. But we shall, in the second section of this chapter, be concerned with a more interesting and powerful device: the *elaborative* ideal-type. As we shall see, elaborative ideal-types are always built up from synoptic constructs, but they are ampliative in that they serve to extend theoretical or factual summaries into hypothetical structures, with definite predictive or descriptive implications. Thus, elaborative ideal-types are usually of direct substantive significance to the social scientist, whereas synoptic ideal-types are usually just analytical artifices, employed at a higher level of abstraction than those associated with elaborative constructs.

As synoptic ideal-types are usually the initiating constructs in a model-building enterprise, we shall start our inquiry with them. We are going to develop a synoptic ideal-type that attempts to explicate the relations between the bases of individual behavior and organizational (collective) behavior. We know now that societal endeavors generally become comprehensible in terms of two sweeping categories: *a posteriori* behaviors and *a priori* behaviors. As a rule, the prelude to the development of a synoptic ideal-type is the generation of a *process* model describing the various behavioral modalities of the subject with which we are concerned. This model sets the stage for the generation of the ideal-type construct, and usually involves the logic by which we attempt to reason from individual components to collective phenomena (irrespective of the specific context in which we are working).

For the case at hand, the process model that suits our purposes is given as Figure 4.1 below. Here we specify the linkages between the various components of our logical development thus far, and thus set out the different modalities of behavior that will have to be accommodated by the synoptic ideal-type construct we shall shortly develop (in Table 4.1). Arguing with respect to the process model, we can summarize some of this volume's key arguments to this point. We have seen, from Chapter One, that there are three bases of human behavior: (1) sense-driven, in its id-level and trial-and-error variants; (2) inductive, with its associative (regressive) and ampliative (inferential) variants; and (3) deductive, incorporating the exegetical and discursive modalities (the latter taking on both functional and pathological forms). In the second chapter we introduced the dialectical engine that serves to "switch" the individual to any of several behavioral engines in the face of some predicate or event demanding a response. Finally, in the last chapter we elaborated on the nature of the most significant engines of societal systems: *a priori* predicates.

The implication of the model is that we have a causal relationship of the following type: (1) Each of the bases of individual behavior is seen as the paramount engine for one or another of the engines of collective behavior. (2) The sense-driven individual behaviors are excepted here because, with very rare exceptions (which we shall later discuss), they do not become collectivized. (3) We must also postulate something of a reflexive, reinforcing process, which finds behaviors of a certain modality tending to improve the probability that that modality will be employed again. This assertion, also, must await discussion in a later section. In general, then, collective behaviors fall into three categories—culturally determined, idiosyncratic and adaptive, with the dynamics effected by the dialectical engine. In this regard, consider our process model, given as figure 4.1.

Now, in our previous chapter, we saw that the entire range of behaviors or activities within any system could be partitioned into subsets, as follows:

$U = A \cup B \cup N \cup M \cup T \cup S$,
A = the set of proscribed behaviors,
B = the set of prescribed behaviors,
N = the set of normative behaviors,
M = the set of non-normative behaviors,
T = the set of tolerable behaviors,
S = the set of sanctionable behaviors.

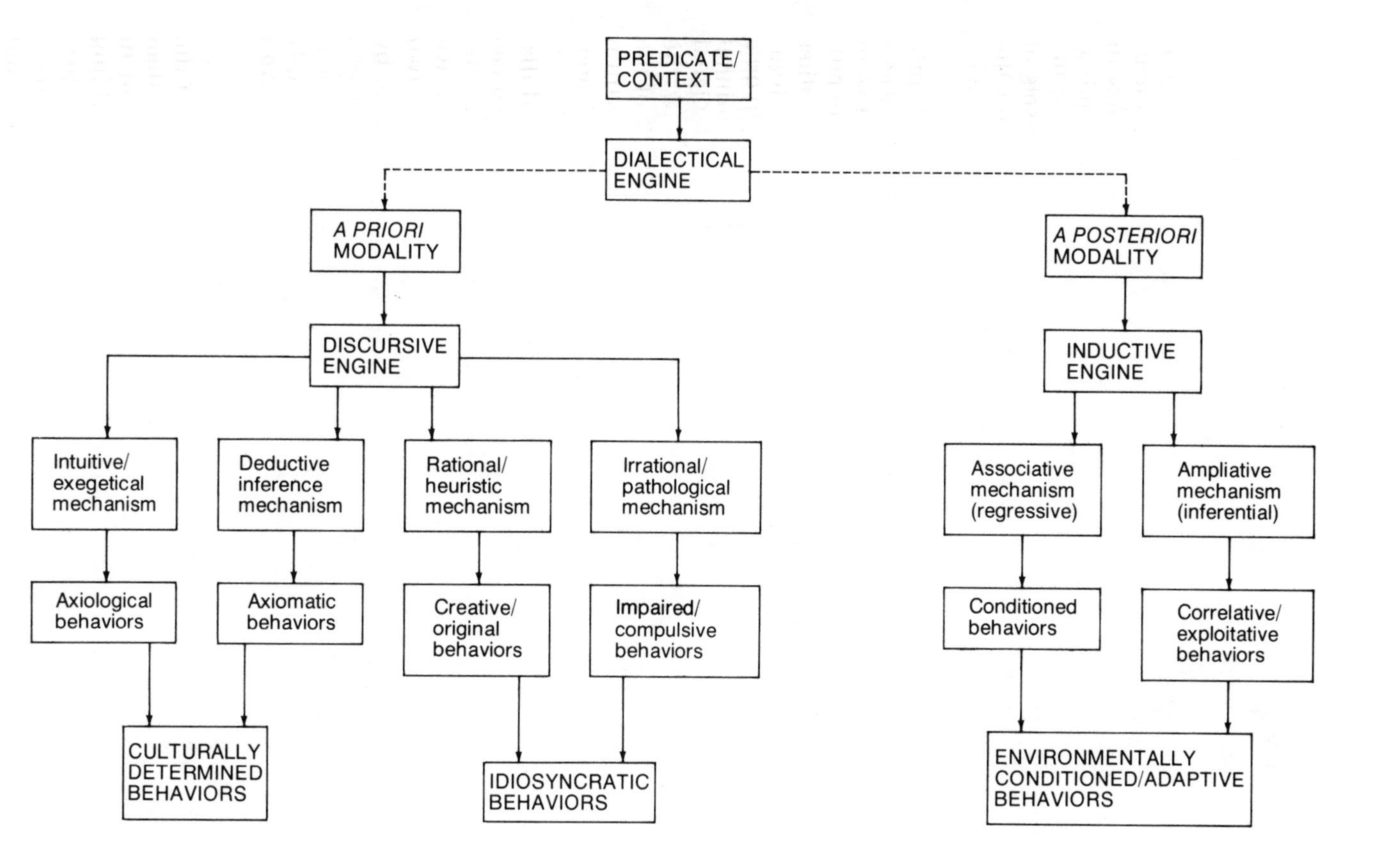

FIGURE 4.1: The modalities of organizational behavior: a generic model.

Our work in the previous chapter was almost entirely devoted to exploration of the behaviors which could be called *culturally conditioned*: the set A, products of axiological proscription; the set B, products of axiological prescription; and the set N, products of axiomatic engination. The implication is that non-cultural engines will account for the remaining behaviors, in the sets M, T, and S. In short, the subsets of non-normative, tolerable and sanctionable behaviors constitute the suprasets of idiosyncratic (intrinsic) and environmentally conditioned (adaptive) behaviors.

As suggested earlier, adaptive (or realistic) behaviors will be products of either association or ampliation. In either case, they serve as a direct link between individuals or collectivities and the properties of the environment. Adaptive behaviors thus exist as alternatives to culturally conditioned behaviors. Idiosyncratic behaviors, on the other hand, introduce significant qualitative differences between environment and behavior. In the pathological extreme, they result in a divorce from reality rather than an elaboration or reconstruction of it. Rational (functional) idiosyncratic behaviors serve the causes earlier associated with hypothetico-deductive or heuristic engination: they allow the individual or collectivity to positively transcend the properties of the environment. In either case, these behaviors have their origin within individuals, and the predicates are therefore idiosyncratic and non-algorithmic rather than codified and cultural in significance.

Again, to the extent that a process model usually precedes and disciplines the development of a proper synoptic ideal-type, we can see that our synoptic construct must be able to accommodate six distinctly different modalities of organized behavior. In effect, then, we will emerge with six unique ideal-type societal referents. Now, two points that we mentioned long ago should be reintroduced here, by way of qulaification. First, we are operating within the framework of a major meta-hypothesis: that societal structures and dynamics are significantly determined by behavioral engines—that they are not simply effects of mechanical adaption to environmental properites. And though this is just a hypothesis, we must proceed to develop its full implications before we are in a position to empirically evaluate its accuracy. So all that follows here is just a deductive elaboration of the meta-hypothesis, and is thus meant to be more of methodological than of substantive significance. In short, I am much more interested in exploring social-science procedure—in illustrating model-building technology—than in trying to change the way working social scientists view the subjects in which they are specialists. The second qualification is this: that at a suitable level of abstraction, social, economic and political properties of societal systems become isomorphic—become, as it were, "effects" of the behavioral engines that primarily underlie societal phenomena. Again, this is just a hypothesis, and the provisions for its validation cannot be fully developed until we have completed our work in Chapter Five. But the crucial implication of this second metahypothesis is the following: to the extent that social, economic and political phenomena are truly isomorphic, then we could have chosen to construct a synoptic ideal-type on any of these dimensions, and would still emerge, at the conclusion of our analysis, with essentially the same models in terms of their basic morphology. But the alert reader will immediately see that the condition of isomorphic relationships between the several sectors of a societal system is not really necessary to our case. Indeed, a much "softer" kind of relationship could lead to essentially the same result. In particular, if the several sectors of the normative societal system are merely correlative rather than isomorphic, then we have the basis for a properly *reflexive* model. Specifically, given the sets of all possible social, economic, political and behavioral attributes, certain members of those sets will tend to hang together in congruent clusters. Thus, the specification of any economic attribute implies the specification of its counterparts on the other dimensions, etc. This concept of reflexivity in societal phenomena has a long tradition in the social sciences, and we are merely broadening its implications in the present context.

For the working social scientist, then, what we are going to be doing in this section will have some familiarity, though some of the terms employed will be defined in ways peculiar to the system scientist. With this in mind, table 4.1 sets out the properties of a synoptic model as it might be applied to the study of complex societal phenomena, arguing essentially from the behavioral base.

Action Implications of the Synoptic Ideal-Type

Our defense of this construct may be rather brief, for most of the implications have already been discussed in previous sections. What is of most critical importance is this: each of the behavioral mechanisms listed in the table would be capable of authoring a defense or rationale for any given behavior. That is, the several behavioral bases are *alternatives*, and as a set are presumed (given our analytical context thus far) to substantially exhaust the alternatives available to societal systems. Let us lend this assertion some illustrative substance by considering the different defenses which might be offered for a common activity, say going to church.

Under a proscriptive axiology, the individual would simply be *pro-*

TABLE 4.1: The Bases of Collective Behavior

	Attributes			
Behavioral Mechanisms	Behavioral Determinations	Basis of Appeal	Psychological Significance	Societal (Collective) Significance
Cultural (aprioristic): (1) Proscriptive axiologies	Dogmatic	Ideological (sacerdotal)	Certainty	Spiritual order; teleological significance
(2) Prescriptive axiologies	Affective	Sentimental (rhetorical)	Spontaneity	Liberalism; permissivism
(3) Axiomatic engination	Rationalized	Logical	Material purposivity	Material order
Adaptive (realistic): (1) Associative engination	Conditioned	Sentient	Mechanisticity (minimal cognitive demands)	Automaticity
(2) Ampliative engination	Correlative	Utilitarian	Instrumental license	Exploitativity
Idiosyncratic (intrinsic): (1) Functional/ rational	Originative	Sapient	Creative release	Aestheticism
(2) Pathological/ irrational	Compulsive	Ab Intra	Existential	Solipsism[a]

[a]Obviously, "systems" composed of "solipsistic" individuals cannot be defined as systems *per se*, but rather as artificial agglomerates. Nevertheless, such agglomerates are important for us to consider, even if they cannot exist within strict system terms of reference.

scribed from not going to church. For example, one may be faced with a proscription that suggests that missing Sunday Mass or one's Easter duty is a mortal sin. In a society ruled by a prescriptive axiology, church becomes a usually pleasant and rewarding adjunct to the social sector. There may be music, entertainment and an emphasis on showing one's *love* rather than one's *fear* of God. The prescriptive emphasis is usually positive. Thus, rather than being proscribed from not attending church, one is offered tangible rewards. In many cases, the benefits are direct, and the experience is an emotionally fulfilling one. In the prescriptive church there is opportunity for companionship and perhaps some element of real religious delirium. Therefore, under the affective behavioral modality, the individual finds emotional release and fulfillment in attending church. We shall later have occasion to take a closer look at the very important differences—on the spiritual dimension of societal systems—between proscriptive and prescriptive axiologies.

When we move into the axiomatic domain, we might find somewhat more complicated reasoning associated with attendance at church. For example, one might "deduce" that going to church builds character. In this view, church is an opportunity to impress upon both ourselves and our children that success in life depends upon being able and willing to exercise self-discipline. Taking another tack, an axiomatic motivation for going to church might be to some such thesis as "the family that prays together, stays together".

What might be a *conditioned* basis for going to church? The predicate for a conditioned behavior usually rests with inertia: the probability of repeating a behavior depends on the number of times the behavior has been displayed previously. Therefore, a conditioned churchgoer is one who has been subject to a "rote" process of some sort, such that church becomes the effectively automatic response to a certain event or stimulus (e.g., Sunday morning; Christmas eve; the birth of an heir to the throne; the first rain of the new planting season). Many activities of societal systems are thus intelligible as products of ritualization (the collective counterpart of rote conditioning), where the origin of the ritual may have once had functional roots that have subsequently become obscured.

There might be innumerable explanations for going to church under the ampliative behavioral mechanism as ampliative behaviors are determined by context or opportunity. Thus, a man might go to church to keep warm for an hour or so, or he might go to impress potential clients or constituents with his piety and humility. Or he might go to a meeting because it is the best way to make friends. Or he might go just to avoid an argument at home, if his wife is more strongly attached to religious display than he. Here there is no aprioristic basis for the activity; it serves ends that are strictly utilitarian and strictly determined by context.

Thus, both associative and ampliative behaviors are adaptive, but in different ways. The associative behaviors become "mechanized" through long periods of repetition and are ultimately conditioned by real-world events (*qua* "lore") rather than by *a priori* predicates. Ampliative behaviors, on the other hand, are adaptive because they serve always to *correlate* the individual and collectivity with the opportunities (or perils) the environment currently presents. This correlation is seldom perfect, for inductive inference processes intervene to extrapolate or project real-world properties. But the original predicates of behavior—the environmental stimuli, events or conditions—will still be recognizable in their fundamental qualitative attributes.

Moving to the idiosyncratic behaviors, the rational individual might go to church despite having no *a priori* convictions about the existence or nature of God. Moreover, he may have no utilitarian stake in going to church. But he might reason as follows: (1) There is a finite (and perhaps even significant) probability that there is a Creator. (2) There is a finite (and perhaps even significant) probability that he is aware of my intentions. (3) I have no other activity which I might perform during this hour that carries with it a greater value than that associated with my effort to appease an intelligent Creator, should he exist. (4) Therefore, going to church carries the minimum opportunity cost for the use of this hour. On any given day, this calculus may lead either to going to church or not, depending on the alternative uses of the time available and the current probabilities the individual ascribes to the existence of God, etc. But in either case, the activity so determined is strictly sapient. It is not rationalistic, for no reference is made to any standing axiomatic or theoretical scheme. Nor is his behavior realistic in the ampliative sense. For, to the strict positivist-empiricist, one cannot generate or manipulate any probabilities that are not drawn from the world of experience and empirical observation. But in the idiosycratic domain, the correlation between rational behaviors and environmental predicates may be very weak (and need not respond to environmental predicates at all in more extreme cases). This is especially true of irrational or pathological behaviors. In the example we have been using, the irrational individual may go to church because he heard a voice telling him to go. Or perhaps he perceives church as a sanctuary against the demons rising up to persecute him from Hell.

It may easily be seen that the attributes for each of the mechanisms on our second dimension of interest—their basis of appeal—derive directly from the nature of the behaviors themselves. Thus, ideological considerations would underlie dogmatic behaviors, along with the instruments of sacerdotal coercion that usually attend proscriptive systems. Ideological motivators thus derive directly from exegesis performed on the underlying axiological prophecy. In short, one does what one does in a proscriptive system because one *must*. In prescriptive systems, on the other hand, motivation generally stems from sentimental sources. One does what one does because one *should*, as determined by the rhetorical bases offered by the prescriptive prophecy. For example, the prescriptive agent may ask alms for the impoverished by exciting the sympathy of the audience rather than citing any transcendental imperative. In short, the key to the integrity of prescriptive systems is making people want to do what they should do. In proscriptive systems, on the other hand, the key is to make people fear to do what they should not do.

We may dispose of the remaining modalities' appeal very quickly. We have already seen that the appeal exercised by rationalized (axiomatic) behaviors is essentially logical. Specifically, one searches for a defense for an activity that follows logically from a set of premises, etc. That is, axiomatic actors are capable of articulating a rationale for their behaviors, and usually that rationale takes the form of a system of simple deductive inferences. The two aspects of realism—associative and ampliative behaviors—have a distinctly different type of appeal. Associative behaviors become virtually automatic or pre-cognitive in engination, and their appeal rests essentially (as earlier suggested) in their minimization of intellectual demands. Ampliative behaviors appeal to those individuals who are products of secularization—who are bereft of any operative *a priori* constraints. Again, then, ampliative or correlative behaviors gain their significance in the complex interplay between individual ambitions and environmental context, and have a distinctly *a posteriori* flavor. Now, originative or functionally rationale behaviors are the essence of sapience. Any apriorisms which enter the behavioral calculus here are (unlike those of the axiological and axiomatic cases) treated as hypotheses or heuristics: something to be tested rather than merely accepted. Sapience, in the rational case, stems from the fact that, where experience or observation can be useful, empirical criteria are employed (whereas these are the sole arbiters of activity for the ampliative realist). Finally, pathological and irrational behavior is motivated by highly individualized, enormously variable psychogenetic factors—by predicates that have their root entirely within the psycho-physiology the individual or, if derived at all from real-world percepts, have been manipulated beyond recognition or relevance.

Thus, as is to be expected, the bases of appeal for the various behavioral mechanisms differ as widely as do the types of behavior they determine. In this very short section we have been able to approach the issue of appeal in only the most rudimentary and preliminary way. In subsequent sections we shall have to be a great deal more precise, for the way in which the behavioral mechanisms attract and discipline their adherents is a critical component in any effort to comprehend societal systems. For the moment, however, we must continue with our defense of the table, turning now to the mechanisms' psychological and societal significance.

Socio-psychological Implications of the Synoptic Ideal-Type

Proscriptive axiologies involve the repression of certain behaviors that can be expected to carry a positive utility for the individual. Now, for this repression to be tolerable, the proscriptive axiology must offer something in return (and the perceived value of that something must exceed the expected value of the repressed behaviors). As a general rule, the abstracted or generic benefit that proscriptive systems offer is certainty, and thereby freedom from anxiety. Often this takes the form of a promise of guaranteed salvation if the individual does not indulge in any of the proscribed behaviors. And, in this context, most proscriptive axiologies also offer a corollary to the salvation theme, the compensatory calculus. This may be stated as: the greater the deprivations one suffers in the temporal world, the greater will be the pleasures he will receive in the transcendental world.

In terms of basic psychological utility, therefore, proscriptive platforms generally propose something like the following: Avoid exercising the non-normative activities, and the probability of salvation increases accordingly. Proscriptive platforms thus bring certainty to what would otherwise be an indeterminate (or at least probabilistic) world. Thus, in essence, proscriptive platforms serve the cause of eliminating anxiety. With regard to the utility of proscriptive platforms with respect to societal or organizational integrity, they may be seen to provide a basis for *order*. For example, without the concept of a compensatory calculus, it would be very difficult to control individuals in the temporal world. Those people who had less—and for whom no ultimate compensation was promised—would find it difficult to decide why they should not try to get more, immediately. But under the compensatory calculus, the sum of delayed gratifications is presumed to be of greater importance than the exercise of immediate gratifications. Now, of course, the cause of social order is served only to the extent that the proscribed activities are *relevant*. Specifically, the proscribed activities should be those which, if given license, would lead to disorder. This is why proscriptive platforms often have a quasi-legislative significance. Thus, to the extent that individuals may be axiologically proscribed from engaging in crimes against persons, property or the state, etc., we have an efficient instrument for maintaining societal or organizational order. Without the axiological proscriptions, more expensive and certainly less reliable means of securing order would have to be developed. For example, the prevention of crimes against property would require police forces and the maintenance of mechanisms for retribution. In the same sense, a government that was not *a priori* assumed to be a partner with God might be challenged on strictly pragmatic or temporal grounds, and have to defend its existence far more rigorously than if it were protected to some extent by the axiological envelope.

But the general problem with any proscriptive axiology is this: those behaviors which are not expressly forbidden are tacitly allowed. In short, any proscriptive system has loopholes, and men of imagination find it no end of entertainment to exploit these loopholes, and in the process contravene the *spirit* of the axiology or law. In part, this is the reason why repressive systems are sometimes found satisfactory to those we would expect to be most offended by proscriptions (as in boot camps and prisons). Thinking up ways to avoid constraints has been one of the ancillary benefits of all dogmatic systems, and has often served to extend their life beyond what would be expected by the strict materialist.

The existence of loopholes or lacunae in proscriptive axiology leads to another characteristic common to most repressive systems—the tendency for the proscriptive base to keep amplifying. The reasons for this constant amplification or elaboration are twofold. First, the loopholes must be plugged as they appear. Thus there is a constant battle between the agents of the proscriptive system (e.g., the legislators or the priesthood) and the constituents. As imagination and arcane readings of the law's letter discover lacunae, an *ad hoc* ammendment or subproscription is added. Eventually, these proscriptive systems become unwieldy, unmanageable and incomprehensible.

Secondly, agents of proscriptive systems can really only make themselves visible—or lay claim to a concrete contribution—to the extent that they are able to elaborate the proscriptive base. Thus, they see their role as *creating* proscriptions, not merely enforcing them. Thus, while some constituents are at work trying to discover lacunae, the priests or legislators or societal executives are at work trying to find new activities to proscribe, and hence get their name on a bill or edict. As a result, the proscriptions begin to apply to more and more activities, both those that are appropriately curtailed, and those that are merely capable of being legislated against. So, if the *ad hoc* characteristic of proscriptive systems eventually makes them incomprehensible, their constant elaboration eventually makes them onerous, and their appeal is diluted in the process.

In short, then, proscriptive axiologies are the ancient and honored "keepers of the kingdom". Up to the point where they become arbitrary, proscriptive axiologies directly serve the cause of linking the temporal world with the transcendental, and in so doing directly serve the cause of societal stability. One misbehaves in this world only at his own peril in the next. Societal stability, of course, means the maintenance of the *status quo*. Now, certainly, this would always serve the cause of the elites in any system who—contrary to the popular thesis that elites most fully embody the value system of a society—generally are much less susceptible to *a priori* arguments than the proletariat or peasantry. For example, the same compensatory calculus that brings comfort to the lower classes could hardly bring much comfort to elites, considering their high order of temporal benefits. Therefore, it is to the elites' benefit not to accept the compensatory calculus themselves, but merely to defend it. Or perhaps they might adopt a distinctly instrumental point of view (e.g., if the compensatory calculus actually works, then we'd better get all we can while we're here, for heaven's going to be a problem for us anyway). To soften this logic, there is the usual concept of charity—a rich man can make the compensatory calculus work in his favor by giving money away. Thus charitable activities cannot really be looked on only as a method of alleviating guilt. They must also be seen as an attempt to modify the operations of the compensatory calculus (even if one is not entirely sure that a heaven or hell exists).

But if axiological proscriptions have usually worked to the advantage of the elite—in that they tend to preserve the status quo and with it the existing distribution of societal prerogatives—we cannot forget their function for the non-elite. For a key factor that makes a deprived temporal existence tolerable is the expectation of a beneficent eternity inherent in proscriptive axiologies. Now (this is the important consideration) the ostensible validity of the expectations—and therefore the degree of determinacy any individual perceives—will in large measure be a reflection of the certainty and determinacy of the temporal world. In short, societal instability and non-stationarity of key societal parameters can only cast doubt on the potency, and therefore the validity, of the axiology itself. Specifically, the best sign of the certainty of the postulated afterlife is the order, stability and continuity of the temporal world. When the temporal present can no longer be predicted because of instability, then the determinacy of the transcendental world also is called into question. Thus, societal disorder and change act to undermine one of our key psychological utilities, certainty. It is therefore no paradox that non-elites are sometimes more vigorous in defending the status quo than those elites who apparently benefit most from it. For if the elites have a material interest in proscriptive axiologies, the non-elites have a psychological stake; and the value of the benefits of certainty and determinacy may often outweigh the incremental material interests of the more "realistic" elites. In many instances, then, the elites may count on the support of the lower socio-economic classes to maintain a system that works clearly to the material disadvantage of the latter. From the somewhat naive materialistic position of the Marxist economists and the empiricist-positivist social scientists, such a situation simply cannot be explained. It is thus not surprising that so many reformers eventually come to resent the very peasantry and proletariat they pretended to serve. But it is the naivety of the Marxist and positivist positions, and not the stupidity, malleability or indifference of the common man, that leads to the reformers' frustration [1]. For they have consistently ignored a fundamental if somewhat subtle law of societal systems:

> In the societal context, there is a constant tradeoff between material and psychological benefits. Societal dynamics—and organizational dynamics in general—cannot be comprehended unless it is first understood that materialistic and cognitive aspects of collective existence are complements.

Thus, a situation that seems to yield little economic or tangible comfort and satisfaction to a group may yield great spiritual or psychological utility, and thus be protected with great vigor.

In terms of gross effect, if proscriptive axiologies serve to introduce order into societal or organizational systems, prescriptive predicates serve to make the system more liberal. The process by which this comes about has already been discussed. First, the prescriptive prophecy will often merely invert the value placed on those activities proscribed by the prior axiology. That is, what is forbidden may, under the prescriptive platform, now take on a positive value. If, for example, impulsiveness is assigned a negative value under the proscriptive platform, it now becomes an imperative under the prescriptive one. The individual is urged to make his behavior *spontaneous*, to give expression to impulses. Thus, society as a whole—once a prescriptive platform has taken hold—tends to become disaggregated. The residual proscriptions (the vestiges of the prior axiology) can be transgressed with increasing impunity. The relationship between law and punishment comes to be mediated by confounding factors such as the "quality of one's intentions". Increasingly (again with reference to the Socratic axiology), society itself begins to take the blame for asocial elements and social casualties. There is usually a shift toward egalitarianism, whereas (as we shall later see) proscriptive systems both institutionalize and defend differentiation.

Now we shall have a great deal more to say about prescriptive systems in a later section, but two points should be mentioned here. First, behavioral latitude is not necessarily the same thing as license, nor is societal liberalism necessarily the same thing as individuation. In a strictly instrumental (realistic) system, the concept of normative behaviors loses significance, for there are no real *a priori* constraints. Under the banner of realism, both exploitiveness and hedonism may reach full bloom, often side by side. The latitude allowed under prescriptive systems, on the other hand, is limited. The imperatives usually cannot be avoided (whereas there are no imperatives in the realistic-instrumental system). As for individuation, this too is constrained in practice. Because of the emphasis on association and reciprocity in prescriptive systems, clusterings tend to form, often around certain specific interests. These clusters may differ radically from one another in terms of ambitions and proprieties, but the members of the clusters are all highly homogeneous. In short, liberalism, under the banner of a prescriptive axiology, usually means the transformation of a once homogeneous society to a cluster-based society, where the clusters differ among themselves, but are internally highly consistent. In practice, then, a prescriptive axiology opens the door to a tolerance greater than that associated with proscriptive systems, but it is by no means the engine of the completely unregulated system sought by anarchists or abject hedonists. Under a prescriptive system, only those impulses attributed to God or the prophets are legitimately to be exercised. Under a hedonistic or anarchic system, any impulse at all becomes an imperative, because there is neither God nor prophet.

The second major point to be made here about prescriptive systems is this: Every latitude allowed by the prescriptive platform entails a cost, at least in aggregate. For societal liberalism and spontaneity implies the loss of certainty and determinacy—and order—which were the benefits derived from acquiescence to a proscriptive system. Beyond a certain point, liberalism thus means *anomie* (just as repression beyond a certain point means frustration). Anomie is a condition where the individual loses teleological purpose and transcendental significance. The proscriptive system entailed an opposite condition. The individual and hence the collectivity became alienated from the temporal and emotional. In both proscriptive and prescriptive systems, then, an element of anxiety emerges.

If the society in which this anxiety emerges is currently a predominantly prescriptive (e.g., permissive) one, then the perceived value of proscriptions will increase; e.g., the benefit expected from the certainty associated with repressions and proscriptions will be perceived to outweigh the frustration associated with a curtailment of permissivism. To a great extent, this is the factor underlying the somewhat anomalous shift of the youth of the affluent nations—after several generations of secularization—to the discipline inherent in reconstituted religions like Yoga and Christian activism (the domain of the so-called Jesus-freak). As perhaps an even more pathetic reaction to the loss of purpose and certainty of the last years, there is the ballooning speculation (or hope) that we are being visited by "gods" from outer space. Thus, we have a "technological" religion emerging to take the place of the traditional forms, complete with visions (e.g., UFOs), superior beings and the presumption that the skills from the skies can solve the problems that man himself has been unable to solve [2]. Here, perhaps, we have the religion of the Space Age, and the somewhat aberrant comfort that many people seem to take in suggesting that "we are being watched". There is, of course, no reason to suppose that there are not UFOs. But I suspect that the increasing number of sightings are being caused by wishful thinking rather than by the fact of an alien presence, just as were the beatific visions of religious enthusiasts in the Middle Ages.

The shift from a proscriptive to a prescriptive system is essentially the reverse of the process described above. At some point in the life of every proscriptive system, the level of immediate frustration will increase to the point where the utility of order and determinacy declines, and the perceived utility of behavioral latitude increases. This alteration in perceived values may be engined by many factors, among them the adhocratic and bureacratic elaboration of the proscriptive base into areas of activity that do not deserve repression. There is also the problem that adherence to the proscriptions in any system is generally not universal. As earlier suggested, elites may depart more and more blatantly from the proscriptions the populace is expected to obey. As any proscriptive system matures, membership in the elite tends to be inherited (ascribed) rather than earned. This leads to the elites' habituation to the fact that their greater privileges are obtainable without corresponding responsibilities. The elites then tend naturally to reinforce their inherited privileges by adopting artificial forms of behavior, serving to more and more distinguish them from the masses. For succeeding generations, it is the distinctive behaviors of the elites that are seen as earning them their privileges, and most distinctive of all are those behaviors that contravene the popular proscriptive base. Eventually, the contradiction between the axiological proscriptions and the behaviors of the leadership leads to a situation where both the societal structure and the axiology are discredited. The reformers invariably condemn the axiology as an instrument of politico-economic repression, and the stage is set for secularization.

We now have some rudimentary concept of the terms of the tradeoff between proscriptive and prescriptive axiologies, and we shall elaborate this considerably in a later section. Here, we may turn briefly to explore the socio-psychological significance of axiomatic engination. Axiomatic platforms usually direct an individual's attention toward the temporal world. They thus provide an individual with a material *purposivity*, whose parameters are rationalistic in origin. This is true of both the great axiomatic engines under which a majority of mankind labors, capitalism and socialism. As we shall later see, both these platforms have axiological predicates (e.g., capitalism has a prescriptive base, while socialism has a sentimentalized base), but these are generally no longer operative except as indirect apologies.

The "purpose" devolving on the individual in the proto-capitalist system is to serve his own welfare, under the rationalization that the welfare of the system as a whole is to be considered as the sum of individual welfares (through the postulation of the Smithian "invisible hand"). But the proscriptive roots of the capitalist platform define certain means which are not to be tolerated: anti-competitive practices, embezzlement, subornation, etc. Therefore, one may pursue one's material ends only within a certain set of normative strictures (which serve to distinguish capitalism from outright instrumentalism, where no normative bounds are defined). The "purpose" of the individual operating in a socialist system is a bit cloudier and less direct: he is asked to serve his own welfare by first considering the welfare of others. This reflects the sentimentalistic origin of most socialist schemes, and subordinates individual interests to "the will of the collective". At any rate both these axiomatic engines urge the individual to seek material gratification, and thus serve to make political and economic properties the *primary* attributes of rationalized systems. Of course, all socio-behavioral systems have political and economic attributes. But as we shall later see, these become subordinate to spiritual or transcendental interests in proscriptive (dogmatic) systems, and subordinate to social considerations in prescriptive (sentimentalized) systems. Thus, the development and maintenance of a politico-economic *order* is the paramount function of axiomatic platforms, whereas it is an ancillary function of axiological predicates.*

We now move on to quickly treat the remaining behavioral contexts. The psychological significance gained by a member of a simple associative system is *mechanisticity*: behavior engined by rote or pre-cognitive conditioning. As we have already mentioned, associative behaviors make the least cognitive demands on the individual, and therefore will tend to dominate in the face of essentially simple or placid milieux. Even within complex systems, many routine behaviors become mechanized, as this is the most efficient means of carrying out highly repetitive or well-precedented tasks. From the standpoint of societal significance, mechanized behaviors tend to introduce *automaticity* into collective behaviors, a property congruent with a set of relatively simple and stable environmental properties [3].

The psychological significance of ampliative behaviors rests obviously with their provision for unbounded gratification—their instrumental license. A societal system exhibiting a predominance of ampliative behaviors will tend to be both exploitative and opportunistic. Moreover, as was earlier explained, there will be no *a priori* constraints on the means which can be employed to achieve ends. Nor will there be any *a priori* specification of legitimate or illegitimate ends. Obviously, ampliative behaviors are vehicles for adaptation to more complex environments than those to which association was the response. In the first place, there may be many ends, some competitive with others. The selection of any specific subset is a matter for discernment (we have already mentioned that being a good hedonist requires a great deal of intelligence). Secondly, there may be many impediments to the achievement of such objectives, especially if other exploiters are also interested in them. Therefore, ampliative inferences are needed for the generation of competitive behaviors. Again, as the environment of a system becomes more complex, adaptivity demands more ampliative behaviors at the direct expense of associative or conditioned activities. In either case, both the goals of society and the trajectories taken toward those goals are determined only by context, freed from the type of *a priori* restraints that operate in axiological or axiomatic systems.

Finally, the idiosyncratic cases may be dealt with rather quickly here

*This thesis is defended in detail in Chapter 6, where we discuss the "corridor concept", whereby social, spiritual and material interests become substitutes rather than complements.

in respect to psychological and societal significance. The rational behavioral base provides the individual with an opportunity to realize what are essentially creative satisfactions. This is true whether the exercise is pure art, engineering, mathematics or whatever. It is a characteristic of originative behaviors that *elegance* becomes an explicit criterion: means and ends become co-significant. For the most part, then, originative-sapient behaviors tend to provide societal systems with their true luxuries: speculative science, serious art and music, educative rather than evocative literature, etc. The pathological (irrational) form of idiosyncratic behavior, on the other hand, permits an existentialistic orientation on the part of the individual, which translates directly into solipsism and structural atomism. As a rule, compulsive individuals act as if nothing except the immediate self is real. Thus, as we shall show in the next section, compulsive-psychogentic behaviors, when collectivized, lead directly to anarchistic societal forms.

In summary, we have now had a look at the collective implications of the several bases of behavior. The synoptic ideal-type model simply tried to order them in a tentative way, such that their implications could be extended and made empirically more significant. Thus, the components of the synoptic ideal-type provide a structured vocabulary of concepts that will now feed into a more operational construct—an elaborative ideal-type.

THE SEVERAL SOCIETAL REFERENTS

As earlier suggested, elaborative ideal-types carry direct substantive (descriptive and/or predictive) significance, and are derived from the sort of synoptic ideal-type that we just developed.The synoptic work thus gives us the basic components of an elaborative construct, which for our present purposes means that we shall set about constructing ideal-type properties for seven different societal referents, as shown in table 4.2.

This is an elaborative construct, because we are now extending the implications of the synoptic construct (Table 4.1) and working on dimensions less abstract than we used there. We are now particularly concerned with socio-behavioral systems on the following dimensions: (1) the process by which the form of behavior becomes the driving system force; (2) the basis of control over the individuals in the system, derived from the behavioral base: (3) the desired behaviors, or those expected to evolve from the particular behavioral base; (4) the range of sanctions available for maintaining behavioral congruence; (5) the

TABLE 4.2: An Array of Societal Ideal-Types

	Attributes					
	Processual Mechanism	Desired Behaviors	Basis of Control	Primary Sanction	Dysfunctional Behaviors	Collective Pathologies
Acculturated:						
(1) Dogmatic	Institution-alization	Temperance/obedience	Sacerdo-talism	Excommuni-cation, etc.	Heresy	Fanaticism
(2) Affective	Sentimental-ization	Sociality/spontaneity	Moral suasion	Isolation/shame	Elitism	Penury
(3) Rational-ized	Indoctrination	Consistency	Principle/procedure	Forfeiture/guilt	Contradic-tion	Inertia
Adaptive:						
(4) Condi-tioned	Programmation	Automaticity/consuetude	Ritualization	Exile	Individua-tion	Atavism
(5) Correl-ative	Secularization	Opportunism	Preemption	Confiscation	Suboptimal-ity	Alienation
Intrinsic:						
(6) Origin-ative	Cerebration	Sapience	Peer review/credentials/ethical codes	Censure	Effeteness	Abulia
(7) Compul-sive	Existential-ization	Hypostatiza-tion	—	—	—	Anomie

types of behavior deemed dysfunctional by the societal system; and (6) the expected pathologies of the systems driven by the several behavioral bases.

These dimensions now allow us to take our abstract (synoptic) ideal-types and lend them some empirical significance. In other words, the societal ideal-types in the elaborative construct are intended to reflect properties of real-world systems, and therefore to serve directly as referents for an emergent societal theory. We may now attempt a brief defense of these seven ideal-type systems, treating first those that are primarily creatures of culture.

The Acculturated Ideal-Types

The acculturated ideal-types will be those driven by the dogmatic, affective or rationalized predicates. Though these three categories have very different properties, all serve to constrain *a priori* the behavioral repertoire of system residents. All initiate and maintain these constraints in rather clear and well-defined ways.

Dogmatic Systems One's first suggestion is that dogmatic behaviors are products of *institutionalization*. From a narrowly functionalistic perspective, institutionalization presumes the existence of a societal mechanism with two missions: (1) the exegesis of rules of conduct from some prophetic base, and (2) the exercise of axiological authority over the population. In all cases, dogmatic institutions seek obedience from their constituents, which in the context of a proscriptive axiology means the avoidance of forbidden behaviors. In our simple set terms from the last chapter, the scope of a proscriptive institution extends over the set of behaviors $A \subset U$.* And, also as suggested earlier, most proscriptive institutions have a tendency to constantly elaborate and extend the set of proscribed behaviors, so that the ratio $[A]/[U]$ constantly increases.

With regard to the basis of control of behavior, institutionalization implies *sacerdotalization*: the summoning of transcendental authority over temporal behavior. Sacerdotalization may be a function of natural societal evolution. What we mean by "natural evolution" in this context is the formalization and coherence of the institution through the natural process of societal differentiation and specialization. We may suppose that in primitive religions, spiritual authority and interest were initially distributed widely throughout the population. That is, such religions are democratic, and the religious functions tend to be al-

*Here, again, U is the universe of all behaviors and A is the subset of explicitly proscribed behaviors.

located more or less symmetrically. Very early in the evolution of societal system, however, religious authority may tend to become concentrated in the hands of a *priesthood*. Concentration continues until the legitimate authority to devise exegetical constructs and enforce behavioral congruence also becomes specialized. When this occurs, the axiological aspects of society become bureaucratized in the same way that governmental and military functions do [4]. In some cases, the proscriptive axiology (with its initially transcendental pretentions) may become either the master of the government, its ally or its adversary, and thereby take on a temporal significance. In short, sacerdotal systems may borrow transcendental authority from the prophets, and then gradually concentrate it and extend it into material arenas. In virtually all cases, sacerdotalism implies *coercion*. When the sacerdotal institution is allied with the temporal governing powers, coercion takes the form of corporal or police action (as with the executions accompanying the Spanish Inquisition or the Salem witch trials). In other cases, sacerdotal authority is restricted to the imposition of penalties carrying transcendental implications, e.g., excommunication, or the array of less severe restraints such as the penances associated with venal as opposed to mortal sins. In virtually all cases, prescriptive institutions most fear heresy—the denial of some dogmatic tenet, or the denial of the authority of some proscription. To the sacerdotal authority, heresy is the equivalent of treason. Finally, if sacerdotal institutions lend societal systems a semblence of order and stability, they also increase the peril of fanaticism. Fanaticism is, of course, the polar opposite of heresy, and implies a condition where the individual becomes the unconstrained agent of dogma. Of course, from the standpoint of the sacerdotal authorities (or of governmental authorities allied with the sacerdotal systems), fanaticism holds both good and bad possibilities. The advantage of fanaticism is that the population becomes easily inflamed and highly susceptible to direction under sacerdotal sponsorship (e.g., in the holy wars of the Mohammedans or, for that matter, the Crusades themselves). But fanaticism eventually serves to weaken the societal system through its dampening of innovation, experimentation and individuation. And it may, in some instances, rebound to the detriment of the elites (who, as we earlier suggested, sometimes pay only lip service to their proscriptions). For fanatics make vicious enemies as well as willing tools.

To ask sacerdotal systems to liberalize—to modify their tenets and constraints in keeping with the times—is to invite them to dilute their significance for the susceptible individual. For the sacerdotal system is the link between past and future, and the "product" that it offers is continuity and hence security. It is the vehicle that removes the individual from his painful environment, even if it also tends to atrophy his discernment and analytical powers. Thus, the integrity of any sacerdotal system depends on its standing against change. In this respect, a great mistake may have been made in the ecumenicalization and liberalization of the Catholic Church—the elimination of certain age-old constraints, the translation of the Mass from the old Latin. In trying to modernize, the Catholic Church may have weakened its fundamental psychological utility. As a result, it has lost not only the affection of some of its members, but also some of the dogmatic authority it was able to exercise over its own functionaries. In summary, then, one of the worst possible policies for a sacerdotal system to follow is adaptivity. Eventually, of course, it may lose authority through the natural process mentioned earlier: the tendency to extend the proscriptive base (the ratio $[A]/[U]$) to the point where the tangible repressions make security and stability less and less attractive. But the attempt to be "modern" or adaptive may hasten the dilution considerably. At some point, in any case—either through the dysfunctional policy of modernization or through the natural process of proscriptive elaboration—the sacerdotal system loses the affection of its constituents. At this point, the way is open for the prescriptive system.

Sentimentalized Systems The process by which prescriptive axiologies gain their moment is *sentimentalization*. The normal implications of sentimentalization are found in the denial of the rectitude (the moral authority) of the sacerdotal system, and in the personalization of the religious relationship. That is—as we suggested in the last section—the Creator becomes transmogrified from a stern, paternal executive into a tolerant and fundamentally benevolent parent.

Instances of unadulterated sentimentalization in the real-world are somewhat rare. The proto-Christian position (perhaps best defined in terms of the Sermon on the Mount) is a relatively pure case. Proto-Communism also has a distinctly sentimentalistic base, especially in its egalitarian implications. And, of course, there are distinctly sentimentalistic overtones in all variants on the Socratic and socialistic schemes, especially in their exhortation to sympathy for the disadvantaged and in their abhorrence of concentration of property or prerogative.

In clinical terms, then, prescriptive axiologies tend to promote *sociality*: the consideration of the welfare of one's fellows. In extreme interpretations, sociality may ask that we consider the other person's wants, feelings, etc., as superior to our own; milder forms ask merely that we at least allow the other person's interests co-consideration. Some prescriptive axiologies have transcendental overtones. But the major instrument of prescriptive control is *moral suasion*, and thus *affective* rhetoric. The emphasis here is on social responsibility, and the primary sanctions are shame and, ultimately, isolation. The presumption of the prescriptive prophet is that the major aspiration of every individual is to be loved. Therefore, the severest punishment one can receive is the collective withholding of love, affection or comradeship.

But moral authority is easily sidestepped, so sentimentalization often gives way to license. This license evolves from the weakening of the association between transgression and recrimination. In proscriptive systems, as we have already suggested, the *letter* of the law is paramount. In prescriptive systems, on the other hand, it is the *spirit* of the law that predominates. Thus, contra-legal activities become more and more subject to defense by way of "extenuating circumstances". In the most extreme prescriptive case, all criminal activity becomes a product of societal rather than individual responsibility: the criminal is a criminal because society made him that way. Thus, as the prescriptive system matures, there are fewer and fewer proscribed behaviors, and more and more latitude. In short, the ratio $[A]/[U]$ gradually diminishes. The legislative process then begins to take on something of the character of the remission syndrome, which we find shared by platforms as diverse as Roman Catholicism and Tibetan Bhuddism. Here, an individual sins, seeks forgiveness, is given a penance to perform and at the completion of the penance is given remission for the sin. He is then free to sin again. In short, guilt is not concatenative. It is the extension of the remission syndrome into temporal affairs that lends prescriptive systems their distinctly "maternal" character. In maternal or "matrist" contexts, as defined by Gordon Rattray Taylor [5], there is an *a priori* denial of the efficacy of the coercive components of "patrist (i.e., proscriptive) systems, and an *a priori* belief that rehabilitation is the *sine qua non* of socialization. Thus, strict rules and regulations—and performance criteria—all become anathema. The emphasis is on "being oneself". And, in the sentimental-matrist system, permissiveness replaces principle, rehabilitation replaces recrimination, subjective performance criteria replace objective standards, results become subordinate to intentions—and intuition gradually begins to drive out analysis.

In terms of the behavioral referents operative in most sentimentalized systems, we have the suggestion that people should be spontaneous, casual, gentle, forgiving, affectionate—more concerned with today than yesterday or tomorrow. It is precisely such attributes that would be proscribed under the prototypical "patrist" system [5]. The patrist is expected to be self-contained, serious, purposive, committed, constrained as to impulse, future-oriented and more concerned with propriety than with motivation. (Of course, to be a patrist one need not be a man. Nor must the matrist be female. Thus, the distinction between matrist and patrist is not sexually determined, as is the distinction between matriarchal and patriarchal cultures.)

The most abhorrent spectre within sentimentalized systems is *elitism*. There are several reasons for this. First, an explicit ambition of prescriptive systems is always to remove (or avoid) the differentiation associated with proscriptive systems, especially that which leads to ascriptive hierarchicalization. That is, one of the temporal appeals of prescriptive axiologies is their egalitarian promise. Secondly, from the psychological perspective, the legitimacy of spontaneity relieves the individual of the anxieties of ambition, self-improvement and sacrifice, etc. Simply "being oneself" is sufficient, and seeking to improve oneself beyond one's fellows would be viewed as antisocial. That is, one may be oneself, usually, only so long as that does not involve being too different. In short, elitism is discouraged because once allowed, it might reinstate the very frustrations that prescriptive systems are supposed to dissolve.

The third reason is a very practical one, and one we shall have a great deal more to say about in the next section. Instances of elitism would operate to dilute the fundamental benefits that are associated with prescriptive systems. When we later explore the social, political and economic corollaries of our behavioral ideal-types, we shall see that prescriptive systems generally imply some economic deprivation. In short, prescriptive systems tend to offer an elaborate and satisfying social sector, in part as compensation for material rigors. Thus, whereas the proscriptive system is able to maintain social order and introduce benefits through the compensatory calculus (which finds temporal rigors traded off against transcendental benefits), the compensatory calculus of prescriptive systems works in a different way. First, it provides the individual with strong social benefits (e.g., affection, extended family association, leisure to enjoy communal relationships, lack of discrimination) which tend to offset poor housing, bland diets, etc. Secondly, the prescriptive axiology itself tends to encourage a symmetry of advantage (or disadvantage), such that the strong and pervasive differences among material possessions and privilege associated with hierarchical, differentiated systems is not in evidence. Therefore, one compensation for any material rigors is that one's neighbors are not demonstrably better off. Thus, the compensatory calculus of prescriptive systems seeks to deter temporal differentiation and discrimination, whereas the compensatory calculus of proscriptive systems may actually reinforce socio-economic asymmetry. In both cases, the ultimate result is societal stability, at least within certain ranges we shall later attempt to define.

As a final note on sentimentalized systems, there are two degenerate conditions to consider. First, there is *penury*. Clearly, where a system demands egalitarianism—and where efforts at differentiation are condemned or discouraged—there will be a tendency toward neglect of the material sector. If this goes too far, then the social benefits of prescriptive systems will begin to pale beside the economic deprivations. There will then likely be a quest for a way to legitimate material ambitions—e.g., a search for an axiological justification of acquisition, such as that associated with the post-Reformational Protestant ethic—and a consequent rejection of the prescriptive axiology and its sentimentalistic correlates. A second pathology to which prescriptive systems tend to become prey is *evisceration*. They lose their ability to defend themselves against either domestic or foreign adversaries. The cause of this is simple: as principle gradually gives way to impulse, essentially principle-based institutions such as law (the defense against domestic chaos) and patriotism (the defense against exogenous intrusions) also become weakened. Lacking a sense of both tradition and purpose, and a practical moral base, permissive systems tend to become defenseless. Their improbability of persistence is testified to by their rarity in the empirical world.

Rationalized Systems Much more frequently encountered are rationalized systems, the last of our acculturated ideal-types. Here we are concerned with the process of indoctrination. Indoctrination becomes operational as logical preemption. That is, indoctrination consists, basically, in the exposure of the individual to *only* a single set of related axiomatic predicates, to the exclusion of all other logical possibilities. To a certain extent, of course, all cultural engines attempt to preempt the attention and affection of the individual. But whereas proscriptive systems reinforce this with sacerdotal authority, and prescriptive systems with moral suasion, axiomatic systems seek logical closure. For example, axiomatic systems secure closure in the early stages of a child's education, by exposing him only to the preferred axiomatic predicates, elaborated by indications of the way in which they produce rewardable behavior. In indoctrination processes, when the prophetic platform takes the form of quasi-abstract principles rather than explicit rules, the emphasis is on the way in which behavior may be made consistent with the preferred principles.* Indoctrination thus

*In proscriptive or dogmatic systems, educational exposure is usually only to the rules and not the reasons for behavior.

may proceed more subtly than either institutionalization or sentimentalization, and often relies heavily on insinuation. For example, within traditional capitalist systems, the children are gradually introduced to the key precepts of the society through continuous contextual exposure to surrogates for the real-world processes. Thus the grading system in school corresponds roughly to the process of economic competition, with differential performances being rewarded differentially. This also introduces the student to the fact that he is not one with his fellows, but in competition with them in an arena of scarce resources. There he is also taught the critical lesson that grades depend primarily on objective performance, and may not always be tempered by emotion or arbitration as they might be at home. In short, the educational system tends to reflect the economic determinants of the broader society, with indoctrination thus taking the form of a pre-living—a pre-experiencing—of the societal forms themselves.

In the axiomatic world, however, there sometimes arises a need to defend one set of doctrines against another. At some point in the indoctrination process, let us say, the individual is exposed to a malignment of any competitive positions. Although this process appears to take place on logical grounds, it is fundamentally a rhetorical process (very reminiscent of the rhetorical pursuasions used to establish a prescriptive axiology). To the capitalist high-school student, for example, the Communist appears as a foreign menace. To the socialist student, in say the fourth or fifth form, capitalists are painted as remorseless blackguards. Seldom does this adversary process involve a comparative examination of the principle base of the competing axiomatic system. For example, our high-school students—usually ignorant of the logical basis of socialism—are at least vaguely familiar with its excesses in the Soviet Union (e.g., the Hungarian repression, the slavish agricultural communes, the murder of the Tsar's family). By the same token, socialist students comprehend nothing of the theory of democratic capitalism, but know all about negro slavery, labor strife and the Vietnamese War, etc. In short, then, indoctrination consists of essentially two phases: (1) the surrogation of the key processes of societal life into the educational or preparational forms (school, compulsory military service, athletic and recreational events, etc.) and (2) the rhetorical condemnation of some adversary system through the judicious use of pejorative anecdotes. To a great extent, then, despite the fact that the various axiomatic systems all have logical pretensions, there is eventual retrenchment to indoctrination via rhetoric.

Continuing now with our defense of Table 4.2, the behavior desired in rationalized systems is *consistency* with the axiomatic base. The instrument by which we seek this consistency has two aspects: (1) first, as frequently mentioned, the axiomatic base provides a set of *principles* from which deductive inferences may be drawn (at least with respect to some subset of the array of societal behaviors, specifically $N \subset U$); (2) for the most common behaviors, there tend to be established sets of *procedures* which serve to guide the individual or collectivity. Procedures, in this instance, are simply predefined "algorithms" that indicate normative or desired responses to well-precedented circumstances or events [6]. It is these procedures that give substance to the institutions of government, commerce and the law, giving an axiomatic base the same kind of tangible expression that sacerdotalism gives proscriptive systems.

There are basically two mechanisms (sanctions) by which an axiomatic society enforces congruence with principle (or obedience to procedure). Adherence to principle is usually maintained through the guilt process: the individual contravening an axiomatic precept is expected to feel guilt (the repugnance of self which accompanies the failure to abide by an internalized code), enforceable much as postulated for the Freudian superego [7]. Obviously, where the bases of desired behavior are specified only in the form of abstract principles, the ability to control behavior externally is highly limited. Some reliance thus has to be had on an intrinsic mechanism, and guilt serves this purpose well. However, in those situations where a procedure has been devised, exogenous control becomes feasible. In this instance, the sanction for contravening a procedure is forfeiture: a tangible loss of privilege, position, freedom, income, etc.

It is important to understand that the viability of any axiomatically predicated system depends greatly on the extent to which principles may be gradually translated into procedures, so that exogenous controls may be substituted for the generally less reliable intrinsic ones. In most axiomatic systems, there is frequently a substantial incongruence between procedure and principle; the translation from principle to procedure is not always a direct or proper one [8]. For example, the Federal Government in the United States sometimes inaugurates procedures which are in contradiction to the "principles" of the Constitution (itself a reification of axiomatic positions popular among certain eighteenth-century European "prophets") [9]. As a general rule, then, we have three levels of concern when dealing with axiomatic systems: (1) the prophetic base, which contains original philosophical predicates; (2) the canonical derivatives, usually in the form of political paradigms, codes or "constitutions" that are products of deductive inference (through exegesis) directed at the prophetic base; and (3) the specific procedures of the system, formalized and codified, which presumably are deduced from the canonical references. Incongruence may appear at either of the interfaces.

Incongruence between the prophetic and canonical axiomatic bases may simply reflect the genuine difficulty in making disciplined deductive inferences. The manipulation of conceptual components is always perilous, so that a proper "mapping" is seldom obtained. On the other hand, incongruence between the canonical axiomatic base and the set of procedures may often be due to another source entirely: the instances of *realistic* (e.g., instrumental) behavior that intrude into axiomatically predicated systems. These behaviors would fall into the non-normative or sanctionable sets, M and S, respectively. Again, the difference is this: Non-normative behaviors represent contraventions of specific procedures, whereas sanctionable behaviors conflict with principles that have not yet been proceduralized. While non-normative behaviors are usually punishable by law, sanctionable behaviors are usually the basis for a new procedure or law. Thus, at the broadest level of analysis, we get some similarities between the operational problems of proscriptive axiological and axiomatic systems. Whatever is not specifically proscribed is considered fair. Proscriptive systems tend to reduce the range of such behaviors by constantly elaborating the axiological base. Axiomatic systems tend to do the same by constantly elaborating the set of proceduralized behaviors, and by constantly evoking new legislative restrictions. Thus, both proscriptions and legislative bases tend to serve the causes of societal control and coherence, by gradually reducing the range of conditions under which judgmental (or intrinsic) predicates may be employed. Thus they both also serve the cause of system simplicity, at least from the standpoint of the system analyst.

In the process of trying to contain the possibility of contradiction, axiomatic systems may, however, generate a collective pathology similar to that of proscriptive axiologies. As the set of procedures expands to cover more and more of the behaviors available within the system—as the ratio $[N]/[U]$ increases—there is a tendency for initiative to decline (leading to *inertia*) and for innovation to give way to *programmation*. In such a situation, the system tends to settle into a behavioral rut. This can sap the vitality of the axiomatic system, and thereby make it less likely to last.

As a final note on the acculturated ideal-types, all three provide the individual with a set of behavioral *expectations*, cast in the form of "paragons" we are urged to emulate. And all cultural systems offer the individual *simplicity*, largely through the successive closing off of the feasible behavioral repertoire. All these benefits act to relieve the indi-

TABLE 4.3: Summary of Properties of the Acculturated Socio-Behavioral Ideal-Types

PROPERTIES	PROSCRIPTIVE	PRESCRIPTIVE	AXIOMATIC
1. Appeal	Ideological	Sentimental	Logical
2. Engine	Institution-alization	Sentimental-ization	Indoctrination
3. Base	Exegesis	Affectivity	Rationalism
4. Utility	Provision of certainty	Release from repression; emotional license	Provision of purpose (structure of legitimate goals)
5. Behavioral modalities	Dogmatic	Impulsive	Rationalized
6. Prophets	Nathan, Elija Mohammed, etc. The Schoolmen Luther, Calvin, etc.	Socrates Rousseau Thoreau Jesus	Augustine, Hobbes Marx, Lenin, Mao Marshall, Smith, etc.
7. Referents (real-world systems)	Sacerdotal (theocratic) and "patrist" societies	Proto-Christian systems; "matrist" societies	Capitalist and socialist societies

vidual of the anxiety of making his own place, structuring his own identity, developing his own expectations and exercising his own cognitive powers and potentials. From the standpoint of the analyst, all these effects combine to produce an essentially simple system—one where behavior may largely be predicted if we know the axiological or axiomatic predicates. As a last construct for this section, then, we have the summary in table 4.3.

The Adaptive and Idiosyncratic Ideal-Types

Still within the confines of the model presented as Table 4.2 (our elaborative ideal-type), we are now going to discuss the referents falling into the second and third societal categories: adaptive and idiosyncratic systems. In all four cases we shall consider in this current section, certain factors intervene to abrogate—or weaken—the determinacy of any *a priori* predicates that might be present. And in three of these cases, this means that the resultant systems will usually be more difficult to describe—and more difficult to predict—than systems responsive to *a priori* predicates.

The simplest system is that which allows simple associative behaviors to prevail. Here, it is the conditions of the environment (the concatenation of exogenous, empirical events and circumstances) that ultimately drive behavior. Whatever behavioral constraints are present tend to take the form of *lore*—an uncodified, unsystematic set of predicates. Now, it is possible to see key relationships emerge between exogenous predicates and responses, for the basic cognitive modality driving such systems is, again, simple regression. But, in sum, the behaviors are products of collective, *ad hoc* learning, and they cannot really be transmitted from generation to generation except by rote or practical demonstration. This is because they lack an encompassing deductive structure (disciplined or otherwise). Thus, again, we want to reserve the term "culture" for those instances where such structure exists, and where transmittal of the societal substance may be accomplished (at least in part) through some sort of argumentation.

Now, there will be very few systems where behavior is totally associative or conditioned in determination. For the most part, as we shall later show, such systems flourished only in the distant past, and even then perhaps only for short times. Today, the few that remain tend to exist at the fringes of civilization, generally in geographical areas that more advanced societies shun. We may find conditioned systems in the deep jungle, or perhaps in the arctic, or in the most inaccessible desert lands. Such geographic conditions are needed because they represent placid environments, all of whose properties (being largely homogeneous and simple in distribution) may be learned in the space of a few generations. Moreover, there are not even significant cyclical or seasonal changes in such areas. As a result, faced with such environmental simplicity, it is efficient for a societal system to allow the environment to *program* its behavior, so that individuals will tend to react with virtual automaticity to the few discrete stimuli or events which such an environment holds. These programmed behaviors probably constitute a very limited repertoire of activities, all more or less directly related to simple survival. Therefore, in addition to automaticity of response, the conditioned societal system would demand *consuetude*, condemning any moves toward either individuation or innovation.

Now, the mechanism by which consuetude and automaticity are introduced and maintained is *ritualization*. As suggested earlier, ritualization becomes comprehensible as the collective counterpart of a simple rote learning [10]. The lore of the conditioned system—acquired through the process of historical accommodation to the environment—now becomes *choreographed*, with every move and response and variant carefully considered with respect to its current or vestigial (historical) survival value.* The major sanction that such systems reserve is exile. Exile, in the environments occupied by the associative systems, is tantamount to death. Indeed, the distribution of villages across a primitive environment may have a special meaning; often, one appears to be just far enough away from another so that an exiled individual is unlikely to survive the journey. As a corollary, primitive systems have always been very fond of elevating the very few immigrants to high positions, this perhaps being a testimony to the special skills or luck that would be presumed associated with such an adventure. Finally, because these autarchic, mechanistic systems are specialists in dealing with a single, essentially invariant environment, they are unlikely to respond well or rapidly to major shifts in the environment. This, again, is testified to by their extreme rarity.

As for our fifth elaborative ideal-type, the system where correlative behaviors predominate, there is again a set of factors which tend to dilute *a priori* determination. Specifically, we already know that correlative behavior allows us to exploit emergent opportunities in a relatively complex environment. Now, by "relatively" complex I mean a situation where there are two conditions that do not prevail in the placid or simple environments that house conditioned–associative systems: (1) environmental properties are distributed in a clustered rather than a random or symmetrical way; and (2) the distribution of these properties changes over time [11]. In such a case, the full effect of ampliative inductive inferences may be brought to bear**. For the future may legitimately be expected to be different than the past (though still some function of it), and one is required to use inference to project properties of the farther milieu from immediately accessible segments.

Again, correlative behaviors are those that take exogenous, environmental predicates, and amplify them through normal inductive inference processes—projection or extrapolation [12]. Behavior here may

*Rituals that have a vestigial quality might, to the casual observer, look like distinctly creative (*ab initio*) exercises; but it is only that the condition to which they once responded—the original predicate for the ritual—no longer exists in the environment. For example, certain African tribes have hunting rituals that portray animals that were thought by early explorers to be creative fabrications. New paleontological studies have shown, however, that the animals did in fact exist—were not mythical or "deduced"—but have since become extinct.

**See, again, Chapter Two.

thus be "explained" largely with reference to the *context* in which the individual or collectivity resides.

Now, the process of *secularization* stimulates correlative instrumentalistic behaviors. For secularization is a process where a societal system is gradually absolved of any axiological or axiomatic predicates that might have once held sway. And, by definition, instrumentalism is merely exploitative license, where neither ends nor means are preconstrained. Therefore, secularized systems are notable in having no cultural predicates, so that no substantive constraints may be passed from one generation to the next. Therefore, secularized systems condone—if only by default—*opportunism*.

The only basis for controlling behavior in a secularized system would have to be both *a posteriori* and *ad hoc*. One that especially comes to mind is *preemption*. It is appropriate because, in a secularized system, relations among individuals may often be ruled by the calculus of the constant-sum game, where whatever one party is able to gain is perceived to be at the direct expense of some other party or parties [13]. Therefore, individuals seek to preempt social, economic or political perogatives, and therefore deny them to others. It is the world of unfettered competition, scarce resources, strategic analysis and "winning". Were there an intellectual model for such a system, it would be a composite of social Darwinism, the coaching philosophy of Vincent Lombardi and an epistemology culled from the more fervid exhortations of Neitzche, Che Guevera and Hobbes. All resolves itself ultimately to power: knowledge is power, cunning is power, etc. In this catch-as-catch-can world, the primary sanction which society holds is the correlation between failure and *deprivation* or *confiscation*. That is, what one cannot hold or gain is lost. For those individuals or subsystems that cannot successfully compete, there is a reduced probability of survival. The only pathological behavior is *suboptimality* *—the failure to be efficient in the pursuit of one's ends. Finally, the rather obvious societal (collective) pathology which attends secularized systems is *alienation*—the effective separation of self from others, from the future and from any socio-spiritual significance [14].

Moving on to our last pair of system types—those where idiosyncratic behaviors predominate—we find again a weakening or absence of direct cultural determination. In originative systems, this weakening is aggravated by cerebration, which leads to a conscious retreat from emotion and ideology, and a deliberate search for opportunities to transcend the environment or experience. If it could, it would develop a society that emphasized *sapience* in its purest form. The sapient indi-

*Where optimality or suboptimality is calculated only with respect to local conditions.

vidual or collectivity is characterized by constancy and quality of discernment. Apriorisms—which become axioms or "givens" in culturally driven systems—take on a hypothetical quality here, and are ultimately judged valuable or fallacious with respect to realities which emerge. Therefore, the sapient system is Janus-faced, with both conceptual and empirical significance. Thus, the sapient system counters complexity with the somewhat idealized "heuristic" modality we described earlier, and is equally at home with moral and with material issues. The behaviors under this rubric are, then, potentially creative in the strictest sense: *original responses to unprecedented situations*.

Generally, control in such systems will be restricted to *peer review*, as peers are the only qualified source of criticism of creative activities. Here, then, as with the system as a whole, there is an attempt to flavor judgements with a strong element of discernment, rather than holding to hard-and-fast rules. This exogenous control modality will usually be accompanied by an intrinsic one, often in the form of some *ethic* which has been internalized, reflecting the "professionalization" of the sapient system's constituency. We also suspect that sapient systems would have to be very exclusive; much emphasis would be placed on evaluation of an individual's *credentials* before he was admitted, and we would expect to find very elaborate recruitment and selection procedures (in part to compensate for diminished control potential).

The usual sanction associated with intellectualistic systems is *censure*, the process of casting objective aspersions on an individual's qualifications. In more extreme cases, there may be an abrogation of the individual's membership in the society, or a revocation of the right to practice his profession or hold some office or title, etc. The pathological behavior most often attending such systems is *effeteness*, the gradual erosion of any pragmatic pretentions or interests. The most likely collective pathology attending cerebral systems is *abulia*: the inability to make decisions or take directed action. Abulia may be traced to the sapient's ability to tolerate a great deal of anxiety and uncertainty in the pursuit of an optimal solution to some problem, and their general unwillingness to tolerate much probability of being wrong. Thus the emphasis tends to be more on study and search than on action, and cerebral collectivities thus become more reflective than reactive.

The last of our behavioral categories—compulsive action—may be seen to emanate from the gradual existentialization of a system. In such a process, the isolated individual becomes the paramount unit of action and analysis, and what is real or valuable or significant is determined only with respect to the individual's idiosyncratic appreciations. As noted in Chapter One, the behavior of compulsive individuals may be considered to be an instance of *hypostatization*—the reification of mentalistic constructs, without the reality feedback that would be associated with heuristic behaviors. If compulsive behaviors are allowed to reach an extreme, they may eventuate in a kind of collective *paranoia*, or perhaps to an anarchistic system. Finally, there is the ever present danger of widespread *anomie* in a compulsive society [15], the somewhat ironic situation where the solipsist loses contact with himself. (We should note, however, that any "collective" components in existentialistic-compulsive systems will be very poorly articulated, given that existentialized systems tend to be inherently atomistic.)

It should be clear that the sapient and compulsive societal forms represent types that have hardly contributed to the history of civilization. They are, more or less, theoretical constructs, representing the best and worst of all possible worlds, respectively. If such systems existed at all, they were generally encapsulated within more traditional societal systems.[Indeed, they may exist as organizations rather than as societal systems *per se*. For example, many university faculties or research agencies would tend to think of themselves as cerebral (sapient) systems. Isolated pockets of the drug culture or religious hermits may be examples of existentialized subsystems.] But, as we shall see, our appreciation of the range of societal alternatives would not be complete without these abstract referents, and we shall visit them again later.

To summarize this section, then, there are at least four conditions under which some non-cultural engine may gain prominance:

1. Where the environment of an individual (or system) is so highly structured that no *a priori* predicates are appropriate—as is the case with essentially primitive (non cognitive) systems operating in the face of essentially placid, non-differentiated natural environments. In such a situation, we would expect that *associative* behaviors would dominate.
2. Where secularization has invalidated all aprioristic predicates, or where the constraints of cultural mechanisms are seen to involve too great a penalty relative to material opportunities. In such a situation, either individual or collective behavior may become susceptible to *correlative* instrumentalism.
3. Where, under a criterion of sapience, it has been determined that cultural predicates are "irrational", or that intellectual and aesthetic constraints represent an unnecessary restriction i.e., they are rationally indefensible. In such a situation, we would expect *originative* behaviors, replacing exogenous formal *a priori* constraints with tentative heuristic referents.
4. Where an individual (or, more rarely, a collectivity) cannot contain

compulsions. In such an instance, cultural constraints are either denied or perforce abrogated.

We restructure the assertions of this section in table 4.4, which summarizes the key properties of the several non-acculturated societal ideal-types.

TABLE 4.4: Summary Properties of the Adaptive and Idiosyncratic Ideal-Types

	INTRINSIC/IDIOSYNCRATIC		REALISTIC/ADAPTIVE	
PROPERTIES	ORIGINATIVE	COMPULSIVE	ASSOCIATIVE	CORRELATIVE
1. Appeal	Sapient	Psychogenetic	Sentient	Instrumental/ utilitarian
2. Engine	Cerebration	Existentialization	Programmation (via conditioning)	Secularization
3. Base	Discernment	Hypostatization	Association/regression	Correlation/ ampliation
4. Utility	Aesthetic release	Eccentric license	Minimization of cognitive demands	Alleviation of guilt, shame repression, etc.
5. Behavioral modalities	Heuristic	Solipsistic	Automatic/ rote	Opportunistic/exploitative
6. Prophets	Natural theologists (e.g., Aquinas) Aristotle Popper, Polanyi, etc.	Heidegger, Kierkegaard, Sartre, etc. Sufis, mystics Kantians, Huxley, Jaspers etc.	Positivists Behaviorists Romantics Functionalists	Epicurus, etc. Instrumentalists Locke, Hume Utilitarians (Mill, Bentham, etc.)
7. Referents	Proto-rational systems (e.g., the "open" society)	Cynical systems, Anarchic societies	Primitive systems	Hedonistic systems

Now, we suggested earlier that any elaborative ideal-type will have direct predictive pretensions. This is important for us here, as societal systems must be thought of as subject to change. The work we have just done should allow us to generate some assertions about the probable nature of the changes that will take place. Thus, our elaborative ideal-types will now be given a dynamic or processual significance.

DYNAMIC IMPLICATIONS OF THE IDEAL-TYPES

In the broadest possible framework, any set of elaborative ideal-types may be thought of as the structural (or state-variable) components of a proper scientific model—one that operates simultaneously on the static and dynamic dimensions. Therefore, our task in this section is to establish the nature of the *processual relationships* among the various elaborative ideal-types. In this regard consider the diagram in figure 4.2.

Basically, this model suggests that any societal subject may be in one of seven conditions. That is, we have now transformed secularization, institutionalization etc., from simple processes into *system states*. We may now speak about a "secularized" system, etc., with the implication that that process dominates. Of necessity, the implications of the model are merely hypothetical. They represent the most logically probable dynamics given our previous analysis. In the next few pages, we shall try to defend this logic and lend the model some empirical appeal.

Initially, our model suggests that societal evolution began with conditioned-programmed systems, driven by simple associative behaviors in response to prevailing environmental conditions. The conditioned system, as the prototypical societal system, is thus an almost "organic" adaptation to its milieu. Such systems make the least cognitive demands on the individual and employ the simplest methods of organization and control (segmentation and ritualization, respectively). Here we have the organizational and behavioral configuration that would be typical of simple subsistence cultures, primitive food-gathering and casual hunting societies. As earlier suggested, such systems have gradually disappeared from the world, with only scattered remnants in the least accessible and generally most undesirable geographic locations (e.g., the high arctic, the deep jungle)[16]. To a certain extent, vestigial nomadic existence reflects this condition of associative predominance, as do the recent attempts to simplify life in some industrial societies by escaping from the socio-economic mainstream. The romantic appeal of a conditioned-associative existence (strong among the youth of affluent materialistic systems) rests generally on the assumption that one may be "one with nature" by de-

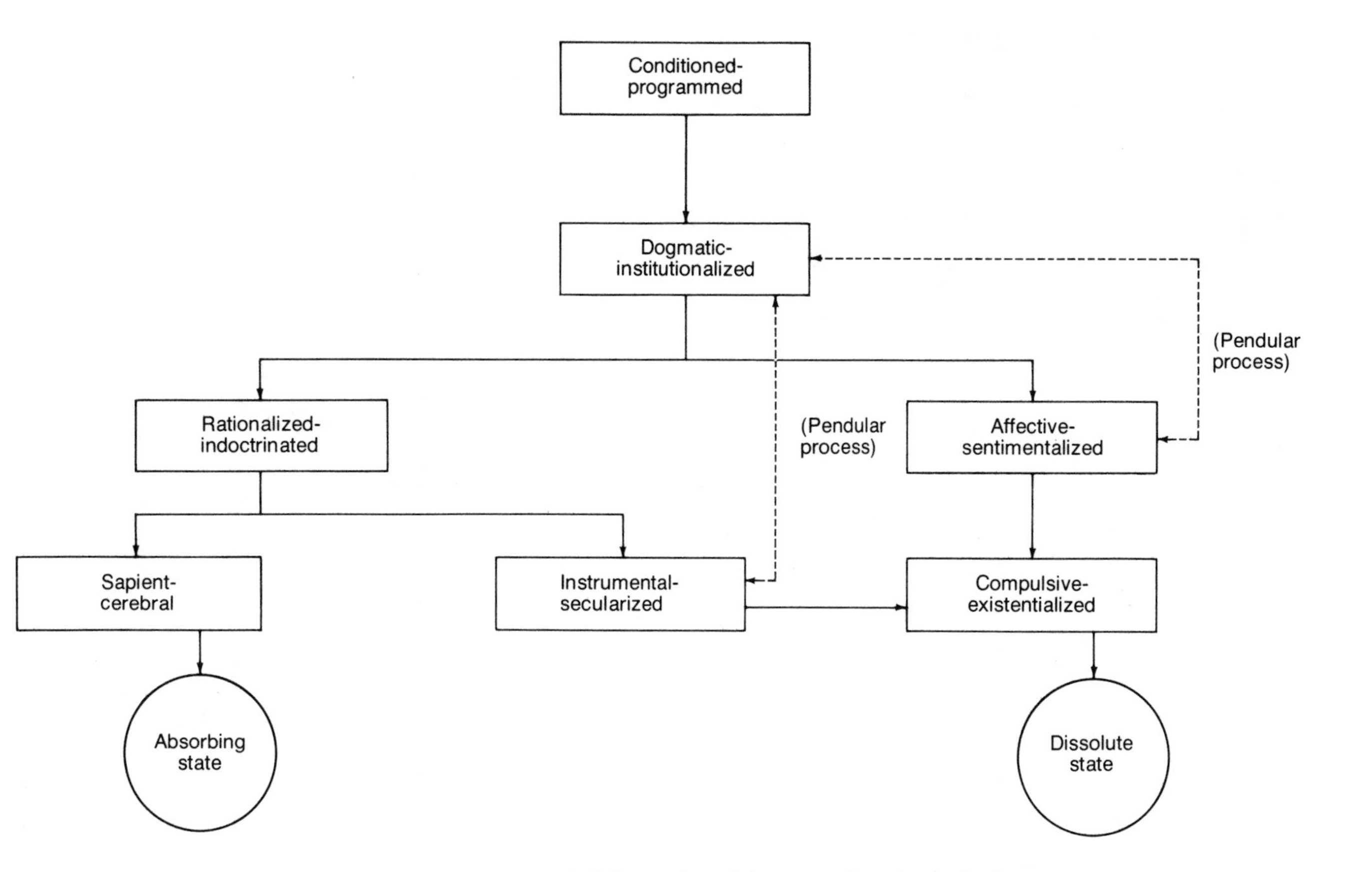

FIGURE 4.2: Normative sequential dynamics of the several societal ideal-types.

priving oneself of ambitions, structure or self-determinations. In a sense, the *Dharma Bums, Bronson, Route 66* genre of popular literature and television portrayed individuals who were content merely to let things happen, and be carried away by what the environment had to offer. All expectations, prejudice, plans and formulae for living were anathema. By some miracle of poetic license, however, all these wandering anti-heroes were able to remain economically solvent and well fed without apparent effort (and were miraculously provided with gasoline for the Jeeps, motorcycle and Corvette sports car, respectively). Such automatic economic security distinguishes them from those who reside in conditioned systems by default. But we shall wait to say more about regressive societal dynamics until a later section.

Now, the implication of our model is that conditioned programmed systems will tend to be transformed into institutionalized systems. In all cases, issues of societal transformation will tend to raise questions on two levels. First, we have to find out why the system did not persist in its original state (why it was transformed at all). Secondly, we have to find out why the transformation led to a particular state rather than some other (e.g., why conditioned systems are transformed preferentially into institutionalized systems). We shall try to answer these questions in order.

Initially, the probability of persistence of any conditioned-programmed system is inherently low. There are several reasons for this, all of which are more or less familiar to us from earlier discussions. First, programmed systems are viable only in a simple, consistent environment, largely because of behavioral rigidity and automaticity due to ritualization. Therefore, any significant shift in environmental properties—within a short period of time—will challenge the system's ability to maintain homeostasis. The simple cognitive base of such systems allows them no capability for strategic decision making or preadaption, nor does it support any major correlative adjustments. Characteristics that make such a system unlikely to survive an environmental perturbation makes it even less likely to be able to stand against a more highly organized, aggressive adversary or competitor. It will either therefore be absorbed into a less primitive system (and gain its institutionalized quality by default, as it were), or gradually be pushed back to the less attractive environment confines. As a rule, then, the probability of persistence of a conditioned-programmed system is in an inverse relationship to the richness or quality of the milieu it occupies. Those prototypical societies in the most fertile and appealing areas were thus the first to dissolve.

A third reason for the conditioned-programmed system's low

probability of persistence is this: being the most simple societal form, it is subject to the full range of pressures for complexification. Some of these are natural, such as an increase in population leading to the demand for new organizational forms. Some are primordial, such as the constant pressures for differentiation, specialization and hierarchicalization.* Others are exogenous, such as the diffusion of technical innovation (e.g., the lesson that settled agriculture is more efficient than hunting or food-gathering). In short, because these systems are so simple, virtually every type of innovation can be accommodated or be found somehow attractive. Thus, once again, the probabilities act against the conditioned-programmed system perpetuating itself.

The second major question can now be answered, perhaps satisfactorily. The conditioned-programmed system is most likely to be transformed into an institutionalized system because the properties of the latter are the natural extension of those of the former. For example, ritualization and sacerdotalism are highly related (the former being simply the less well-organized, spontaneous form of the latter). More specifically, sacerdotalistic institutions are simply the "specialized" form of ritualistic mechanism (where priestly authority is now concentrated rather than diffused). In addition, obedience (as a desired behavior) is a natural extension of automaticity. In terms of sanctions, excommunication derives directly from the concept of exile, with heresy simply the cognitive counterpart of individuation. Therefore, in terms of the basic attributes of the behavioral categories summarized in Table 4.2, the dogmatic-institutionalized system is the *natural successor* to the conditioned-programmed system. In overview, then, conditioned-programmed systems—initially the dominant societal form—gradually become very much reduced, either by being pushed to the marginal environments by more aggressive systems, or by succumbing to broad environmental changes to which they could not adapt (e.g., climatic shifts, migrations of animals or fish). Otherwise, through natural complexification, they transformed themselves into institutionalized systems, which in turn absorbed other primitive systems.

Continuing now with our dynamic model, it is suggested that institutionalized systems will tend to give way either to rationalized systems or to sentimentalized systems. If we identify institutionalized systems with systems susceptible to proscriptive axiology, and sentimentalized systems as products of a prescriptive axiology, then we have already suggested the reasons for the transition from institutionalization to sentimentalization. The discipline, repression and

---

*See Chapter Six.

frustration that accompany the elaboration of the proscriptions tends to make impulsiveness, permissivism and spontaneity marginally more and more attractive. But then the pendular process kicks in, fueled by the disorder of sentimentalized systems, and there is some probability that the sentimentalized system will return to an institutionalized form. This probability, from the standpoint of our socio-psychological appreciations, is quite high, largely because proscriptive and prescriptive axiologies are often merely effective complements or logical inverses of one another.* Thus, the attraction between these two polar opposite states is very strong.

Now, the reason for the improbability of persistence of an institutionalized system has already been given: proscriptions tend to increase to the point where frustration (repression) is perceived to outweigh the benefits of certainty derived from dogmatic context [17]. And the reason for the high probability of transformation to a sentimentalized state rest with the fact that the prophetic process of logical complementation or inversion requires very little imagination or intellectual discipline. But what of the positive probability that an institutionalized system may transform itself (or be transformed into a rationalized system—that the proscriptive axiology will yield to an axiomatic platform of some kind? That is, under what conditions may rationalization follow institutionalization, and sentimentalization be foregone?

The reasons for this alternative are housed in our comparative discussions of axiological and axiomatic platforms in previous sections. There it was seen that the rudimentary properties of proscriptive and axiomatic sytems (as summarized in table 4.2) were very similar. Our point here is that just as institutionalized systems represented the next evolutionary step from acculturated systems, so rationalized systems may represent the next *evolutionary* step beyond institutionalized systems (while sentimentalization represents more of a lateral or horizontal movement). For example, the process of indoctrination is a simple derivative of institutionalization. The transcendental (numinour) authority of the latter are merely replaced with their logical (rationalistic) counterparts. Moreover, the concept of congruence with respect to some set of prior principles is a simple extension of the concept of obedience to a set of dogmatic tenets. To a great extent, the legislative mechanisms that create and defend "procedure" in rationalized systems are the direct descendents of the sacerdotal mechanisms that develop and enforce the rules of institutionalized systems. Even the

*See, again, the discussions on social prophecy in Chapter Three.

sanctions—excommunication and forfeiture, respectively—are similar, while the respective dysfunctional behaviors (heresy and contradiction) are logical cousins. In summary, then, the shift from a proscriptive to a rationalized system represents mainly a shift in predicates—from tanscendent to logical—with the organizational and operational aspects remaining very similar. But the *evolutionary* aspect of the transformation from an institutionalized to a rationalized system consists in the greater abstraction and more elaborate defense of any apriorist predicates, and in the separation of economic and political precepts from the socio-spiritual. Thus, the rationalized system generally represents a more complex form of organization, a point which will be made clear in the coming chapter. This evolutionary trajectory thus exists as an alternative to the essentially pendular shift that we find between institutionalized and sentimentalized systems.

Continuing to the next level of the dynamic model, we see that a rationalized system is expected to be transformed by one of two processes: secularization or cerebration. Of the two transformations, that to the secularized system is the more probable. The reasons are largely those given in the previous section. The principles housed in the axiomatic base serve to define the legitimate (normative) ambitions or goals that individuals should seek to attain: wealth, social significance, self-actualization, political power, status, etc. But the axiomatic base also has proscriptive implications that restrict the means to achieve those ends. Now, the legitimate ends may, in many cases, be more expediently served by illegitimate or non-normative means. To the extent that this becomes possible, there is a shift from a rationalized to a secularized system—from constrained purposivity to unbridled instrumentalism. Thus, the instances of contradiction between principle and practice in axiomatic systems are often merely incursions by instrumentalization. It is a short step, then, from the denial of transcendental authority which leads from institutionalization to rationalization, to the denial of all aprioristic significance which leads to secularization. Now we must consider the other pendular process present in our model—that which leads from secularization back to institutionalization. The rationale for this transformation (an apparently regressive or counterrevolutionary one) is something we have touched on only most obliquely. Secularization ultimately leads to a situation of catch-as-catch-can competition, the Darwinistic drama. Given natural differentiation between individuals in terms of either aspirations or abilities, we get a concentration of prerogatives in the hands of successively smaller groups. To a certain extent, this is the situation with proscriptive and axiomatic systems as well. But the lack of constraints on means associated with secularized systems suggests that the concentration process is likely to proceed more quickly and ruthlessly. Moreover, there is no legal or procedural system to which recourse may be had (as with axiomatic systems), nor is concentration lent any transcendental authority (as in proscriptive systems). Under such conditions, society becomes ripe for a regeneration of discipline and dogma and spiritual significance. The desirability of discipline increases to the extent that the system becomes chaotic. Dogma becomes attractive because it is the polar opposite of collectivized nihilism. And spiritual significance is a relief from the secularized system's lack of moral or transcendental guidelines, a lack which robs individuals not only of psychological comfort and determinacy, but also of the compensatory calculus that compensates temporal deprivations for the noncompetitive. For all these reasons, secularized systems are not likely to persist over long periods. They will either degenerate into existentialistic systems (a possibility we shall explore shortly), or suffer a perhaps rather abrupt and sweeping return to institutionalization. It is significant that institutionalization rather than rationalization seems to be the ultimate destination of this particular pendular process. But in keeping with the concept of pendularity, institutionalization is perhaps more clearly the polar opposite of secularization than is rationalization. Also (with very rare exceptions), rationalization has served as the intermediary between institutionalization and secularization. Therefore, a significant part of the population may blame many of the secular excesses on the axiomatic base rather than on the non-normative behaviors. This tendency is reinforced by the fact that instrumentalists often seek to cloak their behaviors with axiomatic authority. For example, the exploitative monopolies of the late nineteenth century and the collusive counter-competitive industrial giants of our contemporary age both seek to draw authority for their behavior from the capitalist axiomatic platform, claiming to act as agents for free enterprise (even though axiomatic capitalism rejects both monopoly and collusion). Thus, in many instances, the rationalistic predicates are discredited by instrumentalists, and the return to a pure (and transcendental) proscriptive axiology thus becomes the more feasible transformation. It is perhaps thus that many modern Americans are more eager for a return to Jesus or Bhudda than for a return to Adam Smith or Alfred Marshall.

The other transformation for secularized systems—to existentialization—is perhaps less probable than the return to institutionalization. Nevertheless, it can happen. It is an essentially simple and direct evolutionary step. The logical conclusion of secularization is solipsism. In the initial or emergent stages of a secularized system, there is the focus of attention on material ends, and possibly the pursuit of essentially the same objectives as those defined by the prior axiomatic base from which the secularized system evolved. The means may be unconstrained, but the foci of interest are tangible and not inherently condemnable—status, wealth, power, etc., the great societal equilibrators. But where means are unconstrained, there is the tendency for the society to shift its interests from inherently functional ends to psychogenetic pursuits, or at least there is evident freedom to do so without direct sanction. What psychogenetic release means, in effect, is titillation: the excitement of the senses through the immediate gratification of any impulses (with the immediacy of gratification causing the transformation from impulse to compulsion). Thus, it is expected that sadistic, masochistic, sexually aberrative and elaborately perverse sensual experimentations will occur to an increasing extent. Initially, the axiomatic platform's vestigial component may have suggested that basically productive exercises were the most probable route to secular satisfactions—that wealth could ultimately be translated into purchasable pleasures. But in an existentialistic world, titillation becomes the sole end, and does not require any intermediation. In short, existentialism—with its solipsistic and anarchic corollaries—becomes one logical conclusion to the retreat from apriorism which began initially with the transition from dogma to rationalization, and gained both momentum and scope in the subsequent transformation from rationalization to secularization.

But our model also suggests that the existentialistic state may be arrived at via sentimentalization. The reasons for this are already clear from our earlier discussions of prescriptive axiologies and affective behaviors in general. There is, obviously, a very close correlation between the freedom to exercise impulses (indeed, the imperative to do so) and the ability to act out compulsions. To a great extent, then, the line separating prescriptive from compulsive systems is an uncertain one. The difference exists, fundamentally, in two factors: (1) in prescriptive systems, there are *a priori* constraints on the impulses one can legitimately exercise; and (2) in prescriptive systems, there is a sanction (shame) associated with the contravention of the axiological predicates. In the existentialistic system, neither of these constraints exist. No impulses are denied legitimacy, and social disapproval becomes irrelevant. Thus, the existentialistic state is the next evolutionary step beyond sentimentalization (though, for reasons already discussed, the pendular movement back to an institutionalized system is the more likely transformation).

Note that there is only one destination for an existentialized system:

dissolution. The argument for this is essentially the same as in the previous section. Existentialistic contexts eventually just atomize, and therefore lose their identity as systems *per se*. That is, the solopsistic implications of existentialistic systems breed a collective cynicism which (unlike the effective hedonism of secularized systems) eventually robs the system of any possibility of coherence. In short, then, the existential state is one of the two *absorbing states* of our model, one where the system simply ceases to exist. The other is that which evolves from cerebration.

Cerebrated (sapient) systems evolve, in our scheme, from rationalization (as a less probable alternative to secularization). Cerebration is an evolutionary adjunct to rationalization. Rationalization essentially discards the numinous or mysterious aprioristic bases of institutionalized systems, and substitutes a set of axiomatic predicates (purportedly of natural or logical origin). However, like the dogmatic predicates of proscriptive axiologies, the first premises of axiomatic systems are beyond dispute. When these are now opened to examination rather than mere acceptance, we have the basis for sapience. Now, cerebration is a less probable evolutionary alternative than secularization, for the development of a sapient from a rationalized system is a more demanding cognitive process than the simple rejection that produces the secularized alternative.

The eclectic (and hence complex) implications of cerebration stem from a procedure with which we are already familiar. What are indisputable axioms in a rationalized system, now become disputable hypotheses whose utility is heuristic. Thus, in a sapient system, only those *a priori* behavioral constraints that have been logically or empirically validated are operative. The fact that there are any *a priori* predicates at all, of course, distinguishes it from the secularized, acculturated and existentialistic systems. The sapient-cerebral system, may then, combine the best of both worlds: cultural predication and eclectic assimilation. As a result, it should have the highest probability of persistence of all our societal states, and hence be an absorbing state. It is unfortunate that the probability of ever arriving at a sapient state is so low. In the real world such societies have existed only for the briefest times, and then only in adulterated form. But as the sapient state is perhaps the highest form of societal sophistication, its realization perhaps awaits us further down the evolutionary line.

We have now completed a major phase of the hypothetico-deductive model-building process. And from what we have done, we may extract a critical conclusion:

> As we move from the conditioned to the institutionalized system, and from the institutionalized to the sentimentalized system, from the sen-

timentalized to the secularized system, and from there to the cerebrated or existentialistic system, the behavioral latitude allowed individuals increases, and the boundedness and coherence of the system decreases. But (and this is a critical assertion) the net satisfaction available from membership in a societal system need not increase as we move up the evolutionary ladder.

This latter assertion will be the major problem we shall have to face when, in the last two chapters of this book, we turn to matters of societal management. But for the moment, we have a final model-building project to discuss: the generation of linkages between the type of hypothetico-deductive constructs we have been building here and the quantitative-empirical constructs that normal science prefers.

Linking the Qualitative and Quantitative Domains

Our dynamic logic suggests that the future state of any socio-behavioral system is dependent on the current state, and that there are certain definite probabilities of transformation between one state and another. Thus, we should be now in a position to establish a link between a qualitative constructs and the quantitative domain of normal science, and hence generate an apodictical construct. We do this by transforming our directional relational hypotheses into magnitudinal coefficients, as in table 4.5.

With this formulation, societal dynamics now becomes intelligible in terms of a *Markov process* [18]. As may readily be seen by examining the table, there are two explicit premises that concern us: (1) Only the cerebral system is expected to remain in the same state through time; and (2) The probability of transformation from one state to another are not equal—i.e., as we have been saying, some societal transformations are more probable than others. Each of the rows of this matrix may be interpreted as probability vectors (i.e., the sum of the entries for each of

TABLE 4.5: An A Priori Transitional Matrix

Starting States	Concluding States						
	Prog.	Inst.	Rat.	Sent.	Sec.	Cerebral	Exist.
Programmed	0	1	0	0	0	0	0
Institutionalized	0	0	0.4	0.5	0.1	0	0
Rationalized	0	0	0	0.2	0.7	0.1	0
Sentimentalized	0	0.7	0	0	0.2	0	0.1
Secularized	0.1	0.6	0	0.1	0	0	0.2
Cerebral	0	0	0	0	0	1	0
Existentialized	0	0	0	0	0	0	1

the rows is equal to one). The logic that we developed in the last section may now be examined with respect to its empirical implications. It should also be noticed that in moving from the diagrammatic to the numerical form of representing our model, we are able to include several cases here that could not be included in the graph without causing considerable confusion.

Initially, the matrix of transformational probabilities reproduced above is an *a priori, hypothetical* one. That is, the probability values entered there were estimated from the analysis we have conducted thus far in this book, and are therefore subjective (or logical) probabilities. Thus, our array of propositions—suggested verbally before—has now been expressed in terms of numbers. Such a representation might give researchers the basis for an objective evaluation of the validity of our deductively driven conclusions. For example, if scholars with a good knowledge of societal development could review the schedule of transformational probabilities in light of their own specialized knowledge of the history of civilization—and if the resultant transformational configurations were not radically different than those set out here—then our theoretical model of societal dynamics would gain some empirical authority. To the extent that their review of social history and the evolution of civilization generates a significantly different configuration, then our theoretical system may lose conviction, but some superior theoretical system may, in the process, be articulated.

The matrix model may thus serve us in the following way. First, we defined seven possible societal states—programmation, institutionalization, etc. Each of these was associated with certain definite and distinct properties (admittedly, some were more clearly defined than others, but perhaps we can repair this to a certain extent in the next chapter). The transitional logic (expressed verbally) was ultimately expressible as a set of probability statements. Thus, for example, our assertion that, given the conditioned-programmed state, the only significant transformation probability was to the institutionalized state, is expressed by the 1 set at the intersection of the two states in table 4.5. If our theoretical ground holds any merit, then, researchers should not be able to discover too many cases where a system meeting the conditioned-programmed criteria was transformed into something other than a system meeting the institutionalized criteria. The same would apply to any other transformations considered by the model.

Thus, in empiricizing our theoretical propositions, we interpret each row of the transitional matrix as a reference probability distribution. Two such distributions are shown in figure 4.3, where for illustrative purposes we treat what are really discrete events as if they were continuous:

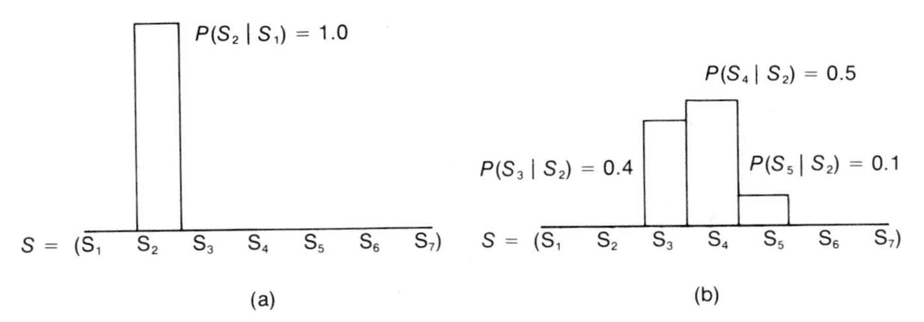

FIGURE 4.3: Reference (*a priori*) probability distributions for transformations from (a) S_1, (b) S_2. [S_1 = conditioned-programmed state, S_2 = institutionalized state, S_3 = rationalized state, S_4 = sentimentalized state, S_5 = secularized state, S_6 = cerebral-sapient state, S_7 = existentialized state]; and $P(S_j|S_i)$ is the probability that the starting state S_i will be transformed into the concluding state S_j.

Now, if empirical analysis tended to justify the probability distributions we presented as the hypothetical set, then the theoretical base's probability validity could be expressed in normal statistical terms—that is, we could establish an *index of confidence* for the theory. But, even if actual empirical research modified the *a priori* distributions (through a Bayesian process, where objective, *a posteriori* probabilities were successively admixed with our *a priori* ones), the great advantage of establishing the transition matrix is this: to the extent that a reasonably valid matrix of transitional probabilities may be established, we may use well-established mathematical procedures to construct dynamic models that will trace the normative course of societal development.

We may, however, offer some very brief (and certainly inadequate) concrete suggestions as to the type of transformation processes the matrix entails. Beginning with the first row, we might suggest that virtually all of the ancient civilizations—Greece, Egypt, Sumer, etc.—developed into institutionalized systems directly from a prototypical conditioned system, usually pursuing simple food gathering, or casual agriculture or hunting. Perhaps the clearest and most compelling example of this first transformation process (from S_1 to S_2) is the story of the rise of the nation of Israel told in the Old Testament. Moses' gathering of the tribes—and the development of institutions through the formalization of an army and priesthood and tribal reconfiguration—is precisely the kind of mechanics that would drive such a transformation.

For the transformation from the institutionalized state, we have some compelling historical examples. Among the most important is the transformation that accompanied the decline of the Holy Roman Empire and the more general derogation of the authority of the Roman Catholic Church in late Medieval Europe. In this transformation, the Protestant Reformation opened the door for the Anglo-Saxon nations to become *rationalized*, mainly through bourgeoise socio-economics and infant capitalism and industrialism. By the same token, many of the countries which remained essentially Catholic—particularly Ireland and the Iberian states and their New World colonies—took the path of sentimentalization. They became strongly matrist, at the expense of general material or economic development. Finally, there were very limited movements from Catholic institutionalization to encapsulated secularization, particularly among certain sub-populations in France and Italy. Therefore, we get the basic qualitative conditions of transformation from the institutionalization state that the model would predict: some significant transitions to rationalization, some to sentimentalization, and a much less significant movement toward secularization.

Moving on to the third row of the matrix, we are concerned about those transformations from a rationalized system state. Only two alternatives are considered. Most probable is the gradual shift to secularization, notable especially among those nations which originally adopted axiomatic capitalism as an alternative to dogmatic Catholicism—the United States, Britain, Germany, etc. The inroads made by secularization may be noted on all societal dimensions. The social sector is being transformed by weakening of family ties and responsibilites, and by an increasing interest in sexual gratification through pornographic titillation, etc. On the economic dimension, the axiomatic predicates of theoretical capitalism are consistently ignored, especially by the larger commercial units. And little need be said here about opportunism in the political sector. There is a *de facto* erosion of the effective authority of the individual voter, and external relations tend to be ruled by a *realpolitik* largely without moral or logical significance.

We have some pressing historical instances of a rationalized system reverting to a secularized state, notable among them the shift of post-Revolutionary France to the Napoleonic regime and perhaps also the displacement of the Weimar Republic by National Socialism in Germany. But National Socialism also had distinctly axiological and axiomatic overtones. It promised a national identity, a national purpose and a provision of order and disciplined growth and expansion where there was politico-economic chaos. To a certain extent, the Napoleonic system promised the same. But the means by which these two processes were given moment and authority—and the eventual concentration of power in a single demogogic individual—mark the processes as distinctly secularized in both emergence and implementation. However, their path to power was eased by their pseudo-axiological aspects, and by their appearance as a "religion without a God". For the French bourgeoisie of the beginning of the nineteenth century, as for the German citizens of the early 1930s, there were numinous overtones to the movements, cast in terms of national destiny, and perhaps even a cosmological mandate about the duty of the French or German people, etc. But for those who actually brought about the transformation—for the vestiges of French aristocracy and the industrial interests of Germany—both these movements offered release from the axiomatic constraints and constitutional proscriptions in the Republican and Weimar legislation. Though there are certainly some equivocations, it is best to consider the Napoleonic accession and the rise of National Socialism as instances of rationalized evolving into secularized systems.

As suggested earlier, there is a slight (0.1) probability of a shift from a rationalized to a sapient system. To a certain extent, the abandonment of rhetorical and partisan interests in the United States—given expression in the popular distaste for the Vietnamese war, industrial pollution and political "realism"—are indications that such a shift is not impossible. But lack of real historical precedent would, again, mark it as unlikely in the immediate future. Finally, there is a somewhat greater chance of a shift from a rationalized to a sentimentalized state. To a certain extent, the growing interest of the youth of the affluent nations in communal, autarchic systems is an expression of their gradual dissatisfaction with the materialistic, proscriptive aspects of capitalism. The moralistic imputations of the Scandanavian socialisms are also indications that a shift toward sentimentalization is possible, given a rationalized system as a starting state.

Moving now to the fourth row of the matrix, we ask about the probable transitions out of the sentimentalized state. As we have already sought to explain, the shift toward an institutionalized system—with its promises of order and certainty and its dogmatic engination—is by far the most probable. But there are at least two other possibilities that deserve some attention. First, there is the chance that a sentimentalized system will shift to a secularized state. Perhaps the best historical evidence for this alternative exists in the totalitarianization of the Iberian nations and their ex-colonies. These systems, following the decline of the politico-economic influence of the Catholic Church, maintained their Catholic tradition but softened considerably. They became strongly "matrist", relatively introspective and generally somewhat dissolute. This left a power vacuum which was eventually filled by the fascistic governments of Salazar in Portugal and Franco in Spain, and the long succession of military despots in Latin America. To a certain extent, the rise of the Bolsheviks in post-Czarist Russia may be explained in similar terms (although initially this appeared to be a transition from an institutionalized, aristocratic system to a rationalized one). Several of the dictatorships in the newly emerged African nations are blatant exercises in secularization and instrumentalism, with virtually no attempt to soften the blow by providing either axiological or axiomatic apologies to the general population.

To a certain extent, all these totalitarian moves may be comprehended as appealing to the public because of their promises of restoring order, discipline and productivity and increasing the general level of material welfare. The people support or tolerate these secularized systems largely in the belief that they are getting a return to an institutionalized form. From the standpoint of the peasantry or proletariat, these totalitarian, demogogic systems often appear to be a restoration of the stability of a monarchy without the disadvantages of a monarch.

For the possibility of a shift from sentimentalization to existentialization we have scant historical evidence, as expected. To a certain extent, the decline of Rome and the gradual introversion of the once-alert medieval Arabian states may provide a clue here. We may, however, have contemporary evidence of a sort available to us, notably in the gradual existentialization of the once sentimentalized hippie and beatnik sub-cultures in the developed nations. China and parts of India, in historic times, also may be used as a reference for this transition. But again, as our valuation suggests, this shift is a less probable one than the shift, along the pendular path, from sentimentalization back to institutionalization.

We now move to the fifth row, concerned with the possible destinations of a secularized system. We have already suggested that there are, in both the United States and the developed countries of western Europe (i.e., quasi-secularized, materialistic systems), two significant movements, each corresponding to one of the two most probable concluding states. First, there is the strong return to what we might call "basic" religion—not encumbered by any rationalistic overtones, but nevertheless offering deterministic doctrine and strong institutional discipline. The "Jesus freaks", the students of Zen Buddhism, etc., are examples of this, as are many other religio-institutional movements now beginning to emerge with increasing frequency. The other direction in which significant (if absolutely small) proportions of the secularized populations are beginning to move is toward the modern versions of existentialism. This, essentially, is the route taken by the drug culture and by increasing numbers of individuals seeking relief from chronic depression, insecurity, etc. To a lesser extent, there is the attempt to return to the simplest societal form—the conditioned-programmed state—as evidenced by the rural communes which have been established in recent years, most of which did not last long. Finally, there is limited movement toward sentimentalization, especially in the form of urban communes, or blocks of generally itinerant artists and intellectuals. At any rate, as the matrix shows, the sentimentalized and acculturated states are expected to occupy only relatively small proportions of the transitional population, with a majority seeking institutionalization, and a lesser number "dropping out" into an existentialistic state.

In summary of this chapter, then, we have tried to show how the model-building technique mentioned at the beginning of Chapter One can be employed: the elaborative ideal-types were equivalent to state variables. We then added a dynamic component in terms of relational functions. Finally, in this section, we introduced quantitative coefficients, and this transformed our hypothetico-deductive constructs into empirically accessible, apodictical form. Thus the implications of properly constructed and manipulated ideal types are very promising indeed from a methodological standpoint. In the next chapter we shall investigate their implications for societal manipulation.

Notes

1. Some of the more obvious lapses of logic or oversights in the Marxian axiomatic system are discussed in Pitirim ·Sorokin's *Contemporary Sociological Theories* (New York: Harper Torchbooks, 1964), especially in Chapter 10.
2. For an interesting analysis of the UFO phenomenon as a religious adjunct, see Carl Jung's "Flying Saucers, A Modern Myth of Things Seen in the Sky" (in *Civilization in Transition*, vol. 4, Princeton: Princeton University Press, 1964).
3. There are, of course, distinct correlations between the environment of an individual or system and the cognitive demands for efficient or effective exploitation of that environment. (For more on this, see the pioneering paper by Emery and Trist: "The Casual Texture of the Environment" *Human Relations* 18, 1965).
4. See, again, W. I. Thompson's *Passages About Earth: An Exploration of the New Planetary Culture* (New York, Harper and Row, 1971, pp. 71–73, 46–50).
5. Gordon Rattray Taylor, *Rethink: A Paraprimitive Solution* (New York: E. P. Dutton, 1973), especially Chapters 2–4.
6. That sociocultural constructions serve to generate *a priori* limits to variety is a well-established concept in the social sciences. See Peter Burger and Thomas Luckmann, *The Social Construction of Realty: A Treatise in the Sociology of Knowledge* (New York: Doubleday, 1966).
7. The senses in which I use the terms "shame" and "guilt" reflect the definitions given by Piers and Singer in *Shame and Guilt* (Springfield, Ill.: C. C. Thomas Publishers, 1953).
8. For example, an excellent study of the way secularized, non-axiomatic economic behaviors have thwarted the promises of capitalism is given in Galbraith's *Economics and the Public Purpose* (Boston: Houghton-Mifflin, 1973). However, as many have suggested, to condemn axiomatic capitalism because of the excesses of secularized behavior is hardly proper. The real benefits of true capitalism cannot be assessed historically, simply because true capitalism has never really existed for any length of time.
9. The construction of boundaries for vested interests and exploitation is a classic development in the sociology of knowledge. See, for example, Karl Manheim's *Ideology and Utopia* (New York: Harcourt, Brace, 1936).

10. For a good study of ritual as an adaptive constraint, see Roy Rappoport's "Sanctity and Adaptation" (*Coevolution Quarterly*, No. 2, June 1974, pp. 54–68).
11. For an explanation of the various environmental types in terms of properties and the algorithms of distribution, see Chapter 4 of my *Systems: Analysis, Administration and Architecture* (New York: Van Nostrand Reinhold, 1975).
12. Note that, in the sense of statistical theory, regression really indicates merely a direction of influence (or more fundamentally whether there is any ordered relationship between two entities). The correlation technique, on the other hand, allows us to suggest or measure magnitudes of the relations, and therefore involves a different level of research and analysis.
13. Most of the original theoretical work on game theory was done by John von Neumann and Oscar Morgenstern, reported in their *Theory of Games and Economic Behavior* (Princeton: Princeton University Press, 1945). Contrasting with the zero-sum game is the nonconstant-sum game, where there is some incentive for the players to cooperate.
14. For a radical treatment of the relations between competition, technocracy and alienation, etc., see A. Wilden's *System and Structure: Essays in Communication and Exchange* (London: Tavistock, 1972). For an elegant interpretation of the alienative influences of modern economic predicates, see Walter A. Weisskopf's *Alienation and Economics* (New York: E. P. Dutton, 1971).
15. A chilling view of existentialized alienation, fishing out the human realities of these two excessively arcane terms, is given by Kenneth Keniston in *The Uncommitted: Alienated Youth in American Society* (New York: Harcourt Brace, 1960).
16. This point has been elegantly defended by Robert Ardrey in his *African Genesis* (New York: Atheneum, 1961).
17. An interesting and frightening alternative—that institutionalized systems can be stuck and thereby generate maladaptive behavior—is considered by Michael Crozier in *The Stalled Society* (New York: Vintage Press, 1973).
18. An excellent treatise on the use of Markov processes is given in Massey et al., *Stochastic Models of Buying Behavior* (Cambridge: M.I.T. Press, 1970).

PART III
MANAGEMENT

One cannot talk about societal management in the same terms that we might speak of the management of some institution or organization. In the first place, we really don't know who—or what—should be responsible for societal management. It is not government, for government can only control institutions, and less directly the array of formal organizations it can identify. A societal system is more than any calculable function of the institutions and organizations within its boundaries. It is also deceptive to speak about the people determining their societal destiny; this, as historians will quickly point out, is at best a vapid assertion. Talk to a budding technocrat and he'll probably tell you that societal management is just a special case of management in general: we simply apply modern cybernetic concepts to make sure that the opportunities of the environment are maximally exploited, and that the system is therefore optimally adaptive. But we have already suggested that adaptation is the least sophisticated of the societal strategies available to us. Without question, societal management is something more than business administration. But what is "more"? And then there is parochialism again: Economists think that they should manage societal systems, because they naturally feel that economic variables are the most significant. Sociologists think they should manage societal systems, because social variables are the most significant. Anthropologists think they should manage societal systems, because cultural variables are most significant.

But it may best be suggested that prophets should manage societal systems. It is they, after all, who largely determine the course of societal events and evolution, largely because it is they who provide the a priori *predicates to which most non-trivial societal systems are susceptible. It is they who have the perspective to see the social, economic and political attributes for what they are: complements potentially, but substitutes traditionally. The function of the last two chapters of our volume is to suggest how prophets might equip themselves with a proper discipline, and how the art of societal management demands perspectives and skills we have hardly begun to think about. For me, then, the question is not so much how we might manage societal systems—a technology exists and can be articulated, however immaturely we may present it here. The far more troublesome question is this: Are there any proper prophets around to exercise it? And here the reader's answer is far more important than any I might give.*

5
The Interdisciplinary Imperative

Even from the simplest possible perspective, the management of societal systems involves staggering complexity. For the societal manager must simultaneously manipulate variables drawn from three different sectors: social, economic and political. Moreover, he must qualify any manipulations with respect to their effect on—or dependence on—the cultural and behavioral predicates operating within the confines of his system. Thus, in this chapter, we must look for every opportunity to create interfaces among the traditional social science subjects. This, after all, is the scientist's task, not the manager's, unless through our own default.

The ambition is to develop typological constructs in which each of our previously defined ideal-types is equipped with implications drawn from sociology, economics, and political science, with our socio-psychological underpinnings already defined and elaborated. The constructive hypothesis that allows us to generate the typologies—and to effect the interdisciplinary interfaces—is that properties of societal systems tend to *cluster* in certain definite ways. This considerably reduces the number of typological cases we have to consider and develop.

The strategy for interdisciplinary research is thus to have social scientists—drawn from various disciplinary or paradigmatic camps—concentrate their empirical efforts and normal hypothesis-testing activities on these clusters. To the extent that real-world systems exhibit the clusteration our hypothetico-deductive analyses propose, then we have at least partial validation of our theoretical artifices. To the extent that real-world systems disappoint our expectations (e.g., to the extent that social, economic and political properties are significantly stratified rather than clustered), then our deductive constructs are in doubt and demand modification or discard. Thus, again, we are trying to operate not only at the interfaces between historically isolated disciplines, but at the interface between conceptual and experimental science.

THE SOCIETAL CORRELATIVES

Anyone seeking to manipulate or manage a societal system—or indeed even comprehend it—must be aware of a possibility that has enormous implications: the possibility that the social, economic, political and behavioral (psycho-social) sectors of any societal system are so closely tied together that they cannot be separated, or meaningfully discussed in isolation. Moreover, not only are the sectors interrelated in a generic sense, but the analysis we have done thus far in this book suggests that the interrelationships occur at another level altogether. For each of the several different behavioral modalities we have associated with societal systems, there will be a unique social, economic and political modality that accompanies it.[1] This, of course, is simply a hypothesis, and the work we will be doing in this chapter will try to defend it. And because societal systems are *reflexive* in nature (at least from the perspective we have been working with), a given social modality would imply the presence of a given economic, political and behavioral modality, etc.

We shall now set out to explore these contentions, but there is one qualification that must be made clear before we proceed. We are going to be concerned here with correlations *not* causality. That is, we are only asserting that certain societal attributes tend to appear together, not that the behavioral causes the social, and so forth. With this qualification in mind, consider the construct in table 5.1, which we shall attempt to elaborate in some detail in the pages that follow.

Because of the rather extensive work we did on the socio-psychological aspects of systems in the last chapter, our elaboration and defense of this typology may be quite explicit. We shall treat the various social, economic and political correlatives in sequence.

TABLE 5.1: Typology of Interdisciplinary Correlatives

	Attributes			
Behavioral/System Category	Primary Social Unit	Economic Mechanism	Political Mechanism	Generic Referent(s)
Dogmatic/ institutionalized	Congregation	Ascription	Oligarchic	Aristocratic/theocratic/ feudalistic
Affective/ sentimentalized	Commune/ extended family	Requisition	Decentralized	Proto-Christian/matrist
Rationalized/ indoctrinated	Nuclear family	Algorithm	Bureaucratized	Capitalist/socialist/ technocratic
Conditioned/ programmed	Tribe	Segmentation	Oracular	Proto-communist/ primitive subsistence
Secularized/ correlative	Bloc	Preemption	Demagogic	Hedonistic/totalitarian
Originative/ sapient	Conference/ association	Mandation	Collegial	Platonic/professionalized/ dilettantish
Compulsive/ existentialistic	Isolated individual/ coven, etc.	Asceticism	Atomistic	Anarchic

The Units of Social Significance

Each of the seven behavioral engines will tend to be accompanied by a specific unit of primary social significance. The unit of primary social significance is simply that which, in a system of social stratification, is the fundamental vehicle for the determination and regulation of relations between individuals.

Our discussion of dogmatic, institutionalized systems in the previous chapter leads us to suggest that the primary unit of social significance there would be the *congregation*. The congregation is a geographic phenomenon, usually representing a local subsystem of the prevailing sacerdotal institution (e.g., a parish). The rules governing the congregation are derived directly from the sacerdotal institution, and therefore carry numinous or ideological authority. The concepts of the family, of different relational groupings and even of distinct socio-economic castes all relate directly to the congregation as a microcosm of the broader society. In short, the socio-economic and socio-political implications of the dogmatic, institutionalized system as a whole find their local expression in the congregation. It is membership in the congregation that lends the individual his "place", in both temporal and transcendental terms.

The important differential aspect of the congregation as a unit of social analysis is this: It is a product of *stratification* rather than clusteration. Here we find rich and poor, promising and penurious, lord and vassal together. The different social status of the membership is reflected in the roles and prerogatives of the congregation. Those of higher rank have greater privileges (e.g., of seating, speaking, advisement). In short, the natural societal differentiations maintain themselves in the congregation as well. This, of course, is the type of social unit familiar in medieval Catholic and feudal Europe, in the Pilgrim colonies in America and to a certain extent in urban Anglican and rural Jewish communities. It is the focus of the life of the Mormon, the Quaker, the Dukhobor, etc. In the case of politically institutionalized systems (e.g., the early nazi party or contemporary Soviet communism), the congregation becomes the agent of government at the local level. For sacerdotal systems, the congregation serves as the basic vehicle for socio-spiritual cohesion and control.

When we move to systems that are dominated by affective behaviors (sentimentalized systems), we generally find the congregation being gradually replaced by one of two other types of primary social units: the *extended family* or the *commune*. The extended family and the commune are the fundamental carriers of the affective, temporal benefits we associated with prescriptive, sentimentalized systems. Neither are particularly efficient as agents of economic development, as we shall see. But both serve to compensate for any material deprivations by giving the individual a closely defined and constant sense of interpersonal substance and security. The hierarchical and algorithmic nature of the relationships found in congregational systems is eased here, and the relations become more casual, perhaps more warm, and certainly less regulated. But, at the same time, the transcendential significance of the congregation is largely lost in the extended family or communal base, the emphasis on immediate context largely replacing numinous referents of the congregation.

Thus, the extended family is the social form we would associate with the prescriptive religious movements. We are not surprised that the Catholic nations of Iberia and Ireland, for example, are organized around the extended family form, whereas the congregational form was once paramount in the more proscriptive Protestant nations (e.g., England, Holland). The commune is a natural extrapolation of the extended family. A commune may be inadvertant in origin, as with the isolated village systems that arose in Europe during and after the feudal period (and persist to this day in many rural areas). Or a commune may be a conscious artifice, as with those that have been established as prescriptive capsules in the midst of industrialized, rationalized nations, or like those encouraged by Maoist (but not Soviet) communism. Finally, communes are generally products of *clusteration*—situations where people of similar socio-economic attributes or potential coalesce (where congregations, as suggested, normally included individuals with heterogeneous socio-economic attributes).

Moving on, we expect that the nuclear family will predominate in systems driven by an axiomatic engine, largely because social and spiritual benefits and emphases are removed from their place of prominence in the transition from axiology to axiomatic platform. The axiomatic engines are purposive in a temporal sense, as was suggested earlier, and generally presuppose the nuclear family as the prèsumedly most "efficient" or natural unit of societal structure. This is especially true of proto-Capitalist theory, and is implicit in materialistic socialism. The nuclear family is thus the social accommodation to industrialization and urbanization [2]. To a great extent, it obviates the congregation, as the spiritual sector becomes merely ancillary to the temporal-material (economic) sector in rationalized systems. Of necessity, it also disaggregates the population. Both the advantages and the disadvantages of the nuclear family are too well known for us to need to spend any time on them here.

For systems whose behavior is a product of conditioning (the simplest form of adaptation), we would expect to find the *tribe* as the basic unit of social structure. The size of such units is generally great enough to provide some security, but not so great as to imperil the casual subsistence economy. They are of a form which permits the society to be, to a certain extent, *portable* (mobile), but also encompass any individual completely. It should be understood that tribal structure is less sophisticated than nuclearization (separation into distinct, quasi-autonomous families), and therefore becomes the prototypical form of societal organization.

When we move to the fifth of our parochial system types, the correlative-secularized system, we consider a very different kind of social unit: the *bloc*. The bloc has no fixed size or composition; it is formed to pursue some opportunity and may very well simply be disbanded at the conclusion of the venture. It is, in a sense, the form of organization that Toffler has popularized as the "adhocracy" [3]. The bloc may consist of any collection of individuals (or systems), and some of them may be adversaries in other contexts. The bloc is thus characterized by its instrumental ambitions, its ephemeral nature, and its contextual orientation.

Systems emphasizing sapient behavior are generally characterized by the use of the *conference* as the primary social unit—a collection of peers who share some common interest (as opposed to the common axiological predicates of the congregation or common objective of the bloc, etc.). Such social forms have their origin in the "advancement" associations of medieval Europe, and have their modern counterparts primarily in academic and professional arenas. The conference is characterized by the demand for some credentials of membership—e.g., professional status—and by the lack of any formal hierarchical structure (the topological map of such a unity being a network rather than a pyramid). Unlike the bloc, the conference is generally permanent, and often an artifice—that is, it is a product of deliberate creation, often delineated by a formal charter of some kind.

Finally, the primary social unit we would associate with existentialized (compulsive) systems is the isolated individual *qua* hermit or solipsist. As should be clear from our discussion of existentialistic mechanisms in the previous chapter, in such "systems" there would be neither the motivation nor the means for establishing any stable social groupings. In short, existentialistic society is an agglomeration of individuals whose probability of contact and interchange (and interdependence) is as small as can be. When any associations do form, they are likely to take the form of *covens* or other incontinent groupings.

TABLE 5.2: Summary of the Units of Social Significance

		Attributes		
Social Unit	Basis of Association	Benefit of Association	Basis of Membership	System Referent
Congregation	Geographic stratification	Spiritual	Coercion	Institutionalized
Commune	Socio-economic clusteration	Emotional/affective	Cooptation	Sentimentalized
Nuclear Family	Biological (Symbiosis)	Functional inter-dependence	Implied contract	Rationalized
Tribe	Autochthonous	Security/survival criteria	Default (no other alternative)	Programmed
Bloc	Common ambition	Expedience	Instrumental criteria	Secularized
Conference	Common interest	Collegiality/synergy	Credentials	Sapient
Individual	Common compulsion	Psychogenetic	Insinuation/initiation	Existentialistic

Table 5.2 summarizes our very brief discussions of the social dimension of systems.

THE ECONOMIC ENGINES

To a certain extent, the units of primary social significance serve to suggest the way in which *status* is distributed across a population. The engines of the economic sector of societal systems serve to suggest the way in which economic prerogatives are distributed: the distribution of real or potential *wealth*. Real wealth is the command over property, or *de facto* ownership. Potential wealth refers to the command over the returns from production (income prerogatives) and credit. As is to be expected, different societal contexts entail very different economic distribution schemes.

As table 5.1 suggests, *ascription* is the modality expected to be associated with most institutionalized systems. In the simplest sense, ascription simply means that economic prerogatives are determined largely by social status—that one's wealth or ability to acquire wealth is related to the social position into which one was born. Usually, because social mobility is highly restricted in institutionalized systems, economic prerogatives normally remain linked almost exclusively to one's birthright.

Associated with the affective/sentimentalized system, we expect to find economic prerogatives distributed according to *requisition*. Under the requisition scheme, one's "requirements" become the basis for the distribution of economic product, as in the Marxian dictate: "From each according to his abilities, to each according to his needs". Thus, all goods become *social goods*, and are allocated on the basis of petitions for sustenance initiated by individuals themselves, or by others on their behalf. As one can readily see, sentimentalization (invoking emotional appeal) is the prerequisite for such an economic mechanism, predicated as it is on sympathy for the afflicted or the uncompetitive. When such appeals become implanted within the society as a whole, as a matter of axiological engination, then the requisitional economic mechanism takes hold [4]. Thus, not surprisingly, there is a distinct similarity between the Sermon on the Mount and the writings of romantic socialism.

Rationalized (indoctrinated) systems employ what may be called an *algorithmic* base for the distribution of economic prerogatives. It is algorithmic in that there are definite rules and procedures for the allocation of income or acquisition of wealth. For the proto-Socialist system, we find the "labor theory" of value suggesting that incomes be correlated with energic investment [5]. In its simplest form, the labor theory of value implies that economic reward should be geared to a weighted index of the effort one expends (e.g., the caloric requirements of the work) and the time involved in a task or productive exercise. In short, the labor theory of value is distinctly empirical, and has great appeal for those who have demanding menial jobs. When we move to the capitalist axiomatic platform in its various versions, we find market-mediated marginal utility operating to determine the distribution of wealth and income. This, fundamentally, is a somewhat more sophisticated form of the capitalist premise that wealth and income should be regulated by free interplay of supply and demand. The algorithm here is that which finds correspondingly greater rewards going to scarcer skills, given some level of demand for services. The return to capital is made algorithmic by simple extensions of this logic. Any asymmetrical distribution of economic prerogatives is assumed merely to reflect the differentiated demand and supply schedules that exist for labor and capital (when viewed as commodities or factors of production). In many cases, especially with the rise of organized labor and with mechanisms such as the civil service, the distribution of economic prerogatives comes to be determined algorithmically, with various grades and seniorities carrying "legislated" returns, similar to the types of algorithmic legislations that regulate distribution in socialist systems.

For the conditioned or programmed system, the simplest kind of distributive mechanism usually prevails: *segmentation*. Whatever material benefits happen to be available to the group are simply distributed equally among all members. All the other system types involve an asymmetrical distribution of some kind (e.g., the proto-socialist position is "to each according to his needs", which is not the same thing as "to each the same amount"). The segmented economic mechanism wins by default, for neither differentiation nor specialization will be strong in the conditioned system.

For secularized (correlative) systems, in keeping with the tendency to play a constant-sum game, the distribution of economic product proceeds on the basis of simple *preemption*. Under a secularized system, any axiomatic constraints on the means by which wealth may be acquired are obviated. Thus, for example, laissez-faire supports the kind of instrumental capital and corporate formulation we saw in the United States during the late nineteenth century, formulations that could not be tolerated under a strictly capitalist axiomatic platform (as they ultimately lead to restraint of competition). The complaints that capitalism doesn't work—because of the kind of concentrations that develop—is a *non sequitur*, for the laissez-faire doctrine is not a necessary corollary to theoretical capitalism, but only an operational alternative. In short, cartels, monopolies and other forms of excessive concentration are injections of realism (instrumentalism) into a system, and not evidence of the axioms of capitalism at work. The same is true of the very serious tendency for certain American and European firms to grow far beyond optimal size of plant. For capitalist axioms demand that firms produce only to the point where profit is maximized [6]. But under a realistic/instrumental abrogation of capitalist theory, many corporations replace the profit index with the calculus of pure growth. As a result, many of the larger corporate enterprises may be distinctly uneconomic, acting more according to instrumentalism than according to proper capitalism. Indeed, much of modern commercial strategy may become intelligible as a deliberate effort to obviate one of the key underpinnings of capitalism— the relationship between risk and profit. Theoretically, in an operative capitalist system, profit should be directly correlated with degree of risk. But collusion, subordination of government agencies, cartels, and many other mechanisms of "mixed capitalism" may be seen to be little more than attempts to achieve profit without incurring the postulated risk. The net result is, again, inefficient use of resources, and a corruption of capitalist precepts [7].

But, as suggested earlier, once a system moves from institutionalization to rationalization, the seeds of secularization are already sown. This is just as true of socialist systems as it is of capitalistic systems; the congruence between axiom and practice is no greater in the former than in the latter.* For example, as we suggested, the labor theory of value would be a theoretically preferred algorithm for the distribution of economic product under axiomatic socialism. But, in fact, there are enormous differences between the rewards accruing to various positions in the Soviet union and the other so-called "people's" republics. Government officials, athletes, certain scientists, etc., have effective incomes and material privileges which in many cases are more extreme than corresponding benefits in the so-called capitalist systems, when we consider the *range* of economic differentiation. In short, the great distinctions that Marx (and later Mao) sought to eliminate—the contrasts between mental and menial work, between town and country, between industry and agriculture—still remain in the socialists' systems as a direct abrogation of axiomatic socialism. They can really be explained only in one of two ways. First, what pass for axiomatic systems are really proscriptive, institutionalized systems (where a repressive political axiology has replaced the repressive religious axiologies of an earlier day); therefore, differential economic prerogatives are really products of ascription rather than realism (instrumentalism). Secondly, there is the normative explanation that we just suggested: that there are always gaps between axiomatic principle and practice, caused by injections of realism (secularization). In short, the process of indoctrination (like the process of institutionalization) is *discontinuous*, affecting certain segments of the population more fully than others. In particular, as suggested earlier, proscriptions (whether axiomatic or axiological) seem to operate more potently among the lower socioeconomic classes of systems than among elites.†

The abrogations of socialist economic principles in the U.S.S.R., for example, are probably due to the former cause (the fact that this system

*Much of the popular criticism of capitalism is thus misdirected, and is due to mistaking the effects of instruments for the dictates of axiomatic capitalism. By the same token, those who defend capitalism by citing the purported benefits of modern industrial concerns are perhaps equally misdirected. In fact, a true capitalistic system has never existed, for all practical purposes.

†Thus elites may not most truly employ the values of a society, but may instead most consistently abrogate them.

is really a dogmatic and institutionalized system rather than an axiomatic one). This may, in large measure, explain the enormously high correlation that we note between position in the political hierarchy and of economic prerogatives. In short, we have the pretense of axiomatic socialism, but the practice of *ascription* (this time using political rather than social referents as the basis for distribution of wealth and income). The second cause of departure between principle and practice is probably more appropriate to capitalist systems. Here we have the pretense of axiomatic obedience (e.g., the evocation of the rhetorical concepts of free enterprise and competition) but distinctly aberrative practices. In short, despite the avowed adherence to the principles of capitalism, some significant proportion of western industrialists are distinctly secularized. As such, business has *de facto* become an exercise in preemption, which is really very different than excercise in competition.

We now discuss a very interesting economic mechanism associated with cerebral-sapient systems. Here we find economic prerogatives being distributed by the process of *mandation*, involving a collective agreement on the benefits that should accrue to certain activities or certain individuals. Generally, mandation involves the voting of prizes or the awarding of annuities, etc., or the provision of stipends on the basis of a generalized appreciation of some performance (or on the basis of collective expectations about the potential benefits from some proposed activity). In short, one here acquires economic benefits through the process of either prior or subsequent appreciation by one's peers, with the precise factors entering into the mandation process highly variable from one judgment to the next.

Finally, there is the ascetic economic mechanism, the engine associated with existentialist systems. Here there is a conscious attempt to avoid acquiring material benefits, driven by philosophical and psychological factors that will be treated in the next section. For the moment, one may conceive of asceticism as being counter-economic in implication (i.e., a situation where, for reasons explained shortly, the normal conditions of "scarce resources" do not pertain).

At any rate, because the several economic mechanisms are so central to our comprehension of societal systems, we must treat them in a bit more detail. In this regard, consider table 5.3, where we look at three attributes of interest: (1) the mechanism by which productive resources are allocated; (2) the mechanism by which economic product is distributed (this column is mainly a restatement of points already made); and (3) the expected configuration of economic benefits, concerned with both the distribution of shares and the aggregate of economic benefits available.

TABLE 5.3: Properties of the Several Economic Mechanisms

	Normative Attributes		
Economic Mechanism	Mechanism for Allocation of Resources	Mechanism for Distribution of Product	Configurational Expectations
Ascription	Largely axiologically determined (e.g., proscriptions against seeking to acquire conspicuous wealth)	Strongly correlated with position in the social (e.g., feudal, aristocratic) hierarchy; also via "just price"	Extremely skewed distribution of product at a relatively low (or stable) aggregate level
Requisition	Investment of resources determined by initiative-referendum processes; some latitude allowed individuals	Ensures a minimal subsistance level for all members (nonconstant-sum game setting)	A "floor" level on individual benefits, with a relatively low aggregate level
Axiomatic capitalism	Investor expectations about rate of return on investment, "invisible hand"	Bargaining and contracting (via supply and demand mechanism, etc.)	Moderately skewed distribution; relatively high aggregate level.
Axiomatic socialism	Rational central planning (supply and demand coextensive)	Legislation, such that economic rewards are set by prior fiat	Symmetrical distribution; possibly high aggregate base
Segmentation	Contextually determined (labor only productive resource); ritualized investments	Simple division of spoils (protocommunism)	Linear (optimally equal) distribution; meager aggregate.
Preemption	Game strategy; short-run, localized utility calculations	Zero-sum game (search for monopoly profits, etc.)	Maximal concentration of benefits from an effectively suboptimal aggregate base

(continued)

TABLE 5.3, continued NORMATIVE ATTRIBUTES

ECONOMIC MECHANISM	MECHANISM FOR ALLOCATION OF RESOURCES	MECHANISM FOR DISTRIBUTION OF PRODUCT	CONFIGURATIONAL EXPECTATIONS
Mandation	Consensus about expected worth of proposed projects (with "worth" criteria possibly complex)	Either coextensive with allocation mechanism, or through a posteriori awards, etc.	Configuration effectively indeterminate in terms of both distribution and base
Asceticism	Independent of utility calculations (e.g., compulsive, idiosyncratic)	Indeterminate (perhaps by simple default)	Shares of a very low aggregate, probably distributed randomly

The several economic mechanisms we have defined differ on each of the three attributes of interest.* The "mechanism for the allocation of resources" considers the procedures and determinants by which investments are made. In ascriptive systems—given their predication in proscriptive axiology—we shall find that there are distinct axiological constraints on the allocation and investment process. For example, medieval Catholic economic dogma argued against attempts at gain. In theocratic, feudal and aristocratic systems, the basis of wealth was land, and not income. Therefore, there was little attempt to optimize the productivity of the land, at least beyond the point where an "adequate" standard of living was secured by the oligarchic elites. Socio-spiritual factors, moreover, argued against diversification into trade or manufacture, a problem still present today in certain regions. The peasantry, for the most part, shared this tendency to underproduce; the axiologies under which they labored bade them be content with subsistence in exchange for grace or transcendental compensation, etc. The ability of the artisan or trade classes to acquire wealth was limited in many cases by the concept of the "just price", which assured them a minimal return on their time and materials, and denied them access to monopoly profits, speculation, etc. In net result, then, we tend to get an economic sector configured as follows: a highly skewed distribution of economic benefits, with the absolute (or aggregate) level of benefits available being limited by the lack of motivation to optimize material

*Note that these economic mechanisms now become component ideal-types—subsystem models.

stock. As we shall later see, this low absolute level of production corresponded to the predoiminantly spiritual or transcendent interests of dogmatic-proscriptive systems; and in fact the ascriptive economic mechanism itself argues against significant development.

The requisitional economic system is interesting in that there are really two domains of allocative interest. First, and paramount, is the concept of social or public investments. These investments may be determined—and hence resources allocated—on the basis of initiatives or referenda instituted by the people themselves, a practice consistent with communal-based societal systems. Alongside this public allocative mechanism there may be a private investment sector, which is not *a priori* denied or legislated against (as it is with communist systems). For various reasons, among them the fact that requisitional systems tend to be preoccupied with other than material interests, this private sector may not be very significant. But what characterizes a requistional system, and what differentiates it from a communist economy, is the fact that a *floor* exists beneath which any individual's economic benefits are not allowed to fall, but that individuals may theoretically rise as far above that floor as they are able. The purpose of the public allocations is to provide the means for maintaining this floor, and not merely to provide the infrastructure (e.g., roads, sewers, harbors) to which public investments are restricted in capitalist systems. Therefore, the commune as a whole represents a "business" when we look at it on the economic dimension, and it is a business where shares are allocated more or less symmetrically. However as with ascriptive systems, there is no strong tendency for the requisitional system to attempt to maximize its return on investments, or to optimize its material stock. Therefore, the aggregate or base-level of economic benefits remains rather low, though the shares of this small pie are more or less equal. Thus, the ascriptive and requisitional systems differ mainly in the distribution function, as they will both tend to have an absolute standard of living or material stock that is suboptimal. The diagrams in figure 5.1 indicate this situation. They clearly show the symmetry in the distribution of economic product associated with the requisitional as in comparison with the ascriptive system. In (a), the more radical curve implies that concentrations are extremely high among the "elites", while the majority of the population has a disproportionately low share of economic prerogatives. The curve for the requisitional system approaches what would be the egalitarian ideal (the dashed line, meaning that prerogatives were distributed evenly across the entire population).

But what the curves do not show is the other side of economic configurations: the base index of aggregate or absolute benefits. This, of

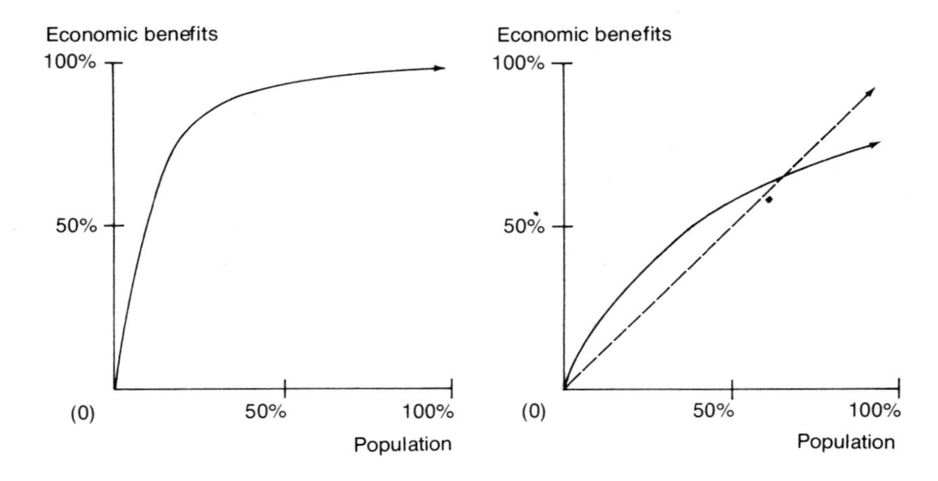

FIGURE 5.1: Comparative economic configurations. (a) Ascriptive society shows more unequal distribution with 80% of benefits going to only 15% of population. (b) Requisitional society shows more equal distribution, with fewer very rich and fewer very poor individuals.

course, would reflect the "size of the pie" from which the various distributed shares are cut. Now, in practice, this base index would have to be a relative or normalized one, so that comparisons between different real-world systems could be made. Such a comparison is suggested in table 5.4. The *base* merely refers to the expected efficiency of the system in exploiting its human, natural and technical resources. It is a measure of the extent to which the system is optimal in terms of its allocation and investment decisions (i.e., the extent to which these decisions all carry a zero opportunity cost, indicating that no more productive alternative exists for the resources at hand). From another point of view entirely, it is also a measure of the extent to which the system is preoccupied with material (economic) benefits, or the extent to which economic pursuits are *a priori* encouraged or discouraged (given the system's behavioral predicates). Thus, the capitalist and socialist variants each pretend to be optimal allocaters and users of available resources. We expect them (in their unadulterated or axiomatically pure form) to most efficiently exploit opportunities for maximizing the material or economic stock of the system, within constraints they cannot alter (such as size of population or quality of natural resources). All of the other system types are expected to be somewhat suboptimal in economic efficiency, with the nadir being reached by the effectively counter-economic ascetic system, and the possibly non-materialistic mandation-based entity.

TABLE 5.4: Differential Expected Base Value for the Several Economic Ideal-Types

Asymmetrical Variants	Base	Symmetrical Variants
Capitalist	Higher	Socialistic
Ascriptive	↑	Requisitional
Preemptive	↓	Segmentational
Mandation-based	Lower	Ascetic

Thus, again, it is possible for two systems that have essentially the same general distribution curve (e.g., capitalist and ascriptive systems are both expected to be fairly asymmetrical) to differ considerably in the base from which the available benefits are to be measured. So the absolute benefits available to an individual at the low end of the distribution in capitalist system—a member of the lower economic class—might still be absolutely greater than a perfectly equal share of the material benefits available within, say, a primitive subsistence (segmented) system. Finally, in a more general sense, the concept of a variable base recalls a point we made much earlier: that two systems with different *a priori* predicates may respond to the same set of environmental properties in different ways.

Getting back to table 5.3, the capitalist and socialist referents are treated here only in terms of their idealized implications (e.g., those consistent with their axiomatic premises). The mechanism by which resources are allocated in the capitalist system is through individual decisions concerning the expected rate of return on investment. Those alternatives with higher expected values are to be preferred over those with lower expected returns. As suggested earlier, economic benefits are distributed through a bargaining process (often formalized into a contractual instrument), where expected demand-supply criteria determine the rewards to labor, capital and management. The net result, theoretically (resulting from the Smithian "invisible hand"), is a system with an optimally high aggregate material base, and with a moderately skewed distribution of shares (indicating the presence of a significant middle class). The socialist system, on the other hand, centralizes its allocation decisions on the basis of expectations about the respective cost-benefit ratios that might pertain to certain *categories* of investment. There is, in the distribution mechanism, a fairly low ceiling on individual benefits (often supported by confiscatory taxes or outright expropriation, etc.). Salaries and wages, like allocation decisions, are generally matters for centralized planning, and are legislated rather than set by free bargaining. Finally, like its capitalist counterpart, the socialist system pretends to be able to maximize the material stock of the system, while forcing more symmetry into the distribution of these benefits, however.

Thus, the idealized distribution algorithms for the capitalist and socialist would be distinctly dissimilar. The more or less "normal" distribution of prerogatives in the protypical capitalist system would reflect its reliance on the postulation of natural differences among individuals (i.e., its essentially Darwinistic roots), whereas the essentially symmetrical distribution of the socialist system would reflect its Socratic-Rousseauean-Marxian referents (with the axiomatic assumption that differences between human beings are artifacts of civilization rather than expressions of nature *per se*). But, in practice, there is considerable skewness in real-world capitalist and socialist systems, due to the tendency for secularized components to intrude into the operational base of the systems. For capitalist systems, this greater skewness would be more apparent were we to attempt to graph the distribution of *real wealth*, which is much more concentrated than income. In socialist nations, on the other hand, there are significant differences in *real income*, which includes perquisites such as automobiles, apartments, chauffeurs, etc., that do not figure into normal income data. If such fringe benefits were included in calculations, the ideal linear distribution would become noticeably skewed, making it more like that for the capitalist system. As a final note, the base index for these two systems is presumed to be very close to the theoretical maximum, reflecting the capitalist and socialist axiomatic schemes as vehicles for supposedly maximizing the productivity of available resources.

Thus far, ascription and requisition, and capitalism and socialism, have been seen to represent sets of polar opposite economic systems, with the two sets operating at fundamentally different base levels. The preemptive and segmented economic mechanisms are also opposites, and as a set operate at a theoretically still lower base level. But the reasons why we suggest this low base level are different for these two system types. The segmentational system may make considerably less than optimal use of whatever resources are available to it for two reasons: (1) there will tend to be a distinct lack of functional differentiation (and consequently specialization) in terms of the way human resources are organized; and (2) only the most primitive means of production will tend to be employed, as indicated by a very small capital-labor ratio. If the system is basically a hunting economy, then there will be no attempt to convert to a herding economy; it will remain a migratory or nomadic system, with low effective productivity per man-hour (as with the Eskimos as opposed to the Laplanders, for example). Or if the system should depend on food-gathering, there will be no advance to settled experimental agriculture, an innovation which may improve yields from virtually any land, irrespective of its initial fertility. It is these two factors—lack of specialization and a low capital-labor ratio—that characterize the productive dimension of primitive, conditioned systems, and that make economic segmentation the most probable distribution alternative in that context.

Segmented mechanisms are proto-Communist in their distribution patterns, i.e., ideally there is equal division of spoils. Now, this does not necessarily mean that every individual in the system will receive the same amount. It may, rather, mean that every hunter, or head of household, will receive an equal share (e.g., among the Eskimos, one goes hunting not a seal or a bear, for example, but a "share"). He may then distribute this to his dependents as he sees fit, so that there may be some complexity in the second-order distribution algorithms. Sons may, by tradition, receive more than daughters, for example. But, nevertheless, the fundamental distribution function, as we have earlier suggested, will be the simplest possible in such systems [8]. The net result of these factors is a configuration of the following type: there is an equal (linear) distribution of economic product, but the base is quite low, so that the equal shares may still represent a penurious situation. In short, the curve looks like the nearly straight-line one for the socialist system, (but the base level is likely to be considerably lower.)

The preemptive economic mechanisms—dominating the economic sector in secularized systems—may result in a lower than optimal base level for two reasons, both perhaps more subtle and indirect than those associated with segmentation. First, there may be a tendency—given the constant-sum-game context in which instrumental systems operate—to be more interested in halting competitors' advances than in optimizing the productivity of available resources. Secondly, there is the tendency to operate on a localized, short-run decisions, which often spells suboptimality or even economic crisis in the long run. But perhaps even more pointedly, the absence of either axiomatic or axiological predicates in the secularized system means that no effective consideration is given to the material welfare of others. Thus, such systems tend to achieve the highest degree of concentration of economic benefits, with one or only a few families perhaps holding virtually all the wealth and commanding virtually all the income and profits. To the extent that they themselves are sufficiently well off, they may be uninterested in expanding the economic base, which would presumably raise the benefit levels available to the general population. Of this, perhaps the most outstanding example has been the traditional Arab systems. Only quite recently (and not to the extent generally supposed) have they shown any interest in the type of economic development sponsored by either of the axiomatic ideal-types (capitalism or socialism). The same may generally be said of some of the Latin American and southeast Asian nations, and several of the newly emerged African states. In all these cases, asymmetry in the distribution of economic product and material prerogative approaches a theoretical maximum.

To a great extent, our academic reference for the preemptive economic system must be monopoly. The constant-sum game calculus, when generalized, becomes a situation where competitors are searching for monopoly profits (a form of economic preemption), but not necessarily through normative (axiomatically responsive) means. Rather, there may be strategies employed that are distinctly diseconomic when viewed in aggregate: unloading, collusion, cartel-building, sabotage, undercutting—even diversification—are all rather subtle techniques for effecting concentration which, while it may maximize the return for a specific set of operators or owners, nevertheless dilutes the productivity of the system as a whole. There are other, more blatant means that might be employed in some cases: bribery, subornation, the granting of exclusive charters to favorites, nepotism, cronyism, seizure of competitors' assets through rigged legal proceedings, etc. All these factors operate to produce an economic system in which there is a maximally skewed distribution of economic prerogative, operating at a possibly suboptimal base level. Thus, the curve for the preemptive system will be even more nonlinear than that for the ascriptive system (in figure 5.1), but may be set at an even lower efficiency level.

The final set of economic mechanisms are asceticism and mandation. For the ascetic mechanism, we know already that there will be the lowest possible base level, for there is likely to be an active lack of interest in exploiting any material opportunities whatsoever. As we shall see in the next section, there is a long and honored intellectual tradition underlying such systems, one enjoying something of a rebirth these days. At any rate, we suggest that the distribution function would be more or less symmetrical, possibly giving each member of the ascetic system a functionally minimal subsistence level, with no attempt made at the generation of a surplus or conspicuous acquisition, etc. Again, then, we have a symmetrical distribution (like those of the requisitional and segmentational and socialist cases), but at the theoretically minimal base level. In some societal contexts, however, a random distribution may result from the lack of material concern.

The mandation mechanism (associated with those systems we earlier defined as cerebrated or sapient) has virtually no historical precedent except in very isolated instances in the arts and sciences. Here, we find the Pulitzer and Nobel prizes, and similar awards for notable performances that were originally undertaken without any contract for return. Here we also cite the ostensibly objective mechanism by which research proposals are selected for support, etc. It is possible, of course, that mandation could be extended to other fields—to industrial functions, for example—but that is highly unlikely. The most we can say is that mandation, as an economic modality, may still be in its emergent phase, and may be better exploited by later generations of systems [9].

The Several Socio-Economic Calculi

As we suggested earlier, axiomatic and axiological predicates sometimes operate in concert. This is particularly true when we look at the *socio-economic* aspect of systems—the interfaces that develop between the social and economic sectors. What is of fundamental interest to us here is the perceived relationship between social and economic benefits. There are three alternatives:

1. There may be a *positive correlation* between social and economic benefits: greater economic (or social) benefits are deemed to imply greater social (or economic) benefits.
2. There may be a *null correlation:* social and economic benefits are seen as effectively independent.
3. Finally, there may be a *negative correlation*: greater economic (or social) benefits are seen as implying lower social (or economic) benefits.

Given the background that we now have in the operation of cultural predicates, we can identify the particular axiological/axiomatic conjunctions which would lead to one or another of these calculi. In particular, a null correlation is seen at the interface between the proscriptive Medieval Catholic dogma and the essentially axiomatic economics of the Schoolmen [10]; positive correlation at the interface between the Protestant ethic (*qua* axiology) and capitalist axioms [11]; and negative correlation, at the interface between proto-Christian or proto-Communist axiology and the essentially axiomatic logic of the Rousseauean or Thoreauvian position [12]. Consider, then, the following construct in table 5.5.

In the most basic sense, social benefits are non-materialistic—that is, they are the whole range of emotional and sensual gratifications avail-

TABLE 5.5: Attributes of the Several Socio-Economic Calculi

Socio-Economic Calculus	Implication	Axiological Predicate	Axiomatic Predicate	Normative Behaviors
1. Compensatory	No necessary correlation between social and economic benefits	Catholic (medieval) dogma	The axioms of the Schoolmen	Acquiscence
2. Amplifying	Positive correlation: higher economic benefits imply higher social benefits	Protestant ethic	Capitalist theory (Smith, etc.)	Acquisition
		Utopian theses (Wycliffe, More, etc.)	Socialist Theory (Engels, Lenin, Marx, Plato, Fabians)	Aggregation
3. Counter-correlative	Negative correlation: social and economic benefits are competitive	The ethico-religious prophets (Amos, Hosea, etc.); proto-Buddhism; Barth's religious existentialism	Stoical, Rousseauean and Thoreauvian theses; secular existentialism (Kierkegaard, Heidegger, etc.)	Asceticism

able within a system. Economic benefits, on the other hand, are those that are either directly or indirectly purchasable, through income, status, power or some other *quid pro quo*. Now, as one can readily see, an ascriptive system would appear, on the surface, to concentrate both social and economic benefits in the hands of the elites. But this is not strictly the case, for one has first to distinguish the notions of social status and social benefits. It is true that in an ascriptive system, economic benefits (both real and potential) accrue to those with social status. But social status is merely a ranking index which serves to stratify society along some structural dimension. Social benefits, on the other hand, are processual. They accrue from a "flow" of affection, emotional release, spiritual peace and sensual appeasement, etc. With this distinction in mind we can briefly elaborate on the several socio-economic schemes.

The Catholic Ascriptive Position The key to the null correlation is, of course, a variation on the compensatory calculus. Recall, from our discussions in Chapter Four, that the compensatory calculus operated primarily to offset material rigors by prophesying an inverse relationship between temporal and transcendental benefits. In its socio-economic guise, the compensatory calculus serves a somewhat different function: it allows those with low economic benefits to still gain an adequate level of satisfaction from system membership by elaborating the social or spiritual benefits available. In short, the greater material benefits of one class are perceived to be offset (at least in part) by the greater socio-spiritual benefits of another. Again, this is not the place to defend the logic completely (that is for a later section), but we can see how the axiological and axiomatic predicates might bring this situation about. Catholic dogma (at least as it emerged in the early Middle Ages) made the compensatory calculus a focal issue. The central thesis, as we already mentioned, was the aphorism: "It is easier for a camel to pass through the eye of a needle than for a rich man to enter the gates of heaven". The axiological corollary was the concept of the virtue of acquiescence: one earned God's favor by uncomplainingly accepting one's own (and of course others') lot. This means that poverty would work to eternal advantage only if it were suffered willingly. In extension, it meant that criticism of the wealth of others would work to the individual's eternal disadvantage (the sin of envy). The distribution of economic product was, in short, God's will, and not to be criticized or questioned by any man (rich or poor). Poverty was to be taken as a sign of God's favor, rather than as a result of one's own competitive shortcomings (as in a rationalized capitalist system), or as a sign of God's displeasure (as in proscriptive Protestant axiologies). This set of apriorisms held benefit for both rich and poor. It protected the property of the former, and obviated anxiety, effort and resentment on the part of the latter.

Now, in order to justify this situation—and elaborate the defense of the very visible and pervasive material asymmetry in ascriptive societies—an attempt had to be made to rationalize ascriptive economics. In short, the axiological engine had to be complemented by an essentially axiomatic one. There had to be a logic behind it, as well as a revelatory (purely axiological) base. The lending of an axiomatic quality to Catholic dogma was, in great measure, the task undertaken by the Schoolmen. The essential axiom of Catholic economics owes its origin essentially to their efforts. It suggests, basically, this: *No man should seek to acquire more wealth than befits his station in life*. To do so was not only a theological contravention (i.e., a mortal sin), but a contravention of logic as well. For the attempt to aggrandize oneself not only was an act of presumption with respect to the Creator's grace, but would actually work to the individual's temporal dissatisfaction as well. For economic differentiation would mean automatic expulsion from the social class into which one was born, and no other would be open to him. But no other social rank was open to him anyway, for the proscription against economic acquisition (backed by sacerdotal threats of excommunication) effectively prevented the rise of any middle class, while the upper class was perforce inaccessible. Thus, while the axiological engine served to deter economic mobility and acquisition (at virtually any level except that of the Papal coterie itself), the axiomatic engine operated to make it infeasible even for those who did seek it.

The result of this conjunction between the Catholic axiology and the axiomatic economics of the Schoolmen was that there was no move toward disruption of the ascriptive system until the Reformation. Freed from the necessity and anxiety of competition and ambition, and supported by elaborate socio-spiritual mechanisms, the Catholic peasants (like many of the present rural inhabitants in Iberia, Italy, Ireland and the Latin American ex-colonies), presented no real demand for broad-based economic development. From the standpoint of the materially deprived individual, no such attempt was necessary, even if it were not *a priori* proscribed. For the array of socio-spiritual benefits gave him an adequate level of satisfaction.* Now, the same situation occurs, to a certain extent, in the ranks of the elite as well. They also, with the exception of the Papacy and the very strong feudal or merchant families, were under the same proscription: they also were discouraged from

*We shall amplify this point in Chapter Six.

seeking material aggrandizement. As a result, no significant effort was put forth to acquire new wealth through investment, trade, etc., even by those of relatively high social status. Now, lacking any legitimate economic outlet, the elites (or the majority of them) turned their attention to social, sensual or spiritual benefits, and operated off the largely unexploited equity of the land and largely "unearned" privileges that went with their aristocratic, feudal or sacerdotal office.

Thus, the satisfactions one derived from membership in the Catholic proscriptive system were restricted—both willing and occasionally by fiat—to the social and spiritual domain, with some sensual license tolerated at certain times and at certain levels. Much the same situation prevailed among certain of the eastern Christian sects (e.g., the Nestorians and the Copts) and the major Buddhist sects until very recently. And in the less accessible areas of Catholic Europe and Latin America, this situation still prevails today. In very simple (and as yet inadequate) form, this is the socio-psychological rationale behind the popular postulate that economic development demanded the demise of Catholic constraints.

More specifically, economic development (and derivatively the development of trade, mercantilism, and even democratic government after a fashion) demanded the shift from Catholic dogma to the Protestant dogma and its ethical implications, *and* the shift from the sacerdotal economics of the Schoolmen to capitalist theory (or socialistic theory).

The Protestant-Capitalist Engine Now, there are great difficulties in actually interpreting the implications of the axiological shift begun with the Reformation. Though it ultimately resulted in a transition from a proscriptive to a prescriptive predication, this was not the immediate result. The problem is that the Catholic axiology had become less dogmatic (more "humanist") after the Aquinian separation of the soul from nature. This eventually led to a dualistic epistemology. On the one hand, the matters of transcendental significance were to continue within the realm of revelatory-exegetical analysis, while natural phenomena could become accessible on a quasi-empiricist platform (reminiscent of the Aristotelean position). In a broader sense, then, the universe of man now became one where "naturalistic" support was sought for the previously invulnerable lessons of the scriptures. Indeed, for the "humanistic" Catholic theologist, the scriptures themselves become somewhat apocryphal. But what perhaps most inspired the Reformation as an axiological phenomenon was this: Following Aquinas, man was allowed some autonomy over his salvation. Christ died for man's sins, true, but it was up to the individual to *merit* salvation by his own acts. This was considered to be a weakening of the doctrine that the scriptures were the actual word of God, and the eternal truth, and that all autonomy rested with God. Thus, the Reformation may be seen on one level as a regression to more intense dogmatism, and a movement away from humanistic realism [13].

But Protestant axiology did not have the retrenching, dogmatic effect that the intellectual sponsors of the Reformation may have intended. Rather, the Reformation ultimately resulted in a series of variants that eliminated the Catholic hierarchy as a mediary between the individual and God, and at the same time broke the back of the proscriptive Catholic economic dogma. Thus, it is much as we suggested when we were speaking of the probable transitions from an institutionalized system in the last section of Chapter Four. We noted, first, that there was the alternative of sentimentalization. We may now see how this could have occurred. It mainly took hold among the Iberian nations and Ireland, and was simply an extension of the humanism and "relativism" of the naturalistic movement begun by Aquinas. The stern God of the scriptures eventually was displaced by the Virgin Mary, who became both a sensual and sacerdotal object in a distinctly "matrist" frame of reference. The church ceased to be a moral marshall and became, instead, the instrument of Mary as Mother. It thus ceased to be primarily the agent of transcendental, spiritual succor, and became instead the focus of a spontaneous and fundamentally affective social sector, where numinous issues were distinctly secondary. Vestiges of the Medieval Catholic economic dogma remained, but their effect was negligible anyway; for the social benefits available to members of sentimentalized systems made it more convenient to ignore impulses toward acquisition or material aggrandizement, along with the demands for discipline, assiduity and sacrifice that economic development would imply. In short, the null calculus maintained itself in the sentimentalized states.

Now the other path from institutionalization, we recall, was the axiomatic route. And despite the originally axiological implications of the Protestant Reformation, its ultimate effect was to fuel secularization, not dogmatic retrenchment. Thus, as ultimately defined and implemented, the Protestant axiology acquired distinct temporal significance. This is not to suggest that there are not small groups of people for whom Protestantism's axiological predicates are paramount. Yet the telling effect of the Reformation was to bring religion into the service of the economic sector, whereas the economic and political sectors had traditionally been subservient to religion under the Catholic proscriptive platform. In the process, the amplifying socio-economic calculus displaced the null correlation.

Amplification suggests that social and economic benefits are positively and causally correlated, so that an increase in the latter automatically implies an increase in the former. In the extreme form of the doctrine, social benefits have no separate identity, but are simply extensions of the set of commodities purchasable through economic means. In the form in which it has stimulated material development, however, the positive correlation may be stated quite simply as the assertion that the wealthier a man is, the happier *and* holier he is. Thus, to the agent of amplification, the pursuit of wealth is identical with the pursuit of happiness and salvation.

Under Catholic dogma, one prepared for heaven—and demonstrated his worth—by the extent of his acquiescense, by accepting either good or bad fortune with equanamity and humility. Under the compensatory calculus, one could even increase his peace of mind through reflection on his misfortunes, knowing that they would be paid for in the hereafter. This logic stemmed essentially from the conception of men as "tenants" rather than owners of the temporal world. Like tenants, men had only to maintain things pretty much as they found them, and to make any significant change (developmental or exploitative) would be presumptuous. But under the prescriptive Protestant axiology, men became the owners of the world, not merely trustees of the Creator. Therefore, the quality of one's earthly tenure could be measured in terms of the improvements and changes he introduced. Those whose earthly activities resulted in significant positive innovations or development, etc., would have tangibly demonstrated their preparation for positions of trust and responsibility in the hereafter. Those who were merely acquiescent demonstrated only their tractability and lack of initiative, and therefore reduced their putative value to heavenly enterprises. In short, the qualifications for salvation now shifted from emotional stability, humility and tractability to energy, assiduity and action orientation.

In this variant of the Protestant platform, then, the prescriptive axiology demanded that one improve one's circumstances through one's own effort and imagination. The prescriptive axiology also seemed to account—through Augustine's concept of predestination—for the different personalities and aspirations that evolved among men. It was noted that even babies at a very young age showed marked personality differences. To the Reformational mind, these differences could only have been deliberately placed there by the creator. Active, energetic and determined youngsters were thus those that could be presumed to have God's blessing and be destined for salvation (to the extent that these characteristics were maintained through later life). Passive, quiescent children were born under a cloud. The significant point, however, was that what now counted was not so much the birthright of the individual—the status into which he was born—as the individual's heaven-endowed personality. The characteristics of the individual, rather than the class, now mattered. Hence the way was open to despise the idle and inept inheritor of a title or position, and to admire the man who rose above his origins. It now became possible to admire the bourgeoisie, and the agents of proscription (and ascription) were intellectually disarmed.

All these essentially axiological (sacerdotal) concepts were fertile ground for the axiomatic seeds that would later grow into the capitalist position, defined successively and concatenatively by Smith, Ricardo, Marshall, etc. They also supported Locke, Mill, etc., in creating their philosophy of "natural law". But for the axioms of capitalism to work, the axiological predicates had to take broad hold. The new prescriptions were well designed to stimulate the material sector. Where men had once been allowed to pursue social, spiritual and sensual benefits without dispute, now they were forced to actively pursue material benefits, with the income statement as the index of piety and the bank account as the measure of worth. Where once men could be contemplative, passive, or even lascivious, they now had to be alert, tenacious and purposive. Where once they could be a bit incontinent, they now had to be temperate. Where once there had been no dictate for self-improvement, this now became an abiding mandate. Where poverty could once be looked at as improving one's probability of salvation, it now became a frightening sign of dismal predestination to damnation. In short, the energies of men were galvanized and directed, and life became a very different affair than it had been under the repressive but undemanding proscriptive axiology where all that was required was restraint.

In time, of course, the concept of predestination was softened. It had to be, or those who found themselves both poor and without prospects would have become intractable. So one was allowed to prove God wrong, so to speak, by working hard and improving one's lot at all times. It thus became the extent to which one was assiduous, temperate and strict with oneself that was the final arbiter of salvation. Any improvement was likely to be viewed as a point in one's favor, perhaps even sufficient to reverse the *a priori* verdict of damnation. Protestantism now became relativistic, much like the humanistic Catholic axiology it originally despised as counter-scriptural. So even those who were never likely to get rich were constantly exhorted to put forth maximum effort. The immediate benefits (consumption) became less important than the delayed benefits (investment), so one accumulated capital through saving and sacrifice, just as one accumulated heavenly capital by exercising the productive virtues.

Now, the positive correlation (the thesis of socio-economic amplification) kicked in to make social and economic benefits virtually indistinguishable. The wealthy man was able to gain peace of mind (through increased probability of salvation) in direct proportion to his economic success. The wealthier a man was, the better the class of his friends (and, presumably, therefore more entertaining). The successful man could presume the love of his wife and children, as they too were supposed to measure worth in terms of his ability to supply them with material goods. Admiration, prestige, companionship, self-respect—all elements of social benefits—were thus assumed to be *purchasable*. If salvation could be measured in dollars as surrogates, then dollars could be safely used as a measure of the value of everything else. That which could not be measured was therefore deemed trivial or mysterious. Eventually, the materialization of society became complete, and Protestant-Capitalist societies thus became systems where economic benefits were the sole element in the perceived store of societal satisfaction. Of course, as others have made amply clear (and as we shall discuss shortly), the release from the *a priori* repressions of the proscriptive Catholic system—with its institutionalization of religion and its ascriptive and hierarchical economic and political mechanisms—did indeed liberalize Europe, and derivatively the colonies of the Northern European nations (particularly the Dutch and English colonies). It powered the industrial revolution and spawned instrumental science, which has resulted in many material advances over prior cultures. But in the process the net satisfaction from societal existence was perhaps not improved. We shall have more to say about this shortly. Here we must devote a few words to a counterpart to the Protestant-capitalist answer to Catholic dogma: the socialist schemes.

The Socialist Schemes Essentially, the capitalist and socialist schemes both got their impetus—or at least their practical meaning—from the advance of the prescriptive Protestant axiologies. Now, capitalism and socialism do not argue about the necessity for an end to ascription, or in their emphasis on material as opposed to socio-spiritual benefits. They differ primarily in their axiomatic preferences for different allocation and distribution algorithms. The capitalist axiomatic position postulated inherent differences between human beings as resulting in different capabilities and aspirations, etc., and draws some axiological defense for this from proto-Protestantism. For theoretical socialism, asymmetry in the distribution of economic and political product is an artifact of capitalist society and a result of a fundamental misreading of the predicate Christian axiology. The socialist searching for an axiological defense discovered that the Reformationists and Catholics had both ignored the axiological justification for egalitarian society that had been laid down in European language as early as John Wycliffe (and later in the works of Thomas More, etc.) [14]. Under this scheme, all men (or all righteous men) had an inherent right to all the blessings of the temporal world. Civil government was of divine origin (made necessary only by original sin) and when combined with socialism would lead to utopia. In short, for the axiological socialist, there is no divine justification for asymmetry. To suggest so merely reflects the corrupt reading (faulty exegesis) of God's will on the part of the traditional religious authorities. In association with the utopian axiological base emerged the axiomatic justifications of socialism, largely through a reworking (or rediscovery) of the propositions of Plato's republic and the works of the Roman romantics (Seneca and Horace, etc.). They eventually found tangible expression in the socio-economic doctrines (axioms) of Marx, Engels, the Fabians, etc. In concept, the axiological and axiomatic aspects of socialism broke down ascription and institutionalization as surely as did the capitalist-Protestant schemes. Both, ultimately, led to a popularization of political authority and to a more symmetrical distribution of socio-economic prerogatives. The significant difference is that while capitalist minions were asked to be assiduous and active in their own interests (trusting to Smith's invisible hand), the members of socialist systems were asked to be assiduous in the interest of society at large. This meant the prescription of self-sacrifice rather than the prescription of economic egoism associated with capitalist systems.

But modern capitalist and socialist systems have not become "prescriptive" in the sense that might originally have been expected. The reason, in large measure, rests with our by now familiar pendular process. Initially, the Reformation axiologies represented a release from the repressions and wide-ranging proscriptions associated with Catholic dogma. It demanded activity instead of passivity; it introduced social and economic mobility instead of the determinism of caste and class. But rapidly the prescriptive components were translated into their proscriptive equivalents. For example, the prescription for assiduity led to *proscriptions* against idleness, sloth, intemperance, emotionalism, as these would disserve the cause of system competitiveness and industrial discipline. In the socialist nations, the significant proscription (naturally enough) was that against economic differentiation—against self-interest. As a result, the prescriptive latitudes of both capitalist and socialist systems were soon transmogrified into constraints. But while the proscriptions in the Protestant-capitalist nation were largely enforcible through the mechanism of sacerdotal sanction and censure (being directed mainly at behavior) those in the socialist systems (the sanctions against differentiation) took the form of tangible government controls, often backed more by coercive than by moral force. As a result, neither socialist nor capitalist systems represent significant advances toward prescription. The axiomatic proscriptions have in large measure subsumed the axiological predicates. Moral force prevents (ostensibly) the emergence of counter-productive forces; in the capitalist systems one is indoctrinated against sloth, etc. In the socialist systems, political forces preempt economic aggrandizement through highly progressive taxation, restrictive licensing and a deliberate attempt to keep old aristocratic (ascriptive) agents from reasserting themselves. And the same instrument, coercion, must often be used to provoke economic efforts and assiduity where one is deprived of the product of one's labor by socialist legislation, as with the collectivized farms in the Soviet Union. Thus, both capitalist and socialist systems (the latter especially in their quasi-communist form) have repressive undertones that in many cases contradict their pretensions to latitude and release.

The Counter-Economic Theses We move, finally, to what we have called the negative correlation—that which sees social and economic benefits in inverse relation, such that an increase in the one implies a decrease in the other. This is at once the most ancient and the most modern of socio-economic theses.

Here it is useful to recall that socialism and communism are different. Socialism is often defined as an intermediate stage between capitalism and communism. In socialism, the private ownership of means of production, and the unregulated distribution of incomes, are modified by some degree of mediation by a centralized authority, presumably representing collective interests. In communism, the concept of private property would disappear altogether, and the community as a whole would own everything. In a communist system, therefore, there would be complete symmetry in the distribution of economic benefits (and presumably symmetry in the distribution of political power and social status as well).

In short, what we have defined as the conditioned-programmed system, with a segmentational economic mechanism, is the only true agent of theoretical communism. The Soviet Union, Communist China, Poland, and many other groups that see themselves as sponsors of international Communism, are merely extremely proscriptive socialist forms (proscriptive in that the rights of the state *qua* institution take precedence over rights and interests of individuals). In actual practice, party-based communism, with its axiological implications and strongly repressive overtones, looks a great deal more like the medieval Catholic states than like any variant of humanistic socialism.

Thus, capitalists, socialists and quasi-communists are equally opposed to the counter-correlative position: that an increase in economic benefits can be gained only at the expense of social benefits. For, to the extent that capitalist, socialist or quasi-communist systems actually achieve their materialistic objectives, the counter-correlationist would see a corresponding reduction in social benefits. Thus, the axiological position of the counter correlationist is distinct from that of his capitalist, socialist or communist counterparts, but also distinct (in a more subtle way) from the advocate of the null correlation.

In this respect, recall that the chief socio-economic point of axiological origin in Catholic dogma was acquiescence: one should accept one's lot, whether rich or poor. Now the axiology of the counter-correlationist suggests that one should actively try to *minimize* one's economic or material demands, for only in this way can true happiness or salvation be gained. The roots of this axiological position go deeper than those of the socialist schemes. They are found in the ethico-religious prophets of the Old Testament (Amos, Hosea, Isaiah, Jeremiah, etc.). The oriental version of the counter-correlative position is found in the teachings of the Buddha and their Tantric and Taoist elaborations.

For the ethico-religious prophet, all material acquisitions beyond bare subsistence would simply lead one further away from the essential purpose of existence: preparation for the hereafter, which is essentially a spiritual domain. Idleness, luxury, sensuality were believed to dull one's piety and dilute one's concentration on spiritual matters. They were all artifices of the devil, designed to reduce the probability of salvation for all those who became susceptible to the sirenic song of materialism and sensual titillation. Therefore, the rich were urged to return their ill-gotten gains, and to distribute their wealth (as were the Florentines more recently under the despotic axiology of Savonarola). To this extent, the axiology of the ethico-religious prophets may be seen as giving root to socialist theory as well as to the counter-correlative position.

Thus, orientalist axiologies are direct progenitors of the counter-correlative position, for they deny the utility of *all* material benefits beyond a bare minimum,* and suggest that virtue is heightened to the extent that one's minimum requirements may be reduced. Thus, for the proto-Buddhist, all material benefits are merely delusions. One's desires can never be satisfied, so why bother at all? Thus, existence in the temporal realm should be utilized for attempts at disembodiment.

*The Hebraic prophets may be argued to have had more of an interest in symmetry of material product than in asceticism *per se*.

Now, if we begin to look for the axiomatic bases of the counter-correlationist thesis, we find an enormously confusing mixture of Stoics, Cynics and religious mystics, culminating in the modern counter-cultural movement [15]. In between is the romanticism of Rousseau (who, like Socrates, thought the pursuit of private property both undignified and ultimately humiliating). And then we have the astute axioms of Thoreau and the aphorisms of Mao. The essential message of Thoreau (and perhaps this is a somewhat idiosyncratic interpretation) is this: Happiness consists in being able to match one's need with one's resources. Both needs and resources are *variables*, not given. Now, one may generally increase one's resources only through directed effort or through undignified pursuit of gain through instrumentalism, appeasement, etc. But one may, on one's own accord, reduce *one's needs*. To the extent that one can reduce one's needs, the same level of satisfaction can be maintained with a correspondingly lower level of economic (material) resources, leaving one more time to pursue the dignified ends of study, introspection, meditation, etc. Therefore, *the key to happiness is in reducing one's material requirements or aspirations*. In short, strive to be poor and learn to fail with dignity. To a great extent, the Maoist aphorisms (if we accept their ultimately materialistic implications) simply suggest that time spent pursuing one's own aggrandizement and differentiation is time denied the pursuit of the social benefits which come with being "of the people" rather than above them. Thus, the mainland Chinese are being exhorted to cast aside their pretensions and seek solace among the classless. They are asked to redirect their aspirations from self to society, and from material to socio-spiritual ends.

One can immediately see, then, that the result of such axiological and axiomatic engines would be a deliberate attempt to deny material benefits the significance that the systems laboring under the positive (or ampliative) calculus grant. Thus, under the negative correlation, *asceticism* replaces the acquiescence of the compensatory calculus, and also the assiduity of the positive correlation. In its modern form, the counter-correlative thesis has become the keystone of what is popularly called the *counter-culture* movement. This, ostensibly, is a reaction by the youth of the more affluent nations against the materialistic preoccupations of their parents. It is intelligible as a simple pendular process merely involving the automatic inversion of prior values, not the creation of any fundamentally new components. The automatic association of virtue with wealth which attends systems operating under the positive correlation is now transformed into the automatic condemnation of success or material achievement. The counter-correlative position therefore ushers in the age of the anti-hero. Poverty means purity of soul, unemployment means dignity, and all truth is visceral or mystical. In the modern literature and cinema, the hero is an economic failure, and the new object of emulation is the itinerant poet rather than the merchant prince. The counter-correlative "culture" condemns patriotism, achievement, ambition and display equally, and is not particularly fond of religion or secular authority either. And, as can easily be seen, such a stance is inimical to the needs of an industrial society, and therefore condemned thoroughly by socialists, quasi-communists and capitalists alike. To a great extent, the rise of the counterculture can be explained in sociological terms as a natural reaction of one generation against the values of its predecessor; in economic terms, the scarcity of "significant" jobs in an increasingly automated and perhaps even contracting industrial domain[16]. In any case, its existentialistic rooting is clear, either as cause or as mediator.

In summary of the economic dimension of societal systems, then, each of the several behavioral modalities that we earlier identified (each associated with a particular form of societal process such as institutionalization, etc.) is seen to be associated with (or as causing) a distinct economic mechanism. The same situation may be found when we turn to the political dimension.

THE GUISES OF GOVERNMENT

We tried earlier to show how the various units of primary social significance (e.g., the congregation, the commune) lead to the distribution of status. Next we saw that the several economic mechanisms determined the distribution of economic or material benefits. Here we must assess the ways in which the several different political mechanisms serve to distribute *power*—the currency of the political "economy" of socio-behavioral systems. These are listed in table 5.6.

We may here provide only the briefest defense of this model. Initially, the *oligarchic* form of political mechanism derives directly from the concept of an institutionalized (sacerdotal) system. Here, the locus of authority may be coextensive with the locus of economic prerogative. Feudal and aristocratic societal systems are characterized by such a clusteration. Virtually all decisions of any import are handled at the very top of the hierachy, with delegation of responsibilities and authorities extremely limited (in contrast with the bureaucratic forms we shall discuss shortly). In short, the distribution of political prerogatives is highly *asymmetrical*, and authority (power) is highly *skewed* toward the highest levels of the sacerdotal hierachy. The reason for this concentration rests in part with our earlier analysis of proscriptive systems

TABLE 5.6: Attributes of the Political Dimension

	Attributes			
Political Modality	Locus of Authority	Structural Form	Engination	Generic System Referents
Oligarchic	Sacerdotal subsystem	Clustered hiearchy	Codes/schedules of mutual responsibility	Theocracy/aristocracy/feudal system
Decentralized	Ad hoc leadership	Reticular	Consensus	Participatory democracy
Bureaucratic	Professional agents	Stratified hierarchy	Consent	Technocracy
Oracular	Exogenous	Extensive/random	Ritual/revelation	Primitive subsistence systems
Totalitarian	Despot/fascistic bloc	Absorptive	Gaming	Demagogic/tyrannical systems
Collegial	Committee	Modular	By-laws/charter	Platonic system
Atomistic	Indeterminate	Entropic	Default	Anarchic system

as very conservative. But some theocratic, feudal and aristocratic systems are prevented from being capricious despotisms by a set of *codes* that define the mutual responsibilities of governors and governed, despite the strong concentration of political prerogatives. When these are present, the populace is not an instrument to be exploited and feared, but a foster child for whom the elite has a paternal responsibility. It is to be guided, succored and patronized as well as disciplined. For a member of the lumpen proletariat this is good news indeed. The whip is replaced with wit, the chain with the catechism and the sword with suasion. Society now becomes an exercise in reciprocity of obligations rather than the unilateral containment, coercion and resigned resentment associated with despotic or Fascistic regimes.

These codes differ widely in their net effect. Some have resulted in highly restrained exploitation (e.g., the Saxon, Latin and Japanese aristocracies). Others have been only a little less vicious than out-and-out tyrannies (e.g., the Turkish and Teutonic feudalisms). Nevertheless all such forms have this in common: the relationship between governor and governed is dictated entirely by the governor, and whether it is benevolent or tyrannical seems to be largely a matter of axiological accident. Those code-based systems, where ascribed privilege is in part paid for by assuming the paternal burdens, are perhaps more insidious than those systems where privilege carries no compensatory responsibilities, and where the exercise of class prerogative becomes identical with the exercise of power. But from a practical standpoint, I suspect that most of us would rather be constrained by a kiss than a cudgel.

The *decentralized* system of government—where authority and power are distributed more or less symmetrically across the population—is of course entirely congruent with the sentimentalized, prescriptive base of communalized systems. Here, authority is exercised in an *ad hoc* manner, largely through individuals lent political prerogatives for brief periods of time and in the face of exceptional circumstances. When the circumstance passes, so does their authority. In theory, then, authority rests ultimately and completely with the members of the commune; and it rests there directly (whereas it is indirect in bureaucratic systems). In terms of the political structure, to model it we would have to develop *reticular* referents—models reflecting the fact that each individual has a power potential more or less equal to that of every other. Such a system is distinctly non-hierarchical, so that government by consensus takes on its purest expression. In short, we have here the prototypical version of participatory democracy [17].

Associated with the simplest of our societal ideal-types (the conditioned, programmed system) is what we shall call the *oracular* politic. Here, authority rests outside the system, in the form of the purportedly implacable and irresistable forces of the environment. In short, it is not useful, when dealing with conditioned systems, to consider the possibility of deliberately initiated action. Such systems are, rather, primarily reactive. Now, when an occasion arises demanding a response—and when no programmed, ritualistic reaction exists in the repertoire—then the tribe may search for a sign from their gallery of gods. Such signs may appear in the form of tribal members being possessed and serving as a medium for the ultimate authority. Any member of the tribe may be so possessed, which makes the actual politics somewhat random so far as the outside observer is concerned. In some cases, the sign is an unequivocal one, such that the directed action is clear and comprehensible. In other cases, the sign may be parabolic or ambiguous, in which case the tribal council (usually a casual collection formed on the basis of simple seniority) deliberates an action that appears to be consistent with the oracular message. Much of the ritual of primitive systems seems actually designed to bring a medium to light, usually in response to the frenzy or trance which accompanies the ritualistic dances, chants, etc. For the routine activities of the tribe—the normal hunting or fishing or food-gathering function, etc.—there may be simple deference to the demonstrably best hunter or the eldest fisherman. Even so, there may be a deliberate attempt to avoid authoritarian or hierarchical overtones, so that authority appears manifestly diffuse. However, because conditions demanding a response probably arise infrequently, issues of power and authority may not generate the same basic concern here as they do in more sophisticated societal systems.

Little need be said about the totalitarian form of government, which essentially is merely the exercise of caprice and coercion. As it is associated with secularization (in its pure form), there are no rules or *a priori* constraints on governors, and therefore power tends to be entirely absorbed by a single dominant individual (a despot) and his sycophants. Thus, power—like economic prerogative—becomes primarily an exercise in preemption within the secularized, instrumental system, with authority far more concentrated than in either the oligarchic or bureaucratic constructs.

We can easily suggest the logic behind the tendency to associate totalitarian (tyrannical) forms of government with secularized systems. The fundamental philosophy of the secularized system (or the prophets of realism who underlie such societal movements) finds all differentiation among men to be dictated by nature. That is, asymmetry of ability leads to asymmetry of social status and economic benefits, and derivatively to asymmetry of political prerogatives. In the realistic-instrumental world, the responsibility of those in power is to stay in power, and power is a legitimate end in its own right. Under such a scheme, the people become instruments of class privilege or party purpose, to be exploited at the same time they are feared. The government under this perspective, becomes a processual rather than a procedural phenomenon. The governors bear no responsibility to the people that the people are not capable of imposing on them. Of course, no opportunity is lost to reduce the probability of mandates being effectively formulated and passed up from the people. Instrumental politics thus becomes a constant-sum game played between governors and governed, reminiscent of the unbounded competition on the economic dimension of secularized systems. Such a situation is relatively easy to maintain. So long as the enormously wide tolerance limits of the governed are not exceeded, a degeneration of the game into open conflict can be avoided.

In collegial government, authority rests with the set of credentialed individuals comprising the "college". As a rule, authority here will be *modular*, with certain exercises in decision or policy making being more or less the province of certain committees, and with the right of challenge exercised only infrequently. The nature of the modular powers are usually dictated by some set of *by laws* or rules of procedure, usually complemented by certain protocols, which serve to restrict decisions to those best qualified to make them.* For example, scientists may be given authority over matters of technical import, clergymen over matters of morality, etc. Usually there are opportunities for contravention, and certain procedures for overriding the tacit distribution of prerogatives. As a general rule, however, authority may be seen to rest with presumed or demonstrated expertise.

Finally, a majority of the world's population is governed by some sort of *axiomatically* predicated politic. The three great political referents of our day are, as we have seen, all essentially axiomatic in origin: democratic capitalism, democratic socialism and quasi-communism (or proscriptive socialism). In all these contexts, authority is presumed to rest ultimately with the people (in contrast with tyrannical forms), but not so completely or directly as in real participatory (decentralized) democracy. For, in practice, all axiomatic systems evolve a class of professional political functionaries who become the effective locus of power. These governmental functionaries operate with the *implied*

*Note that this precludes simple democracy, where qualifications are usually not functional prerequisites for degree of influence etc. The collegial polity is thus a complex or "weighted" democracy.

consent of the people, though the mechanisms by which this consent is sought and determined vary greatly between the various bureaucratic forms. Finally, if the political structure of theocratic federal and aristocratic bureaucratized systems may be defined as a cluster hierarchy, that of rationalized (technocratic) states is a stratified hierarchy. The difference is that authority—though still basically hierarchical—is delegated to different levels of government (state, federal, local, etc.) and there are also distinct functional bases (e.g., defense, commerce, propaganda). Therefore, authority is spread more widely throughout the government itself, though still concentrated with respect to the system as a whole.

Government under the bureaucratic modality may take basically two forms. First, there is the limited democratic system (or indirect democracy), where the presumption is that the governors act as agents of the public will, gaining their significance through the execution of policies presumed to resonate from the popular base. The mechanisms of this resonation are the franchise system, limited tenure of public office, and the institutions of initiative and referendum. As we well know, however, the "public will" is difficult to define in practice, and open to many abuses of interpretation and execution.

The second bureaucratic form adopts what we might call the *didactic* posture. Here, the governors are presumed to be servants of the people in that they embody some (ostensibly) commonly held philosophy. The absence of a franchise system is thought to be no impediment, for the ideological dictates under which the societal system operates are presumed to be held equally in awe by all the people. The philosophical base specifies the "correct" response for every event, either directly or via exegesis and interpretation. Under such a situation, it is the governor's responsibility to set policy, and policy is theoretically a matter of inferring the correct response from the Marxian manifestoes, the "little red book" of Mao or some other aprioristic base. Governors are thus thought to serve both an educative and an executive function.

In the western industrial democracies, the Soviet-style proscriptive socialisms and the dogma-driven Maoist state—the major bureaucratic referents of our age—rationality is restricted for the most part to the pursuit of material ends. Specifically, the central concern is the raising of the material stock, on the assumption (somewhat inaccurate, as we shall later see) that the aspirations of the people are themselves restricted to the material domain. The foci of rationality are thus production, distribution and the defense of that which is produced and distributed. This means that the bases of governmental action are essentially empirical. Governors are *a priori* absolved from responsibility for rationality with respect to moral or normative issues except in so far as they directly affect the causes of production, distribution, internal security or defense. Thus, Mao preached involvement and commitment, the necessary predicates for the development of a distributed cottage industry, for increasing agricultural productivity and for the erosion of the last vestiges of the traditional self-effacement and disinvolvement of the vast rural proletariat. The Soviet Politburo preaches self-sacrifice, the behavior necessary to offset the historical tendency toward "bourgeois" behaviors which have affected both the peasant and urban classes in Russia and its satellites. Moral pronouncements in the industrial democracies on the other hand, are typically about self-reliance and assiduity, the engines of capitalism. Even those anomalies—the European democratic socialisms—are returning to the Smithian fold after long spins at the wheel of nationalization, paternalization and secularization.

In summary, then, the mandate is for maximization of the material stock of the state, irrespective or whether the vehicle of production is socialist or capitalist. As we have seen, it is the algorithms for the allocation of resources and the distribution of realized national products that mainly differ between these vehicles, though the degree of actual difference is not so great as the theoretical distinctions would lead us to believe. For government in both capitalistic and socialist (and quasi-communist) states takes on a distinctly technocratic flavor. Particularly in the quasi-communist and some socialist states, the public sector is more or less coextensive with the economic. The system's governors, in the name of centralization, are responsible for the significant allocation and investment decisions, and ultimately for the distribution decisions as well. Policy exercises are directed at maintaining the integrity of the context within which economic activities are performed, the socialist policy constraints usually being softer and less specific than the quasi-communistic. In the industrial democracies, theoretically at least, we expect to find the government's role restricted entirely to policies aimed at protecting the integrity of the "natural" market mechanism that is the essential underpinning of capitalist theory. However, with the advent of Keyneseanism, and the increasing proportion of national product accounted for by government services, the distinction between the effective responsibilities of the democratic and socialist-communist executives is gradually being obscured. This dampening of differences is also due, in large measure, to the realization that there are severe limits on the functionality of economic centralization. Thus, there is some rewakened interest among socialist-communist nations in developing "bourgeois" mechanisms.

THE INTERDISCIPLINARY INTERFACES: A TYPOLOGICAL CONSTRUCT

The past several sections have merely sought to provide some sort of initial defense for the model we presented as table 5.1. There, we set our seven ideal-type socio-behavioral systems, each with unique social, economic and political properties. Through our subsequent argumentation, we sought to show that these ideal-types are of some theoretical merit because social, economic and political properties appear in *clusters*. That is, socio-behavioral systems in the real world tend to be such that certain attributes appear in the company of certain others.[18].

Underlying our development of Table 5.1—an elaborative ideal-type construct—was thus a *typology*. In the sense we use the term in this volume, a typology sets out the theoretically most probable relationships from among a universe of alternatives. That is, it is a construct disciplined by subjective (or judgmental or logical) probabilities, and thus represents a proper hypothetico-deductive artifice. It tends to focus subsequent investigation on a subset of relationships, and thus serves to initiate inquiry into a very broad, complex and protean subject area. The typology which underlay the ideal-types we have been defending is shown in table 5.7.

The use to which such a typology might be put is obvious to us now: various interdisciplinary scholars (or teams) would take it into the field and develop a frequency distribution around various real-world socio-behavioral or societal systems. The empirical frequency distributions, if they are to lend support to our hypothetico-deductive exercises, should show a significant majority of real-world cases having properties clustered along the main diagonal, with sparse entries in the other cells of the typology. To the extent that our logical (qualitative) front-end analysis is flawed, then we would expect to generate real-world "maps" that entail significant variance, such that the postulated clustering is very weak. Thus our typology presents, in the most definite terms, a qualitative template that feeds into the testing procedures of normal science. (In actual practice, we would probably attempt to validate it by conducting some sort of Bayesian investigation, in which empirical probabilities are constantly fed back to modify the *a priori* probabilities on which the typology is predicated.) In terms of general procedural logic, then, we have a *vector of congruence* model similar to those we developed in the context of the dialectic paradigm in Chapter Two. The components of this current typology are simply less abstract, and deliberately designed to develop productive interfaces among the various social-science disciplines.

TABLE 5.7: The Typology of the Interdisciplinary Interfaces

Economic Mechanism	The Primary Units of the Social Sector						
	Tribe	Congregation	Commune	Nuclear Family	Bloc	Conference	Individual
Segmentation	Oracular						
Ascription		Oligarchic					
Requisition			Decentral-ized				
Algorithmic				Bureaucra-tized			
Preemption					Demagogic total-itarian		
Mandation						Collegial	
Asceticism							Atomistic/anarchic

From another standpoint, this entire chapter may be said to be primarily useful as a vehicle for laying down a single, critical dictate for societal management:

> A congruent system will be one whose social, economic and political sectors obey the normative dictates laid out in table 5.7. Only in such a case may the societal system be said to be synergistic—more than the sum of its parts. For it is only along the major diagonal that the criteria for social, economic and political complementation are met. Any other combinations, arguing from the theoretical perspective we have adopted, will result in an anomalous and hence unstable system, one where the net benefits of membership are lower than they could be were congruence maintained.

Thus, from the simplest and most direct perspective, the problem of societal management is to make sure that the several dimensions are complementary—that the social, economic, political and behavioral dimensions act to complement and reinforce each other. But there is a contradiction here. For what we have been suggesting is that these congruences occur *naturally*, and that our typology simple reflects the conditions we expect to find among a majority of real-world systems. In this regard then, the managerial problem may be stated not in a positive sense, but as a negative: the societal manager should refrain from introducing any change agents that might act to unbalance the system.

Well, now we're in trouble. For from the perspective we've been operating on, this may be translated to mean that there should be no attempt at societal management whatsoever. We can, however, resolve this apparent paradox by suggesting the following: no attempt at societal management can be successful if directed at the level of the sectors we have treated in this chapter: at the social, economic and political institutions. Rather, societal management must take place on an entirely different dimension, the one we are concerned with in the final chapter of this book.

Notes

1. These may be viewed as intricate psychological *packages*, elaborating on the use of those terms given us by Berger, Berger and Kellner in *The Homeless Mind: Modernization and Consciousness* (New York: Vintage, 1973).
2. That is, the nuclear family is the primary unit of social significance within even the most adamantly socialist states, the commune being mainly an artifice for certain types of productive activities.
3. See Chapter Seven of his *Future Shock* (New York: Random House, 1970).
4. For a note on the practical policy implications of requisitional economics, see Chapter One of my *Managing Social Service Systems* (New York: Petrocelli Books, Inc. 1977).
5. One of the most important discussions of the concepts of just price, utility and supply and demand—all related to the axiological issues of economics—is given by Walter Weisskopf in his *Alienation and Economics* (New York: E. P. Dutton, 1971).
6. One of the most interesting phenomena in this regard is the multinational corporation, with its characteristically obscure profit portraits. See, for example; Barnet and Muller's *Global Reach: The Power of the Multinational Corporation* (New York: Simon and Schuster, 1974). For a more detailed, semi-classical analysis of the implications of optimal size, see Chapter 5 of my *Administrative Decision Making: Extending the Bounds of Rationality* (New York; Van Nostrand Reinhold, 1977). When we consider the enormous unsecured debt that private corporations support—along with government units—we get an even more frightening picture of what it means to consistently exceed economic limits. In this regard, I recommend Paul E. Erdman's remarkable book *The Crash of '79*.
7. One of the prime examples of this, if perhaps one of the most subtle, is the restricted, parochial use of retained earnings by industrial corporations. Under axiomatic capitalism, even retained earnings should be allowed to seek the optimal return, and not be automatically dedicated to the growth of an individual firm generating it. For more on this, see Chapter 5 of my *Administrative Decision Making* [6].
8. Not that these systems are without their complications. Stuart Piddocke, for example, cites some of the complexities of redistribution in "The Potlatch System of the Southern Kwakiutil: A New Perspective" (*Southwestern Journal of Anthropology* 21, 1965).

9. But such systems may be becoming increasingly important, as W. I. Thompson explains in pp. 7–11 of his *Evil and World Order* (New York: Harper and Row, 1976).
10. A fine exposition of medieval Catholic economic dogma is given in R. H. Tawney's classic *Religion and the Rise of Capitalism* (London: John Murrary, 1926).
11. Cf. Talcott Parsons's translation of Max Weber's book *The Protestant Ethic and the Spirit of Capitalism* (New York: Scribner's, 1930).
12. Thoreau's position is of course give in his *Walden*. A fine look into formative concepts is given by Mario Einaudi *The Early Rousseau* (Ithaca: Cornell University Press, 1967).
13. See Francis Schaeffer's interesting, if somewhat sacerdotal, *Escape from Reason* (Downer's Grove, Ill.: Inter-Varsity Press, 1968).
14. For more on the axiological bases for axiomatic socialism, see Chapter 1 of Harry Laidler's impressive *History of Socialism* (New York: Crowell, 1968 edition).
15. For the properties of this latter, see Theordore Roszak, *The Making of a Counterculture* (New York: Anchor, 1969).
16. Socioeconomic interpretation of the counterculture does not end here. For an extended analysis, see Berger and Berger's "The Blueing of America" (*The New Republic*, April 13, 1971).
17. A fascinating, if somewhat strained, set of defenses for the implementation of participatory democracy is made by the articles in *The Case for Participatory Democracy* (edited by Benello and Roussopoulos; New York: Grossman, 1971).
18. It is difficult to find empirically predicated studies that point out the interconnectedness between social, economic and political properties of operating systems. The most exciting of such studies (sadly ignored) is Jacob Fried and Paul Molnar's "A General Model for Culture and Technology" (*Technological Forecasting and Social Change* 7 no. 1, March 1975).

6
The Criteria of Societal Quality

Alfred North Whitehead once suggested: "The problem is not how to produce great men, but how to produce great societies". From the analytical exercises we have gone through thus far, we should be able to see something of the magnitude of this problem. For to change society in any meaningful way, one must also change the bases of behavior. And to change the bases of behavior, one must change the cultural predicates to which the system's residents respond. And to make predicate changes effective, we must change people's minds. Now the management of societal systems takes us beyond these simple statements. For one does not manage a system by inducing change; rather, one *manages change*. Change, as we saw from our discussion of societal dynamics in Chapter Four, is the natural state of society. Managing society means breaking the hold of the natural or spontaneous dynamics, and replacing them with reasoned or desired processes. But why would we want to do this?

The answer is that the natural processes have not resulted in much substantial increase in the quality of life offered by societal systems. There is a societal evolution of a sort, true; but this mainly results in parametric rather than basic alterations. Therefore we suggested that societal systems have not so much evolved as *revolved*, with one form giving way to another that, from the standpoint of the system constituent, merely replaces one set of contraints and benefits with another of more or less equal magnitude. To this extent, the history of civilization is a sadly unimpressive one.

And then there is the matter of the prophets. First, in Chapter Three, we criticized them; then, in the last chapter, we suggested that they were the only salvation for us—that they would have to be the true societal managers. This contradiction must be cleared away before the book is finished. I shall try to do this by pointing out that *proper* prophecy bears little resemblance to the axiological and axiomatic prophetic processes we discussed earlier. Secondly, I shall try to use our previous hypothetico-deductive exercises to point out those trajectories of change that the prophets might lead us along—the trajectories that do promise genuine societal evolution. And then, in the last section, I shall have to apologize for being able to offer these prophets so little insight.

THE MANDATE OF THE MASSES: THE SEARCH FOR ISOMORPHISMS

We may search for isomorphisms on either the strucural or the dynamic dimension of societal systems. In this section we shall be concerned mainly with the former, leaving the latter to the next section. On the structural dimension, we are concerned with the possibility that from the universe of structural alternatives available to societal systems, only a few are given significant expression. To the extent that this is true, many otherwise very different real-world systems may become intelligible in terms of a few generic structural referents. The implication for the social sciences is thus that we need not wake up in a new world every morning: these generic referents may be used to yield a comprehension of large blocks of societal systems, so that the structural properties of many subjects may be subsumed *a priori* in a much smaller number of theoretical constructs. In this way, isomorphisms directly serve the cause of scientific efficiency.

Within the framework of societal systems analysis, we are quite fortunate. For of all the basic societal alternatives available to man, he has consistently sought to exercise just two: institutionalization and rationalization. The other societal forms have had their moments, but none has captured significant populations for sustained periods of time. Now, if we propose that *societal substance is ultimately either what men want or what they will tolerate*, then there should be a strong lesson for us in the predominance of the institutionalized and rationalized societal forms. We know that both these societal forms offer the individual perhaps the strongest *a priori* discipline and direction; both imply well-defined behavioral constraints. But what is also clear is that institutionalized and rationalized systems will all tend to be characterized by: (1) a certain degree of *functional differentiation*, exhibiting itself in terms of specialization, division of labor, etc.; (2) some form of *structural differentiation*, usually taking the form of a hierarchical configuration; and (3) some *asymmetry* in the distribution of social, economic and political prerogatives.

Some readers may find these three attributes offensive. For many modern social-science schemes suggest that egalitarian systems are not only to be generally preferred to differentiated, asymmetrical ones, but that the former also represent the "natural" and prototypical societal form. Following the Socratic, Rousseauean and Marxian arguments, one sees egalitarian innocence at the dawn of history, and a gradual corruption of this innocence to the point where differentiation and asymmetry enter and elaborate themselves. At the base of such arguments—or perhaps as a quasi-axiomatic first premise—is the suggestion that egalitarian systems are the *most natural* form. This would imply, from the reflective standpoint, that sentimentalized or conditioned-programmed systems would dominate social history. But, as we suggested, they had only brief moments. Yet it is easy to understand the axiological preference for such systems. For example, most academics and social critics are themselves products of quasi-egalitarian, collegial systems, and they may have a very difficult time understanding why an individual would want, or even tolerate, that which they themselves find abusive or unsatisfactory. The axiological predicates of science are again at work.

But the sentimentalized sapient and conditioned systems—with their symmetrical distribution of social, economic and political prerogatives—must remain primarily of normative significance,* for history has not given them much play. This is not hard to understand, at least when we consider what the opposites of functional differentiation, hierarchical structure and asymmetry would be. The opposite of functional differentiations would be a system where there was segmentation of roles—in short, no specialization. The survival tasks would be distributed symmetrically throughout the system, with each man a rep-

*Or they may be viewed as *prototypical* but not, from our perspective, as *archetypical* societal referents.

lica of every other in terms of skills and responsibilities. The hierarchical structural form would be opposed by an extensive configuration—one where both actual and potential authority was distributed equally among the members of society with respect to all affairs and substances. Finally, a symmetrical system would be one where social, economic and political prerogatives were distributed equally, so that there was no differentiation of status, wealth or power. Now, again, our argument is that the primary properties—differentiation, hierarchy and asymmetry—are more probable than their opposites. The historical or empirical evidence is, once more, the relative infrequency of the appearance of egalitarian systems. But the theoretical evidence is no less telling. For the Socratic, Rousseauean and Marxian positions are all vulnerable to a case that can be made in terms of logical (judgmental) probabilities.

The case begins with the conditional assertion that attributes of societal systems tend to be clustered, so that there is a high probability that differentiation, hierarchy and asymmetry will appear together. Thus, given any one, the others have an extremely high probability of following. To make this case somewhat crudely, as we must do in the limited space available, let us try the simplest and most straightforward sequence: functional differentiation leading to hierarchicalization leading to asymmetry in the distribution of societal prerogatives. We must first search the logical arguments for differentiation, trying to show that it is a logically probable occurrence. Then we must set a case for subsequent hierarchicalization and asymmetry, and the basis for their persistence and replication. And what I shall try to show is that there are indeed myriad opportunities for the realization of functional diferentiation—so many, in fact, that if the symmetrical system (postulated, for example, by Marx,) existed at all, it was but for an instant. It must have succumbed to the laws of probability shortly after the dawn of collective civilization, its vestiges remaining in only the most remote regions of either memory or geography.

All we need to make the logical case for functional differentiation is inequality of human attributes on any of three dimensions: physical, mental or motivational. The role of physical and mental differences in producing functional differentiation is obvious. But when we consider the motivational base, it is possible to let men look alike and have the same mental apparatus and still move directly toward specialization. For wherever we can postulate motivational differences, societal differentiation becomes a result of a natural and ineluctable bargaining process. Imagine two individuals, A and B, who have two different utility functions, X and Y respectively. Now let us suppose that a set of tasks $(a_1\, a_2 . . . a_n)$ devolves on each member of the group. Now, so long as the utility functions of A and B differ with respect to any of these tasks, there is an opportunity for bargaining which will not be foregone. For example, let us suppose that A has a high utility for task a_i, whereas B has a high utility for task a_k. There will be a natural tendency for A and B to reach an agreement where A concentrates on the performance of the task a_i aspect of B's repertoire, while B assumes some proportion of A's responsibility for task a_k. In such a way, when other members of the collectivity enter into the bargaining process, a differentiated functional structure will emerge, with performance *prerogatives* distributed asymmetrically, as a reflection of the asymmetric distribution of performance *predelictions*.

To the extent that such specialization intensifies, we have the basis for transgenerational differentiation, in that special skills tend to be handed down from father to son, etc. As functional differentiation intensifies still further, there will be a tendency for structural differentiation to follow. Certain tasks become the province of certain "lines" or qualifications, e.g., age, ritualistic assignments (the first male child born at the turn of the winter moon), physical attributes, courage. And where monopolistic pressures may be exercised, there will be the basis for asymmetry in income, then wealth, and eventually in social and political prerogatives as well. Asymmetry on any of these dimensions can then be the basis for the institutionalization of privilege, and more important, the rigidification of utility functions and expectations. It is now possible to differentiate a societal system not on "natural" differences, but on ascriptive contrivances, where individual differences become subordinate to class-based differences (e.g., demonstration is replaced by inference, and latitude by predetermination). But it was "real" differences and not merely artifices that led to this situation, a point that often eludes apologists for the Socratic-Rousseauean schemes.

As it is so simple to develop a model that can mimic the transition from a fully segmented society to a highly differentiated one simply on the predication of minimal differences in motivational (utility) functions, it is no trick at all to develop a model based on more tangible differences. Where we recognize physical differences, we simply allow another basis for differentiation, and also allow the possibility that the bargaining process will be displaced by coercion: B will perform task a_k for A or A will simply punish him. Another basis for differentiation—to complement the physical and motivational—has its roots more deeply in the psychological dimension. It is possible that conceptions of self and worth have always been somewhat different among different individuals. There is no particular reason to suggest that prototypical social men did not have a capacity for generating conceptual sets that differed from one to another. Thus, it is possible to conceive of one man having a high utility, say, for self-assertion, and another a high utility for placidity. Under such circumstances, the latter may be willing to trade some quantum of autonomy for a release from anxiety and responsibility. Or one may defer to another simply because his tolerance for dissonance is inherently lower, or because he prefers tending his grapes or flock to the challenge of leadership, etc. The possibilities for differentiation on the cognitive dimension are vast in richness and variation and possibly in intensity as well.

Now, clearly, any of these three bases—motivational, physical, cognitive—is sufficient to cause the rapid transition from initial segmentation to functional differentiation which in its turn causes a shift from symmetrical to asymmetrical distributions of social, economic and political prerogatives. But when we allow the possibility that two or even all of these engines of differentiation are operating simultaneously the force and potential rate of differentiation becomes stunning indeed. So it is small wonder that even the societal systems which appear at the very dawn of recorded history have been highly differentiated and elaborate artifices, and susceptible to interpretation as institutionalized or rationalized (rather than sentimentalized or segmentalized) variants.

On balance, then, asymmetry looks to be a most "natural", if not applaudable, state of society. That is, men have been either willing to tolerate, or actually appreciative of, the unequal distribution of social benefits. The dominant social systems are those where the marginal utility of system membership is a variable, not a constant. And to the extent that lack of symmetry—lack of apparent equity—is in evidence, then we have a prima facie case for dissension, alienation and dissatisfaction. But we don't necessarily have a cause for conflict, reform or revolution. There is an intermediate variable that is seldom made explicit, in even the most erudite social and political theorizing. This is the phenomenon of *toleration*. Men apparently have a great capacity for suffering insult, hardship and abuse—and contradiction. This toleration is surely as much a component of our fundamental behavioral repertoire as any other attribute of human nature.

The question for us here, then, is this: why have so many people apparently been content to reside in differentiated, hierarchical and asymmetrical systems? The trivial and popular answer is one I think we can dispose of: it is not necessarily because for thousands of years the vast majority of people in the world have been consistently and completely cowed by authority or "conditioned" to acquiescence by clever

TABLE 6.1: Properties of Generic Societal Benefits

Benefit Category	Engination	Maximizes	Minimizes
1. Placidity	Spiritual	Resignation	Anxiety/cognitive dissonance
2. Gratification	Material	Titillation	Guilt/ frustration
3. Individuation (identity)	Social	Ego differ- entiation	Redundancy/ anomie

concept-mongers. At special times and for special groups this might have been the case. But to suggest that it is true of all men at all times—or virtually so—does great disservice to the fundamental dignity (or perhaps perversity) of the human race. There is a better reason, derived from our earlier discussions: that differentiation, hierarchicalization and asymmetry serve definite psychological functions.

A Schedule of Societal Benefits

There is another level at which we may be interested in structural isomorphisms. We may here ask about the structure of societal benefits. This would suggest that we look at the various societal forms (our several societal ideal-types) in terms of their *output characteristics*. This question extends—within our framework of analysis—to the following question: are there any *classes* of benefits that can be identified, such that many different systems may be made comprehensible in terms of a few generic referents? Table 6.1 suggests that perhaps there are.

Though any attempt to categorize societal benefits is bound to suffer from oversimplification, a taxonomic construct is initially the preferred method for trying to gain some tentative appreciation of complex phenomena and to search out isomorphic properties. Therefore, we shall proceed to develop our generic benefit sets without apology, but with the full realization that few systems will be found in the real world that offer precisely the benefits defined here, or produce them in precisely the way suggested. In general, however, we expect to find *placidity* being available to those individuals who have resigned themselves to some superior power. In absolving themselves of the responsibility for their own welfare, performance or destiny, these individuals may be expected to gain release from the anxiety and the cognitive demands (and insecurity) that self-responsibility and realism imply. Gratification is gained through the satisfaction of one's appetites, which becomes generalized as *titillation*. The individual operating under the calculus of titillation may be expected to lose placidity, but to gain release from repression, frustration or guilt. Finally, there is the benefit of *identity*, the appreciation of self and place and purpose. In the most basic terms, one gains a concept of identity through ego-differentiation. This, in turn, may be had only with respect to the properties, interests and ambitions of others. Hence, individuation depends on the *other* individuals with whom a subject has interchange, or of whom he has knowledge. And, as we shall show, these other individuals enter into two distinctly different types of relational-identificational process: symbiosis and synergy.

We may now display the focal model for this chapter, as below in table 6.2.

Placidity is a generic type of spiritual benefit and, again, derives from the process of resignation. Now, resignation may be engined by any one of three vehicles: .

IMMANENCE — The thesis that the spirit of some creatorial force is present in all things (e.g., pantheism), and that man is equally susceptible to determination with the animals, seasons and other natural phenomena.

THEONOMY — The thesis that the creator takes on anthropomorphic qualities, and becomes "intelligent". He also becomes especially the author of human events, to whom each individual is known as a person and a personality. Here, then, the creatorial force is first humanized, and then personalized to a less or greater extent.

ABSORPTION — The thesis that the creator is coextensive with natural and human phenomena, and that man is therefore completely enveloped in a pervasive and unanalyzable creatorial engine. In its more moderate forms (e.g., proto-Buddhism), the creator is a disembodied intellect, with no recognizable corporeal or emotional human qualities; he is "one with the soul".

It would take a dissertation in its own right to set forth adequately the subtle differences between these three positions; a very brief treatment of their implications will have to serve us here. Intially, as table 6.2

TABLE 6.2: The Typology of Societal Benefits

System Type	Behavioral Benefits	Mechanisms for Achieving		
		Placidity	Gratification	Individuation
Conditioned/ programmed	Mechanisticity (lack of cognitive demands)	Immanence	—	—
Institutionalized	Certainty/determinism	Theonomy	—	—
Sentimentalized	Emotional/affective release	—	—	Symbiotic relationships
Rationalized	Purposivity/directivity	—	Constrained optimization	—
Secularized	Instrumental license	—	Instrumen-talization	—
Sapient/cerebral	Creative/originative license	—	—	Synergistic relationships
Existential	Psychogenetic/com-pulsive release	Absorption	—	—

suggests, the immanence thesis is expected to be associated with conditioned systems. Indeed, as anthropological evidence indicates, primitive systems are characterized, to a great extent, by undifferentiated pantheisms. As is consonant with the simple (associative) adaptivity of such systems, the primitive collectivity sees itself at the mercy of capricious forces that it cannot comprehend or control. Immanence thus correlates well with the psychological benefits of conditioned systems; it simply releases the individuals from any cognitive commitment [1]. The individual becomes the tool of his environment and its immutable pantheistic agents, and thereby exists with minimal self-determination. In short, he becomes the personification of behaviorist presumptions.

The theonomous thesis is more sophisticated than immanence. Here, the capricious exogenous forces that cannot be rationally comprehended give rise to a purposive creative force in the form of an implicitly *embodied* creator. He may be endowed with both intelligence and quasi-human aspirations. The anthropomorphism of the institutionalized religions expresses this implication. The comfort or benefits derived on the spiritual dimension are essentially those of a child relating to a strict but essentially interested (if not loving and affectionate) parent. In short, the individual persuaded of the theonomous position is neither orphan nor victim; he is capable of assuaging or modifying the Creator's will through certain actions (e.g., prayer, sacrifice). The implications of the theonomous position, then, correlate nicely with the terms of the type of proscriptive axiologies that tend to underlie institutionalized systems, and offer complacency in exchange for obedience and adoration. It is no accident, then, that proscription, institutionalization and theonomy have tended to appear as a societal cluster throughout recorded history.

Finally, the absorption thesis postulates that spirits, man, nature, mind and matter may all be viewed as coextensive in some numinous creatorial force. Cause and effect disappear as meaningful terms, as do distinctions between subject and object. Individuals thus become *absorbed* to the extent that they deny any real differences between themselves and others, or between themselves and nature. In the absorptive domain, one gains spiritual benefits by becoming successively disembodied—through meditation, asceticism and even starvation and attempts to lower one's respiration rate, etc. The thought is, as we have already explained, that one approaches a consciousness of fundamental reality by shutting off the cognitive mechanisms, a concept that has found much favor among the drug cultures of both our own and ancient times. For the "acid head" or the mesquite-eater, for example, truth and reality are thought to emerge only through pre-cognitive hallucinations, with the (drug-induced) denial of one's own separate identity as a prerequisite. It is clear, then, that existentialization is a useful precondition for the realization of spiritual benefits of the absorptive type, and that the conditions of asceticism, anarchy and hermitism we associated with existentialized systems directly serve this purpose. In summary, then, all systems that emphasize spiritual benefits—through any of the three mechanisms—may be seen to *disinvolve* the individual from the empirical, material world. For the seeking of societal satisfaction through spiritual benefits discourages the individual (for reasons we shall explore shortly) from attempting to gain any significant satisfaction on the other two dimensions [2].

Gratification, as a source of societal satisfaction, is an easier condition to analyze. Here we are concerned mainly with the processes by which material benefits may be realized. Fundamentally, material benefits become comprehensible in terms of their ability either to satisfy appetites or to minimize corporal discomforts (and, in some cases, psychological discomforts as well, although this we must explore at some length in a later section). Thus, material benefits really extend to anything that may be *purchased* through any of the three commanding societal currencies: status, wealth or power. And appetites may therefore run the gamut from the exotic to the essential. Now, there are fundamentally two processes by which material benefits are sought, both of which have some familiarity for us:

CONSTRAINED OPTIMIZATION	The ability to pursue certain legitimate material ends within procedural constraints imposed by the system or established by principle.
INSTRUMENTALIZATION	The license to pursue any ends by any means.

Constrained optimization is the mechanism for societal satisfaction associated especially with rationalized systems. The axiomatic bases of such systems license certain forms of gratification (the set N), while condemning others (the set M).* Further, as we saw in the last chapter, the three dominant forms of axiomatic societal engination—capitalist, socialist and quasi-communist systems—all ultimately confine themselves to the pursuit of material ends. Therefore, the normative behaviors (the set N) generally take on distinctly materialistic overtones, while the non-normative behaviors (the set M) generally serve to circumscribe the means by which material ends may be sought or main-

*See again Chapter Three.

tained. Contrasting with this situation is the secularized system, allowing (if not actively encouraging) strictly instrumental behavior. Here, as we have said so often, all *a priori* constraints disappear, which allows any end to be pursued by any means.

The last of the three generic benefits, individuation *qua* identity, is the process of gaining a concept of self, either actual or perceived. That is, uniqueness may be real—in tangible terms—or simply assumed by the individual. In terms of basic satisfaction, either mode of self-conception is functional, at least up to a point, and may only be obtained in relation to other individuals [3]. For individuation always involves the concept of distinction, and therefore requires a reference. In more specific terms, identity may result in either of two ways:

SYMBIOTIC REFERENCE	The individual's identity is secured through his more or less unqualified acceptance into a situation of emotional or biological interdependence, the intensity of the resultant individuation being related to the intensity of the interdependence. The relationships here, however, are merely additive.
SYNERGISTIC REFERENCE	The individual's identity is dependent upon the specialized functional role that he can fulfill in the company of other functional specialists, the intensity of individuation being determined by his perception of the rarity of his particular talents, skills, etc., relative to the demand for them (or relative to the density of their distribution among the population as a whole) [4].

In sum, then, the symbiotic and synergistic contexts allow the individual to escape from the ultimate societal horror: abject redundancy [5]. For those for whom neither material nor spiritual differentiation has much appeal, this is a saving grace indeed.

Now this admittedly simplistic treatment of societal benefits should not be interpreted as an attempt to rewrite social psychology. Rather, it is simply intended to provide us with a basic vocabulary of concepts to carry us into an area of far more immediate concern to this volume—a discussion of the apparent terms of the tradeoff between classes of societal benefits, and derivatively between social, economic and political endeavors.

TERMS OF THE SOCIETAL TRADEOFF: THE CORRIDOR CONCEPT

That the archetypical societal form (differentiated, hierarchical and asymmetrical) has persisted for so long, and replicated itself so widely, is testimony to the fact that it must provide its constituents an adequate level of satisfaction. In the generic terms we have been using, this means that these systems deliver a threshold level of benefits on one or more of the three dimensions: placidity, gratification and individuation. That is, a societal system must provide its members with peace of mind, or with emotional or functional significance (non-redundancy), or with the opportunity to satisfy appetites—or with some combination. Though our basic benefit dimensions were perhaps not the best that could be defined (and readers may indeed have better-schooled ideas about generic societal benefits), the essential point is this: an adequate (threshold) level of satisfaction may be gained by the societal resident on any of several benefit dimensions, so that these dimensions become intelligible to us as either substitutes or complements.

The working proposition that we shall be exploring here, then, is that there is a constant (and dialectically driven) *tradeoff* among the several dimensions, at least potentially. A corollary is this: most empirical and historical societal systems have tended to exploit the tradeoff potential only incompletely, and have instead tended to concentrate societal benefits in one dimension (or "corridor") at the direct expense of the others. Here is the crux of the societal problem and also the seed of its solution. Now, in the general case, this concentration on a single corridor may be effected in several ways:

1. It may be that the concentration on a single benefit set reflects the situation where neither of the others is available.
2. It may be that individuals simply find sufficient (threshold) satisfaction in one corridor, so that there is no great interest in seeking satisfaction through the others.
3. It may be—for the short run only—that individuals have been coerced or otherwise conditioned to concentrate on one benefit set and ignore the other dimensions of satisfaction.

We shall now briefly explore the implications of these alternatives, first with respect to our three dominantly spiritualized societal referents.

Systems of Spiritual Significance

Initially, conditioned-programmed systems represent a situation where spiritual benefits (*qua* placidity) predominate primarily by default, in that gratification is limited by factors beyond the collectivity's control, and where other essentially exogenous factors set a definite limit on individuation. From our previous discussions of the societal implications of conditioning and programmation, the reaons for this configuration should be quite clear.

In the conditioned system, spiritual benefits predominate for the simple reason that such systems are found consistently in unattractive (e.g., marginally productive) locations, usually in company with undifferentiated organizational structures in all three sectors (economic, social, political). Because of these factors, the material benefits of association with such a system are small. Subsistence *per se* implies a lack of any surplus of production (i.e., basic survival needs exhaust the productive supply). Now a material surplus is to a certain extent a prerequisite for societal elaboration, and it is definitely a prerequisite for cultural sophistication. Conditioned systems lack both. Especially, they lack the collective cognitive sophistication to develop the kind of axiological base required to produce affective behavioral referents (and to maintain and elaborate these through sentimentalization). As a result, societal relationships will tend to take the most rudimentary form, and there is likely to be little concept of the worth of the individual. For there is neither stimulus nor tolerance for differentiation and also very little display of properly solicitous or duty-driven behavior. Rather, social relationships will tend to be damaged by the material rigors and deprivation. They will not be the affectionate and dignified interchanges posited by Rousseau for his "noble savages".* Rather they are likely to be the expedient, abrupt and often cruel counterrelationships that ethologists find in hard-pressed animal collectivities. In short, then, material and social benefits are contextually denied the members of primitive conditioned systems, leaving the realm of spiritual benefits as the primary source of societal satisfaction.

To the individual whose behavioral repertoire is exhausted by con-

---

*The society that Rousseau visited was not "primitive" in the sense that we are using the term here. Nor are the exuberant, free-loving Pacific island societies that have long been favorites of many anthropologists.

ditioned or associative activities, there is no distinction between sequence and causality. For example, for the primitive intellect, if a cat crosses his path and later that day he becomes ill, the cat may be seen as the "cause" of the illness by virtue of the placement of the events in linear time. If he later recovers, after perhaps having seen a rat on the floor of his hut, he may then attribute the recovery to the rat's presence. The cat thus becomes a hostile agent, to be appeased, and the rat becomes a guardian to be thanked. Or the cat may be thought to be held at bay by the rat. In this case, the man might produce icons of the rat and the cat, and keep the former always between him and the latter. Eventually, when enough such cause-effect events have run their course—with causality always generalized to sequence—he may have a very large repertoire of "gods" or spirits to either propitiate or court. Moreover, he may have devised specific procedures for propitiation or appeasement, etc., for each of the gods or sets of spirits. These rituals may precede routine events such the season's hunt, crop sowing, fishing, etc., or they may be invoked by some extraordinary event (e.g., death, war, earthquake). In this way (and there are certainly many other scenarios we could have developed), we eventually arrive at a very elaborate spiritual sector associated with otherwise very simple and primitive systems. And in virtually all cases, these rituals may be thought of as influencing events that are effectively beyond the full control or comprehension of the system members. When we note that the terms *influence* and *control* are not synonymous, then there is really no contradiction here. Nor, in reality, do the primitive magic makers have pretensions to anything beyond probabilistic suasion of the gods and forces whose victim or instrument they have become, or to which they have resigned themselves. Unlike King Lear, the primitive collectivity does not rage against the elements, but courts them.

The ritualistic base of primitive, conditioned systems is simply the collection of all these assuaging (courting) techniques, cast in a form capable of being transmitted from generation to generation by rote. These techniques become imperatives for the individual or group. When they are accomplished (e.g., when the rain dance is done, when the parrot's feather has been burnt, when the sacrifice is made), then the primitive mentality is at peace, for there is nothing more that he can do. He is fundamentally a reactive creative who feels that both cause and control always rest in some other realm. Hence resignation, and hence cognitive placidity. Someone resigned to his destiny, and conceding himself to be non-autonomous, is less likely to suffer anxiety than someone accepting responsibility not only for himself, but quite presumptuously for nature as well. But the limited purpose of ritual must again be stressed. It is to influence events, not to control them. Thus, non-autonomous man does not lack influence over his life (or ritual would be purposeless and hence non-existent); all he lacks is final authority should it ever come to a test of wills with some superior spirit. And as a note on the theology of pantheism, it must be mentioned that in the face of capricious if somewhat mutable forces, there is a natural tendency for the society to dwell entirely on the present. Thus, both reflection and projection, both strategy and self-determination, are largely gratuitous in conditioned contexts. To men overburdened with conscience, confused by intellectual demands and harassed by competition and self-doubt, it is small wonder that so-called primitive systems hold such fascination.

When we come to institutionalized systems, the concentration on spiritual benefits (at least for the majority of the population) occurs for reasons somewhat different than those we associated with the conditioned societal type. For the most part, we have already explored these: (1) the tendency for ascription to strictly limit material benefits for all but the system's elites; (2) the tendency to discourage entrepreneurism and economic development (for both the lower and the upper classes); (3) the tendency for institutionalized systems to associate themselves with agrarian culture, where the agricultural technology is usually inefficient [6]. As a result, the material stock of institutionalized systems may not be great to begin with, and the heavily skewed distributions leave most of the population rather poor. But there is, given the spiritual emphasis of such systems, considerable expenditure on sacerdotal display, particularly in the form of cathedrals, ornaments, etc., and other artifacts that represent collective "consumption" rather than investment. Finally, the congregationalized social sector of most institutionalized systems generally constrains the spontaneous human interchange we find in sentimentalized systems, and dogmatic prescriptions are almost always at work to repress sensual impulses.

Thus, the individual becomes close to God, as it were, at the expense of intimate ties to either family or environment. Now, as an ancillary point that we made much of in an earlier chapter, it must be expected that as one rises through the hierarchy of institutionalized systems, material benefits (in the form of gratifications via status, wealth or power) begin to increase relative to spiritual benefits. As a result, it is common for piety of the lower classes to coexist with (often veiled) venality of the elites. Thus, with very rare exceptions the distribution of spiritual benefits across a population is uneven, with material benefits beginning to dominate more and more fully as one approaches the apex of the sacerdotal hierarchy.

Here then we have the source of the "sentimental" view of the lower-classes expressed occasionally by elites. For, indeed, religion is the "opiate of the masses", but not quite in the sense that Marx suggested. For the modern Marxian, the organized religions are seen performing a dysfunction only: the constraint of the lower-classes from the pursuit of economic advantage. But the Marxian generally does not comprehend the fact that proscriptive axiologies offer the individual palpable psychological benefits (viz. placidity), for he himself is rooted in the material domain. Thus, he is often confused when his rhetoric fails to arouse the proletariat.* When this occurs, he tends to blame either the people (for their lack of interest or indignity, etc.), or the church for its supposedly conscious and deliberate reduction of men's ambitions. To a certain extent, then, this is why the Marxist hates the church, and also why he often feels such distinct disdain for the very people it is his mission to save—the lower classes.

But those who have heard the siren song of resignation—who live lives free of striving and strategy—do not share the materialists' motivations, and therefore do not appreciate their message. For the plea of the materialists ultimately comes down to this: abandon the benefits of placidity in exchange for potential material gratification. Thus, both the capitalist and Marxist apologists are really asking that residents of institutionalized systems merely rob Peter to pay Paul (with the materialists fundamentally unaware of Peter's existence). Thus, what appears to be an unqualifiedly advantageous move to the advocates of materialism may often appear to be a dubious tradeoff to the person enjoying the benefits of placidity. And the end result is the truly pathetic question that the capitalist and socialist evangelists so often ask: How can these people be so happy when they are so poor? And, in their turn, the people might ask this question: How can these people be so rich when they are so stupid?

Thus, much of what we said about the causes and effects of resignation in the conditioned system applies to the institutionalized system as well. The major difference is that whereas the primitive collectivity resigns itself to a multitude of spirits and numinous forces (the pantheistic gallery), the member of an institutionalized system resigns himself to the will of some anthropomorphic God (a Jesus, a Buddha, Allah, Jehovah, etc.). The "rituals" of sacerdotal religion are thus comprehensible in essentially the same terms as the conditioned rituals of the

*That many Marxian apologists have these problems of perspectives does not mean that Marx himself had them. His "materialism" was an analytical tool, and very much broader than the prescriptive and dogmatic materialism that some of his more ardent followers espouse.

primitive: both are attempts to appease or influence an autonomous but mutable exogenous force. In the primitive system, spiritual benefits were victorious by default, there being no real opportunity to elaborate material benefits or individuation. In the institutionalized system, however, spiritual benefits alone may yield sufficient satisfaction for the individual, so that his interest in pursuing gratification or individuation is dampened.

To comprehend the attraction of existentialistic systems, we have to extend our logic somewhat and consider the process of absorption. Absorption yields placidity in that the individual gradually is removed from reality, and gradually comes to the point where his needs are both generated and met in isolation. In short, the individual has resigned himself to *himself*, rather than to some exogenous force. Existentialized individuals, as we earlier explained, become ascetic by default, and may deliberately ignore material benefits and actively seek to minimize gratification (through adherence to the negative socio-economic correlation we discussed in the previous chapter). By the same token, association with individuals—the source of social benefits—becomes contradictory to the mission of existentialism. The logic here is clear, if intricate. In virtually all cases, existentialism urges that the individual rid himself of the concept of "self" as opposed to others. In the theophilosophical forms, the "I-it" distinction—the subject-object distinction—is condemned [7]. The individual is asked to consider himself as one with God, nature and his fellows. Thus, somewhat paradoxically, solipsism does not lead to individuation, but away from it. In short, the individual absorbed in himself loses himself, for there is no opportunity to gain identity through reflected differentiation.

In virtually all cases, then, existentialistic societies become loosely articulated collectivities where individuals become absorbed by highly personalized, idiosyncratic "religions". The ascetic monks of Buddhism retreat not only from the material world, but from the social world as well. The Protestant mystics do the same, as do the Sufis and certain incommunicado and penurious orders of the Catholic Church. The psychedelic syndrome accomplishes the same thing. Through the medium of drugs, the individual's concept of self is removed and he is left awash in a sea of basic sensations. There is no attempt to communicate these sensations or generalize them, as they are fundamentally non-cognitive and therefore incapable of articulation. The drug "culture" thus becomes a kind of parabolic phenomenon, perforce turning the individual inward even when it promises associations. But the mystic, psychotic and acid-head all have the benefits of cognitive placidity to offset the material and associational deprivations, at least while the spell lasts. For they, like their counterparts in the conditioned context, exist in a world perceivedly beyond their control or comprehension and therefore obviating both anxiety and analysis. The difference is simply this: members of existentialized systems become absorbed by (and therefore resigned to) *ab intra* psychogenetic engines, while the forces driving conditioned systems at least have names and symbolic form.

In summary, then, the conditioned, institutionalized and existentialistic systems gain their appeal for the individual essentially through some form of resignation. All offer the individual the opportunity to absolve himself from responsibility for his own welfare, and therefore from cognitive demands and directed intellectual effort. The processes by which this is accomplished are different, but the result is essentially the same: a societal system with a relatively low order of material benefits (weak productive sectors, few amenities) and limited social benefits (highly proscribed emotional and sensual relationships).

Systems of Social (Associational) Significance

The sentimentalized and cerebral-sapient systems (see Table 6.2) are postulated as offering the individual strong opportunity for individuation, but limiting his pursuit of material gratifications and placidity. In short, the primary benefits of membership in a sentimentalized or intellectualized system stem from association with one's fellows. Both these systems offer the individual the benefit of conceiving of himself as *non-redundant*, of having a tangible if contingent identity.

Initially, it is necessary to talk about non-redundancy in the broader context of association. To a great extent, spiritual benefits (placidity) may be obtained by the individual operating in isolation. So, as we shall later see, may certain forms of material gratification. But the benefits of individuation always imply the presence of other human beings. In the case of sentimentalized systems, these other human beings offer the individual a sense of identity through the exercise of emotionally significant symbiotic relationships. In cerebral systems, the sense of identity proceeds through the process of successive differentiation, i.e., specialization. Both these processes are very complex and both deserve some treatment here.

The process of specialization—which operates in cerebral systems—acts to make the individual different from others with whom he might have contact or of whom he might have awareness. Often, this process becomes comprehensible in terms of reflective complementation. Complementation means that individuals search for an unoccupied *functional niche*, and thereby move the system toward the completion of some set of skills or interests. Suppose, for example, that an individual is aware of a subsystem exhibiting the following set of skills: (A,B,—,D,E). Under the complementary calculus, that individual would seek to occupy the empty element (C). The complementary strategy thus seeks always to complete a sequence. The utility of such completed sequences lies in its synergistic potential, where a collection of complementary parts allows the whole to achieve a value greater than the sum of the values of its parts, or to "complete" itself. Complementation is the key to such systems, and in cerebral-sapient contexts the opportunities for specialization increase constantly; that is, the areas of legitimate functional specialization tend always to increase in number, becoming ever finer in distinction.

Within the specialized system—be it intellectual, athletic, industrial, etc.—the individual's identity is in most cases coextensive with his functional position, as in the Platonic societal scheme. That is, one comes to know himself (both directly and through reflection) by what he does. One is, for example, the second-string tailback on the football team; one is the orthopedic surgeon in a medical group; one is the statistician on a corporate research team; one is the tail-gunner on a B-17 bomber. In such cases, the cause of system efficiency is served directly to the extent that the individual's identity is coextensive with his function, and the depth of identity is dependent on the perceived importance or worth of his contribution. And, of course, the value of a functionally dependent existence for the individual depends directly on the extent to which he feels that he is non-redundant.* For this reason, functional specialization serves better in more restricted, highly technical areas than, say, on an assembly line or even in an army platoon. If one knows that there are many others who could replace him with little disruption, then the psycho-social utility of specialization quickly erodes, and there must be an attempt to motivate the individual's performance through the use of either emotional or material leverage. [8].

There is another modality through which identity becomes available via specialization. This is the process of *pure differentiation*, a vehicle entirely consistent with the orignative behaviors we expect to find at the fringes of the cerebral system. Complementation, as we have seen, was predicated essentially on the individual's searching for a functional niche, *given* an uncompleted functional sequence existing somewhere. Differentiation, on the other hand, does not seek complementation or synergy, but uniqueness. This guarantees non-redundancy, and thus may often be an end in itself, independent of any

*Professionalization, as a collective phenomenon, becomes comprehensible in these terms.

functional utility. Artists, for example, often deliberately seek to evolve an unprecedented style, irrespective of any other criteria. Some artists thus gain their identity (and to a certain extent this is true of speculative intellectuals as well) to the extent that their product is sufficiently unlike what has been produced elsewhere.

In addition to the processes of differentiation and complementation, which lead to individuation, there is individuation via *symbiotic clustering*, which we associate with sentimentalized systems. Individuation also revolves around the concepts of assocation and non-redundancy, but in a very different way. Note that association in the arena of specialization (through either complementation or differentiation) demanded that the individual be somehow different than others within his purview. Thus, we do not look here for societal satisfaction stemming from emotional ties with one's compatriots or colleagues. The associations are mainly functional, largely because of the substantive (and self-) preoccupation we associate with individuation via specialization. In short, a collegial system does not necessarily imply congeniality among the members. Indeed, the relationships between colleagues bound in a functionally synergistic relationship may run the gamut from benign indifference to jealousy or outright contempt. It is a perhaps damaging myth that functional efficiency among a group of specialized colleagues—be they academics or athletes or surgeons—depends on their being friends as well as operative complements.

But, as is clear from our earlier and quite brief definition of symbiotic clustering, we are here concerned about benefits of individuation deriving from the development and maintenance of emotional ties. The key, again, is the concept of non-redundancy. Non-redundancy in the sentimentalistic system is available largely without great sacrifice, and is an essential appeal of such systems. Recall that sentimentalized systems emphasize affective behaviors, and applaud spontaneity within the limits provided by the prescriptive axiology. Now, it is this latitude that generally gives individuals the opportunity to cluster in small units, and to find a *localized* identity. The calculus that operates here is not the objective or reflective functionality of the specialized system, but an emotional engine. We thus replace functional utility with affection, love or mutuality as the basis for association. One thus comes to know oneself as part of an association predicated on emotional significance: the commune, the family, the love affair, the friendship, the comradeship.* Such associations may exist independent of any objec-

*Many of what I have referred to as "emotional" relationships may in fact have distinctly biological or "instinctual" roots. The sociobiologists, as their field matures should be able to provide significant definition in this area.

tive or purpose, and simply yield the intra-species contacts that virtually all cognizant life values, at least to some extent. Thus, individuals who have no particular distinguishing skills or attributes—who are functionally redundant in an objective, purposive sense—now have the opportunity to become individuated by virtue of an ostensibly unique bio-emotional relationship. To some extent, both emotional and sensual (sentient) license operates within the confines of the sentimentalized system. As a result, the individual would appear to be able to exercise satisfactions on two dimensions simultaneously: he can gain the benefits of identity through his emotional relationships, and to a lesser extent gain gratification (release of frustration) through the exercise of limited sexual or other sensual urges.† Yet, to some extent, sentient (sensual) and emotional release may represent essentially competitive benefit sets. In short, it is suggested that the exercise of sensual impulses may act to the detriment of the individuation which emerges from emotional release. In fact, as we suggested earlier, it is the sentient license that, when exercised too freely, turns the sentimentalized system into an existentialized one.

It is important to understand the difference between emotion and sentient release. Emotional release always has an *a priori* referent, usually some component of the prescriptive axiology. It "prescribes" those relationships from which the individual may derive legitimate satisfaction, or at least specifies the imperatives of the societal system [9]. These imperatives may be very sweeping, such as the dictate that one should love the members of one's family. Or they may be ritualized and formal, such as the rule that demands that members of the high school social club always come to the aid of a fellow member. These prescriptions all serve to give substance to the symbiotic associations that predominate in sentimentalized systems, and their observance lends the individual a sense of both psychological and emotional satisfaction. In short, emotions are always axiologically conditioned, whereas sentient impulses may be thought of as developing without cognitive mediation or reference (i.e. they are equivalent to the Freudian id-level drives).

It is thus easy to see that life in a sentimentalized system may be marked by some conflict between sentient and emotional impulses, and thus there is an opportunity for the same type of "conflict of conscience" we might expect in institutionalized or rationalized systems. Social satisfaction via individuation, in the symbiotic context, may be amplified as the union persists. Amplification here means merely an

†Note that it is these characteristics of sentimentalization that field anthropologists find so compelling among the island cultures that they visit so frequently, and that we tend to think of as primitive. From our perspective, most "primitive" systems anthropologists have studied are really sentimentalistic.

increase in the intensity of the interdependence, and therefore an increase in the probability of the association's persistence. But when the prescription to form emotional unions occurs in company with sentient license, there is ample opportunity for conflict and for reduction of purely social benefits. For if the calculus by which one gains emotional satisfaction is predicated on the observance of the prescriptive axiology and on the duration of some union, the calculus of sentient license is *titillation*. Titillation is an instrumental form of gratification, and belongs therefore within the context of secularized systems. But its entry into sentimentalized systems has been a recurrent historical fact. It might happen, for example, like this. One is told to marry for love, and expects thereby to be granted identity and emotional satisfaction. But in a secularized society, love may often come to mean nothing more than a sensual (e.g., sexual) phenomenon, lacking in any axiological overtones. To the extent that marital relationships are actually predicated on titillation, their probability of persistence is very low. For titillation is a non-concatenative process: for a given level of gratification to be obtained, there is the need for constant variation and elaboration. Therefore, relationshps based on gratification of appetite generally *decrease* in ardor and satisfaction with time, whereas relationships predicated on bio-emotional symbiosis are expected to generally hold greater satisfaction the longer they persist.

Beyond a certain point, then, titillation can be had only by way of successively more exotic experiences, which eventually may become transmogrified into perversions (as the sentimentalized system begins to give way to the secularized system, and from there to the compulsive-existentialistic state). Well before, however, the axiological and therefore the emotional significance of relationships has disappeared, and so have the societal benefits by way of individuation. What remains is only the benefits of absorption (in the existentialistic state) or the material benefits via titillation. Therefore, the emergence of sentient license may lead to a dissolution of individuation rather than its elaboration. In particular, it is improbable that marriage partners can exhaust each other *both* emotionally and sensually. The attempt to do this may simply lead to a sub-satisfactory return on both the emotionally and sensually. Thus, the failure to distinguish carefully between social and material benefits (in this case between emotional and sensual satisfaction) may often lead to a situation where there is both guilt *and* repression, an anomaly indeed within the context of the sentimentalized system.

In summary, then, the sentimentalized system offers the individual a strong set of social benefits, among them succor, sustenance, affection, abetment, tolerance, consideration, companionship. All stem from the tendency of individuals within sentimentalized systems to form associations that give expression to emotional (or sociobiological) urges, and therefore lend a sense of societal satisfaction even in the absence of material sophistication or spiritual comfort. Indeed, it is possible that the strongest symbiotic associations emerge in the face of quite harsh material conditions (as, for example, among men in army boot camp or prison blocks, or among refugees), and also when spiritual benefits are at their lowest (as when, for example, a family or group suffers the death of one of its members and therefore has reason to doubt divine interest in its welfare).

On the other hand, individuation in the cerebral (professionalized) system will tend to stem from either differentiation or complementation. In either case (as above), the requisite for concept of self is the existence of a social reference in the form of other individuals. Complementation usually leads to synergistic associations. Differentiation, on the other hand, leads,to situations where the individual's concept of self (his identity) is obtained by nurturing some uniqueness, in behavioral properties, achievements or interests, etc. But here also, the concept of self is reflective, and employs other members of the society as a reference [10]. Thus, the very popular distinction between inner-directed and other-directed individuals is deceptively simple when we try to apply it within a formal societal context. [11].

As a final note, the emphasis on social benefits—via the mechanism of individuation—tends to reduce opportunities for placidity. Particularly, the maintenance of either emotional symbiotic or complementary-synergistic relationships demands that the individual remain alert to the shades and nuances of the temporal, tangible environment. In short, resignation is the surest way to defeat individuation in either the sentimentalized or the cerebral-sapient context (though, when emotional relationships do collapse—or when his career collapses—the individual might rather quickly turn to the placidity which resignation permits; both jaded lovers and professional failures are ready candidates for religion or existentialism). And it should be noted, also, that neither sentimentalized nor cerebral-sapient systems offer great opportunities (or motivation) for material advancement or differentiation, a point that should be clear from our discussions of the socio-economic calculi in the previous chapter. We must rather, turn to our last category of systems to find an emphasis on material concerns.

Systems of Materialistic Significance

As we have shown, it is no accident that rationalized and secularized systems are essentially materialistic. In the case of rationalized systems, there was the conscious effort to displace transcendental apriorisms (the elements of the proscriptive axiologies) with purportedly rational or "natural" axiomatic platforms, where the numinous or moralistic arguments were displaced by arguments of logical persuasion. This transformation reduces the opportunity for enjoying the placidity that attends essentially spiritualistic (de-empiricalized, proscriptive) contexts. Indeed, in our earlier analysis of the capitalist and socialist systems, we found the axiomatic dictate that the individual be *assiduous* in the pursuit of material ends. The difference was that the capitalist did so for his own direct benefit, whereas the member of the socialist system was expected to direct his efforts toward collective ambitions of some kind.

Again, assiduity demands attention to the generation or exploitation of contextual opportunities, and hence the rationalized individual is *a priori* directed toward the temporal and material as opposed to the spiritual or social domain. In the most basic sense, historical rationalizations of the Western world deliberately sought to drive the individual away from the placidity and acquiescence associated with the ascriptive, enervating Catholic socio-economic dogma. The net result is that both capitalist and socialist sympathizers were edged toward an existence defined almost totally in terms of material parameters. And in both these socio-economic contexts (though more thoroughly in secularized systems) the material dynamics become more or less comprehensible in terms of some competitive algorithm.

A competitive context obviates the social benefits we associated with the sentimentalized system. Specifically, in a materialistic world, emotional symbiosis becomes a luxury that societal members can hardly afford, or that they may exercise only at the expense of their material welfare. In most rationalized systems, there are usually prohibitions against certain types of emotional attachments (e.g., against homosexual or bestial sexual liasons; against miscegenation). Even a parent's relationship with his children may be one of didactic rather than emotional significance. The parent's duty may be to prepare the child to compete in the material arena, and the giving or acceptance of affection may often be predicated on certain performance criteria (e.g., getting good grades in school). In certain socialist systems, the social context is sometimes even more severe. For to stress the good of the collectivity beyond the good of the individual (or the good of the family) often means that natural emotional ties must be consciously and formally weakened. This was carried to its logical conclusion in National Socialism, where children were recruited as agents against their parents and family, and even removed from the family for socialization by the state. To some extent, the radical reform movements in the United States and the developed western nations also try to drive a wedge between parents and children as a step toward communalization, as do some of the regressive attempts at institutionalization (e.g., the Unification Church of the Rev. Sun Myung Moon).

If materialism disrupts the possibility of emotionally symbiotic unions within the family, extra-familial associations are also disrupted, further dampening the social benefits available to members of rationalized systems. For example, the lesson of scarce resources is lost on no one in the capitalist world: children are taught that only one team in any contest can be a winner; members of companies are informed that they are in competition with their colleagues for promotions; labor sees itself in competition with managers and owners for a proportion of the firm's total return. College professors are in competition for full professorships, military men are in competition for rank, etc. Even love and sex become a matter of competition, as in the somewhat pathetic song that announces an engagement with the lyric: "You belong to me". Neighborhoods become collectivities of geographic significance only, for there is competition there for the largest house, the greenest lawn, the most awesome automobile. Parties become opportunities to do business rather than just socialize; we go to church to be seen, or perhaps to convince potential clients of our piety. Wives become social assets in the most material sense— as an adjunct to the corporation's welfare. All readers, by now, should be familiar with such plaints, as they are as ancient as the ethico-religious prophets, and as recent as the American hippies. Yet a vast human population continues to dwell in systems that offer neither significant spiritual nor social benefits. The immediately significant question is, of course, this: what types of gratification are sought on the material dimension? There are as many attempts to answer this question as there are philosophers or social scientists, from Epicurus to Seneca to Augustine to Hume to Smith to Maslow, and on and on. We cannot approach exhausting the list, nor would it be useful for us here to attempt to produce another list of fundamental human demands or attempt to restructure the popular hierarchy of needs. But gratifications—or categories of material benefits—fall into at least three general categories: (1) survival *requirements* (adequate nutrition, water, etc.); (2) *appetities* (lust, rapture, deference, etc.); and (3) *licenses* (curiosity, mobility, expression, etc.).

Within each of these categories there is unlimited possibility for elaboration. For example, shelter may be met very simply, by a portable tent, or very grandly as with the palace of Versailles or the Hearst castle at San Simeon or the mansions of the old Eastern wealth in Newport or the Hamptons. The requirement for nutrition, likewise, permits almost limitless elaboration, from a meal of stream water and corn cakes to a sumptuous feast at the Four Seasons or the Cafe Ritz. Appetites and licenses also exist on broad continua of elaboration, from the primitive to the exotic.

Knowing that there is a wide range of material benefits available to societal systems, the next question is how these benefits may be obtained by individuals or by collectivities. We already know the answer in essence—material benefits are distributed (or allocated) through the three great equilibrators of materialistic society: *status, wealth* and *power*. These are the commanding currencies by which material benefits may be acquired, and hence the primary vehicles for gratification. To a great extent (and this is perhaps a critical point), most material benefits may be obtained through *any* of the three means. What can be had through status can also be had through wealth or power.

The mechanism by which status acts to secure material benefits is imputational ascription.* For example, one may gain material advantages (e.g., credit, gifts, privileges) by virtue of those with whom one has social contacts, by virtue of one's manner or by virtue of one's personality or charm. In short, through the cultivation of a real or supposed social status, one may claim certain material benefits. In some cases this may mean cashing in status for tangible goods and services (the gift of an automobile from an admirer, a legacy or annuity or an income-granting position). The melodramatic literature of England and the United States, particularly in the nineteenth century, is filled with references to this process. In certain historical periods and to some extent even today, the pursuit of status (as an end in itself) has preoccupied whole societal subsystems or collectivities. Among the subaristocracy in eighteenth and nineteenth-century Britain, for example, the end of all existence was to become fixed in a particular social circle, and such ends were pursued to the complete abandonment of economic or emotional attachments. To a certain extent, much the same was true in certain oriental societies, particularly that of medieval India. The point is that status may serve as an end in itself as well as a vehicle for the securement of tangible material benefits. It was a mark of some distinction in Britain of the period in which we are speaking to be poor but cultivated, and to make a great show of remaining free of the taint of business or commercial interests. Patches on one's clothes and a certain air of respectable malnutrition are still requisites for entry into certain cliques, and in many cases the badge of dignity in a supposedly otherwise crass world.

*But of a casual kind, not the legislated ascription we earlier associated with institutionalized systems.

The relationship between wealth and the acquisition of material benefits is more direct and less interesting: one simply purchases what one wishes, through credit or cash. Goods and services imply a standard consumer transaction. Certain privileges may be purchased through bribery or subordination. Power involves the even simpler medium of preemption: one acquires one's ends through direct or implied coercion. But neither the issue of what constitutes material benefits nor that of the means of acquisition is as compelling as the issue of their perceived value and the elaboration process itself.

Why people pursue material benefits is a question we have already partially answered: because, within the confines of rationalized and secularized systems, spiritual and social benefits are scarce or highly diluted. Therefore, material benefits predominate by default. We also know how this situation might have originated in the first place. The emphasis on spiritual benefits associated with institutionalized systems (which usually are the predecessors of rationalized systems) was dampened in the transition from a proscriptive axiology to an axiomatic platform. In the same way, the social benefits associated with sentimentalized systems disappear in the transition from a prescriptive axiology to either an axiomatic platform or an instrumentalistic (apriorism-free) context. In both cases, what remains for the societal members is the material corridor, the social and spiritual sectors having been closed off.

Now, once the spiritual and social benefits have been eliminated from serious consideration, the question we must answer is this: why do individuals seek to raise their material benefits beyond the level of adequacy? From the standpoint of our analysis, there must be three answers to this question, each of which is a plausible explanation within certain limits.

The first answer pertains to rationalized systems. There may be some *a priori* predicate that argues against the individual's ceasing to pursue material benefits beyond some certain point. We have already suggested both an axiological and an axiomatic argument of this type. For example, it was a dictate of the early Protestant axiology that one could not legitimately cease striving to obtain wealth except at the peril of one's salvation. In capitalist axiomatic theory, a similar exhortation exists against being prematurely satisfied with one's lot, suggesting rather that the welfare of the economic sector as a whole depends on the constancy and depth of acquisition and persistence of individual self-interest. And caveats against acquiescence and self-satisfaction are frequent in the literature of socialist economic theory and Maoist ideo-economics. In particular, the historical tendency of the peasantry to be satisfied with a subsistence economic standard, and to use what remains of their time to pursue social or spiritual activities, has long been a thorn in the side of the socialist and communist reformers and prophets.

A second motive for the pursuit of material benefits beyond any given level of adequacy pertains especially to secularized systems, and may also reinforce the axiological and axiomatic motives found in rationalized systems. This has to do with titillation as the calculus of material benefits in the appetite category particularly. Titillation—and the gratification that results from appeasing one's appetite—is a highly elastic function. That is, the very act of ceasing to elaborate one's activities or advance one's ambitions attenuates gratification. The two requirements of the titillation process are thus *more* and *different*.

Finally, the third reason material benefits tend to be pursued with constancy recalls the logic of the positive socio-economic correlation. There we saw that economic (i.e., material) benefits were the currency for purchasing other categories of benefits: placidity, social-emotional benefits, etc. To a great extent, this logic has direct relevance for us here. Status, wealth and power may be seen to be media for the attainment of identity, solace (cognitive security) or some suprabenefit. For example, power may be perceived as lending the individual placidity because he is able to control events, and therefore cannot be surprised by them. Status may be seen as lending a tenuous but tangible individuation, assuming that prestige or ascribed position is a corollary of the symbiotic and synergistic processes we discussed earlier. Wealth, also, may be seen as a buffer against the vagaries of the future, and therefore as a path to peace of mind.

To a great extent, then, one works to develop a stock of wealth that is expected to be sufficient to give a reasonable prospect of security in one's standard of living, despite illness, old age, infirmity, natural disaster (fire, flood, etc.), or even revolution or war. The fallacy, however, is this: anxiety, real or imagined, now becomes the motivator of efforts in the materialistic arena. Therefore, through material acquisition alone, one cannot gain the placidity that is directly available through resignation. But if one is *a priori* proscribed from being resigned—if one has been cautioned that acquiescence or acceptance is a sin, unmanly or otherwise unvirtuous—then the choice is only between anxiety, shame and guilt. One must avoid resignation and continue to strive, accepting the anxiety and expending the energy, or one must court shame through the "copping out" process that has recently become more and more popular. Thus, even to this end, the dismal dialectic continues to haunt us. Similar equivocations attend status and power. The maintenance of prestige generally demands the same assiduity as does the maintenance of wealth or authority. Those seeking social advancement or maintenance of position in the societal hierarchy must continually sharpen their manner, hone their wit or elaborate their repertoire of amusements, knowledge, or talents. Charm, like any scarce commodity, is not constant; nor of course, is power.

Again, then, we are forced back to the position we first took: that the properties of rationalized and secularized systems argue against acquiescence and for assiduity, against complacency and for constant alertness. For material benefits cannot merely be husbanded if the benefits of gratification—in any of their forms—are to be maintained at parity. Rather, it appears that the maintenance of satisfaction via material benefits demands their constant extension. So, in the domains of rationalization and secularization, we all join Alice and run as fast as we can just to stay in the same place.

In summary, then, granting that placidity and emotional satisfactions are not likely to be found in association with gratification, material benefits become ends in themselves, and life becomes a contest, mostly tribulation with rare moments of elation. Here the dialectic is most at home, for in the material world the engine of behavior is aspiration, and the motivation is the gap between aspiration and achievement.

As a more general point, we have seen that virtually all societal systems become *isomorphic* with respect to the corridor concept, i.e., all seek to provide a threshold satisfaction for societal members through concentration on one of the available benefit sectors. To this extent, we should be able to view our different ideal-types as *analogous*, at least in abstract. And to the extent that they are analogous, we should be able to develop some sort of model to describe the behavior of all our ideal-type referents in the same essential terms. This is the task of the next section, where we work toward making a *generic production function* as the operational point of analogy between otherwise very disparate phenomena.

THE MANAGERIAL METHODOLOGY: ANALOGY AND PRESCRIPTION

The essence of the corridor concept is this: societal systems have tended to emphasize only one set of benefits, to the direct exclusion of the others. Thus, were we to attempt to analyze the range of real-world systems within the framework of the corridor logic, we would expect to find systems of spiritual significance, systems of social significance and

TABLE 6.3: The Generic Benefit Typology

	Spiritual Benefits			
	High		Low	
Material Benefits:	Social Benefits		Social Benefits	
	High	Low	High	Low
High	[Certain expatriate societies]	[Feudalistic elites]	[Annuitied systems]	**Industrialized systems** (rationalized and secularized)
Low	[Proto-Christian communities]	**Patrist and subsistence systems** (conditioned, institutionalized and existential)	**Proto-communist and matrist societies** (cerebral-sapient and sentimentalized)	**Refugee systems**

systems of material significance. But we would not expect to find many systems of synthetic significance, where all three generic benefit categories were more or less equally accessible and elaborate. Were we to make an empirical test of this proposition, we would do so within the context of a simple typology, arranged so that the possible permutations of benefit configurations are set out as in table 6.3.

The point to the typology is this: it suggests that real-world systems will either tend to be normative within the confines of the corridor concept, anomalous or hybrid. The normative cases are the three major referents in boldface. It is here that we find agents of the ideal-types we developed in our previous chapters. These, as has been suggested, become targets of analogic interest, because each is restrictive in terms of the benefits it yields, and hence consonant with the corridor logic. The anomalous and hybrid systems cannot be explained by the corridor concept. For example, the "refugee" referent is anomalous because it offers adequate benefits on no dimension, while the other four "hybrids" all offer more than one exclusive benefit set. But the inference is this: the hybrid systems have accommodated only insignificant populations, and persisted for only short periods. The vast majority of mankind lives within the three major referents. These represent the situations where the behavioral base is the least complex. In short, the major (or normative) systems are explicit extensions of the various ideal-type constructs we developed in earlier chapters. But the other members of the typology (the five non-normative cases) represent exceptions to our ideal-type logic and thus represent the best and the worst possible worlds. Each of these deserves a bit of explanation.

History yields very few examples of systems enjoying simultaneously high (or even adequate) levels of spiritual, social and material benefits. The total proportion of the world's population within such systems is miniscule, yet they do exist, and show what is possible (if not what is probable). The best examples of such systems are probably a very limited collection of *expatriate* systems of the following type:

1. All gain their spiritual benefits through specific axiological predicates that differ from those prevailing within their context of residence (e.g., the Parsees).
2. All gain their material benefits through the mechanism of an extended and centripetal economy where each member of the system puts his resources at the disposal of all others, and where capital stock or interest is confined to a limited set of individuals (as with, for example, some of the Chinese communal families in the Western United States or Canada,
3. The social benefits occur because, being different in culture than the majority of the population in their region of residence—and often being subject to exogenous harassment—emotional-interactional ties intensify (as with the Basques in America, for example).

Now the fact of forced closure—due to ethnic or axiological differentiation—does not necessarily (or even usually) lead to the situation we have described. It may lead more easily to the worst of all worlds, where there are no adequate benefits whatsoever, as in certain refugee systems. The key to the situation where social and material benefits may be combined seems to rest in the conjunction of a materialistic axiology (or at least one permitting acquisition, if not enabling it), coupled with the fact of expatriation the deliberate decision to move elsewhere, often explicitly for economic advantage. It may be doubted that there is much spiritual placidity in such situations, yet there is often some resignation associated with the religious life of expatriate communities, if only to buffer them against their political defenselessness (as in the expulsion of the Jews from Portugal or the more recent expulsion of the Indian merchants from Uganda). As I am sure the reader is aware—at least if he is at all familiar with the enclaves of the Parsees, the Basques, and the merchant Jews and Chinese in both East and West—the simultaneity of all three benefit categories may be short-lived. Yet while they lasted, these expatriate systems have provided us with instances of man's most interesting societal achievements.

As an example of the situation where relatively high spiritual and high material benefits may go hand in hand—where there is some opportunity for both placidity and gratification—we might look to certain of the feudal elites (in locations as diverse as the Middle Ages in Europe, nineteenth-century Tibet and the several great African kingdoms of the thirteenth, fourteenth and fifteenth centuries). Such conjunctions of high spiritual and high material benefits are very rare, and are usually based on the inheritance of material privilege (through a deep-seated and relatively impervious ascription mechanism) coupled with a strong and dogmatic religious envelope. Then (as we explained briefly in discussing medieval Catholic economic dogma) inherited wealth becomes a privilege completely sanctioned by the axiology, and there is therefore no conflict between conscience and concentration. Placidity stems from the fact that wealth and privilege under such a system are supposedly God-ordained, and it is therefore almost a duty to accept the gratifications such wealth entails. In short, the material world offers gratifications that may be accepted without guilt, and the ascription mechanisms (e.g., inheritance, entailment) allow the individual to maintain his material benefits largely without effort. Thus, to some extent, we have the logic behind those few situations where men's consciences were at peace (placid) *and* where their milieu was commodious.

The case where we find high spiritual and high social benefits—in company with relatively low material benefits—is again rare, perhaps restricted to the very few communities which practiced the proto-Christian axiology in its purest form. As a general rule, these are Edenistic and Utopian ventures that lasted only a short time (as with, for example, the early Pilgrim communities in the United States and Leyden). We would *not* include here the more popular Christian communal societies. For example, in Quaker society, the relationships on the social dimension are somewhat strained and formalized, with the benefits of spontaneity (e.g., emotional symbiosis) largely proscribed. Essentially the same thing must be said of the Mormon system, where the spiritual sector often becomes significant primarily as a subordinate to the material sector, and where social relations are highly formalized. Some of the very modern counter-cultural communities illustrate a situation where high social and spiritual benefits are available—at least in rhetoric. The Hari Krishna movement and the communities of "reborn" Christians exemplify this pattern, and are perhaps as close to the case we are discussing as any modern societal systems.

The fourth of our non-normative system referents may be found in the theory if not the fact of certain types of professionalized systems. For example, with the enormous increase in the pay scales of academics, many individuals are able to enjoy a fairly high standard of living and yet still maintain fairly satisfying relationships with their family, colleagues and students, etc. However, when incremental material benefits become dependent on objective performance (e.g., publication of books and papers, consultancies), then material benefits may only be obtained at the expense of the social benefits. To a certain extent, though, the institution of tenure puts many professors into the enviable situation that existed (and to some extent still exists) in western Europe among the *annuitied* set. Here, the basic material sustenance of the individual is assured through no effort of his own, and he is largely free to pursue social benefits or "identity" through emotional symbiosis or dilettantism (where he has the leisure to pursue aesthetic or intellectual pleasures without paying an economic price). In annuitied systems, the individual may receive *adequate* material benefits without the necessity of competition or directed effort, and therefore is largely free to elaborate whatever benefits may exist on the social dimension.

Finally, there is the anomaly: the case where individuals are deprived simultaneously on all dimensions. They are not materially comfortable, they are not spiritually placid, and they do not enjoy strong, deep or spontaneous social relationships. To a great extent, this situation is restricted to refugee and certain migrant groups. It is perhaps best represented currently by the communities of urban Negroes in the northern United States cities, many of them first- or second-generation immigrants from the agrarian South. Now, when neither social, spiritual nor material benefits are adequately high, there will be considerable motivation to shift to another context, perhaps along one of the trajectories indicated in figure 6.1.

Through the mechanism of Black Power (a distinctly secular movement predicated on the strategy of martial, political or economic preemption) there is the attempt to develop a black business sector, increase transfer payments from the developed sector or otherwise secure increased *material benefits*. This process, in a sense, is merely an attempt to emulate the perceivedly instrumentalistic rules by which white America operates. And indeed, to the extent that the axiomatic bases of capitalism and democracy are declining in real influence (and thereby transforming U.S. society from an essentially rationalized into a secularized system), the strategy is an appropriate one. But it should be noted that while secularization is a superior position in that it at least provides the promise of material benefits, it will not lead to a Uto-

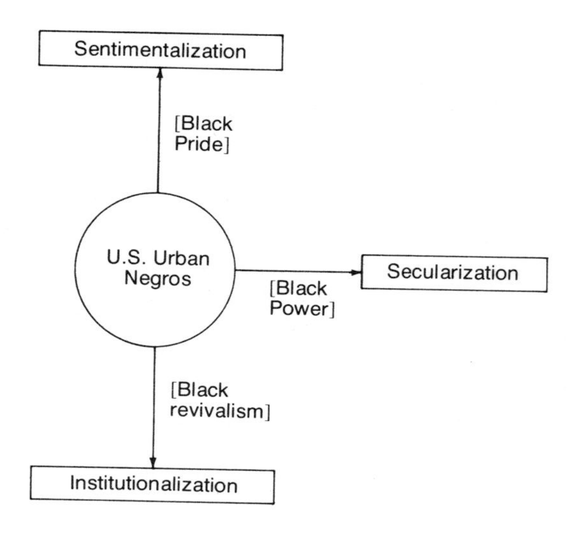

FIGURE 6.1: The alternative trajectories from the deprived system state.

pian state. For, as we have already sought to suggest, the price one pays for material benefits via secularization is a substantial denial of both placidity and social benefits. But, as some street-wise commentators have suggested, it is much better to be rich and unhappy than poor and unhappy.

Black Pride, on the other hand, seeks to engender the *social benefits* that were largely discarded in the migration north. Here, inspired by slogans such as "black is beautiful" and the concept of black "brotherhood", active attempts are being made to induce emotional symbiosis among members of the Negro community, and to forge a situation in which sexual, social and artistic references become available.

Black revivalism seeks to provide an institutional basis for placidity via theonomy, and in the process to offset attempts to seek solace through dysfunctional absorption (through drug addiction or alcoholism, etc.). Here we find, from the pulpit, a resonation of a proscriptive axiology very similar to that of the more vociferous Protestant sects. The result is a turning away from a hostile and deprived environment and a deliberate attempt to endow the susceptible population with the benefits of placidity via transcendental resignation.

Finally, there are dysfunctional surrogates for every benefit category. Gang membership yields a surrogate for social benefits; drug stupors provide an ephemeral sense of placidity; thievery and muggings, etc., provide a somewhat unsure but temporally satisfying source of material benefits. Proper societal management would try to discourage these surrogates through means we shall now explore.

An Analogic Construct: A Generic Calculus of Satisfaction

It may be inferred from our earlier arguments that the hybrid system states are less probable, less stable and less persistent than the normative referents (those where a single benefit category pretty well exhausts the system's repertoire). But it should also be clear that, with the exception of the "refugee" context, the *hybrid states hold more potential for satisfaction than do the three major or normative states*. For, while each of the latter provides *adequate* satisfaction through one or another of the mechanisms we have defined, our logic suggests that there is a *ceiling of satisfaction* associated with the exploitation of any single benefit category. In particular, each incremental addition to a cumulative benefit stock (of placidity, individuation or gratification) will yield successively less incremental satisfaction. In short, all our major system referents (which means all of the seven elaborative ideal-types we have worked with since the fifth chapter) become victims of decreasing returns to scale. Hence, all are *analogous* in terms of their calculus of satisfaction, as all may be modeled in terms of a generic production function like the ones in figures 6.2 and 6.3.

The horizontal axis in both diagrams refers to an incremental increase in one of our generic benefit categories (placidity, gratification or identity). Now, as we have shown, each of the generic benefit categories eventually transforms itself into a level of satisfaction for the individual, the satisfaction (like economic utility) being a dimensionless variable. Thus, marginal satisfaction refers to the net increase or decrease in satisfaction associated with the incremental increase in benefits during the last interval.

Now, the implication of these two curves is fairly obvious. The properties of Curve 6.2 are essentially those of the "learning curve" model we used in an earlier chapter. Initially, the satisfaction from the exploitation of a particular benefit builds quite slowly (in the interval from the origin to *a*). This may be thought of as "acquiring a taste". Beyond this point, in the interval from *a* to *b*, satisfaction accelerates, and eventually becomes roughly constant over the interval from *b* to *c*. Beyond *c*, decreasing marginal returns set in, and each additional increment of the benefit yields less satisfaction. The same sequence is reflected in the cumulative curve (figure 6.3), which is the integral of figure 6.2. Initially, cumulative benefits build slowly; then they gradually accelerate and eventually approach an asymptote. Beyond *c*, the radically decreasing marginal returns means that addition of more benefit incre-

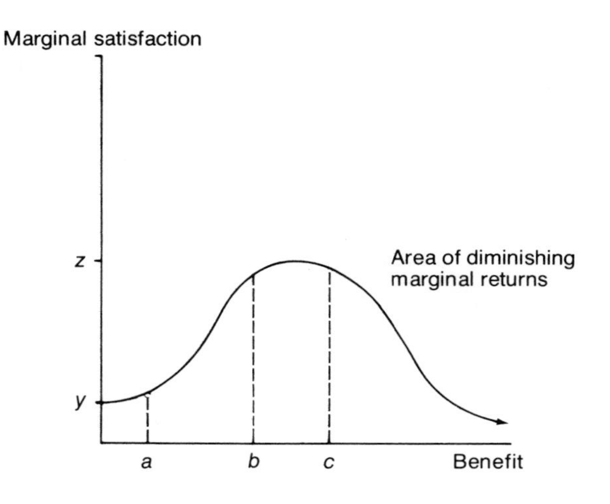

FIGURE 6.2: The marginal-benefit curve.

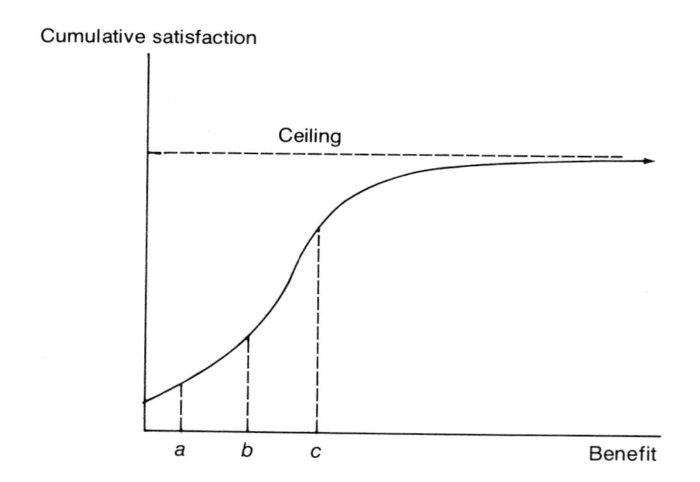

FIGURE 6.3: The cumulative-benefit curve.

ments yields very little increase, if any, in net satisfaction. That is, there is a definite limit to the level of placidity, gratification or identity one can accommodate.

We must also consider why one might continue to seek a particular benefit even when a ceiling has been reached. Certain types of behaviors fall rather neatly into this category, and we generally refer to them as *addictive*. In some cases, the addictive behaviors are driven directly by *a priori* predicates. For example, the tendency to acquire unlimited wealth was, in part, dictated by the axiological predicates of proto-Protestantism, and also finds some *a priori* justification in the axiomatic predicates of theoretical capitalism. Here, the behavior becomes in part independent of material satisfaction (the income cannot be spent as rapidly as it is generated), and is exercised because of the fundamental lack of alternatives [12]. For example, the man who has spent the major portion of his life accumulating wealth through assiduity and concentration in business affairs has not, in all likelihood, developed other sources of satisfaction. Again, then, *a priori* predicates tend to restrict behavior *by default*. A simpler explanation for addictive behavior is, however, the *treadmill* model. For, as the marginal satisfaction gained from each additional increment of a benefit declines, one must accumulate more and more of that benefit to maintain a past level of satisfaction. The alcoholic must, through time, consume a great deal more alcohol to get a glow on, having been inured to smaller quantities. The man addicted to the production of wealth must make ever grander and more sweeping deals—and gain ever higher returns in absolute terms—just to maintain his previous level of gratification. Thus, from our standpoint, addictive, compulsive and repetitive behavior may indeed be looked at from the standpoint of the model of decreasing returns. Only in the simplest societal contexts (e.g., among conditioned systems) can iterative behaviors be looked at merely as products of operant conditioning.

Of course, though we have spoken mainly of the case of material benefits, the concepts of decreasing marginal satisfaction and the treadmill model also pertain to spiritual benefits (placidity) or to satisfactions that might emerge in the social dimension through ego-differentiation or emotional symbiosis, etc. Yet we are not really finished with our analysis yet, for the corridor concept may be extended to levels other than that of the generic benefits. By way of example, we may look deeper into the situation of materialistic systems, and show that our model may operate at several different levels of analysis. The lower-level formulation of the corridor concept may be stated as follows: *given that the primary source of societal satisfaction is gratification*—so that a societal system is essentially materialistic in character—*there will be a tendency to concentrate on only one of the three subcorridors for obtaining gratification*: status, wealth, or power. That is, systems will tend to offer their members gratification through only one of the several societal sectors, the social, economic or political.

This proposition explains (1) why some societal systems, as aggregates, are predominantly social, economic or political in significance, and also (2) at least in part, a special kind of correlative concentration we find in most materialistic systems. Let us briefly explore these implications.

Once again, it is to be recalled that satisfaction within a materialistic system proceeds through gratification, and gratification may be obtained through the accumulation or exercise of status, wealth or power. But interest in, or opportunity for, exploiting any of these three vehicles is not equally distributed among systems. There have been situations where each of these has tended to dominate at the expense of the others. For example, in the United States, wealth is the predominant currency. To a great extent, men who seek status or prestige rather than financial gain are looked at somewhat askance, perhaps as effete. Similarly, American politicians are usually regarded with a very jaundiced eye, being presumed to be either crude or corrupt. Even for the American proletariat, politicians seldom carry prestige or even real presence. For the most part, like film stars and athletes, they are mainly a source of entertainment. As has been said, "the business of America is business."

In other systems, status is the predominant calculus, and may often by pursued at the deliberate expense of economic gain. For example, as we have already mentioned, there has long been a bias—often bordering on real antipathy—against business among certain of the minor British aristocracy. Entrance into commerce was regarded as rude, as were the often conspicuous displays of the nouveaux riches, the popcorn kings or the aspiring bourgeoisie. To a certain extent, some oriental cultures also regard the concerns of commerce or finance as less attractive than the world of manners, protocol and "dignified poverty". And in other nations, it is the pursuit of power that has prestige. For example, in Ireland, Portugal and Spain—the prescriptive Catholic nations—political interests monopolize attention, and one's political posture may be of more concern than either wealth or manners. To a certain extent, the same is true of some nations of the Middle East, and of some developing countries. The man of affairs in America might rightly ask why it's so important who rules a nation where there is really no industry, infrastructure or significant wealth. But the Irish patriot might ask how a man could be so shallow to pursue his business rather than the political interests of some party. And the Indian or English sub-aristocrat might wonder why anybody would want to "control" anything at all. In Ireland, and Portugal (and now in Lebanon and Angola, etc.), political interests and advantages are fought over at the direct expense of economic development. But for an Irish patriot, to gain rank within the IRA is no more or less important than a promotion is to an executive of an American commercial corporation, and perhaps no more or less satisfying than entrance into a more exclusive social set would be to a young lion in the British or Indian social hierarchy. In short, satisfaction is available through either the social, the economic or the politic corridor, and there is (in terms of the net satisfactions they yield) perhaps really very little difference whether one seeks gratification through attention to business, to politics or to refinement of one's airs and acquaintances.

In some societies, there will be more than one corridor available to an individual. Yet, as with collectivities, the individual may tend to occupy himself exclusively with one or another. For example, members of academic or clerical communities generally tend to have high prestige but somewhat modest economic benefits; that is, status is traded off against wealth. Most businessmen may be somewhat deprived in terms of prestige or admiration from the population at large, but may compensate for this by the size of their bank account and ability to support a level of conspicuous consumption usually beyond the professor or minister. Again, politicians, though generally lacking in prestige or great wealth, may enjoy the benefits of power. For example, though the chairman of the Securities Exchange Commission in the United States (the agency which regulates the sale and offering of corporate shares) may have a salary that is only a fraction of that of the corporate executives with whom he deals, the power he wields may make these high-paid executives quake, and there must be some satisfaction for the civil servant in that.

The corridor concept operates at a still lower level: that of the individual firm or organization. For example, some positions within an industrial firm lend the individual a certain power, but may not carry great prestige or economic reward—e.g., the private secretary of the corporate director or president may often determine which subordinate gets to see him, and may also influence the reception the subordinate will get. Other positions may carry high economic rewards but little real prestige or power, such as those occupied by public relations, advertising or marketing executives. Finally, there will be some positions of great prestige but little real power or income, such as perhaps the scientific ones in the corporation's research and development offices. Though their income will usually be well below that of the line executives, and they may really have no direct authority, they nevertheless may be held in some awe because of their reputed knowledge and expertise.

Well, the point of all this is simple: We showed earlier that most societal systems will tend to concentrate on a specific generic benefit category (placidity, identity or gratification). We then showed, using the material dimension as an illustration, that most societies will tend to concentrate on a single vehicle for gaining gratification. Finally, we showed that the same type of closure may operate at a still lower collective level, the individual organization. Thus (and this is the pivotal proposition) there is a tendency for virtually all collective phenomena to become susceptible to the engine of decreasing marginal satisfaction.

The Managerial Challenge: Optimizing at the Margin

But we are not irrevocably condemned to the treadmill (decreasing-returns) model. There are at least two options available to us—to be exercised in concert—which should improve the societal systems available to us: (1) to reduce the tendency for material benefits (social, economic and political) to concentrate at the highest level of societal systems, and (2) to develop "hybrid" benefit mixes by deliberately sophisticating our societal systems.

The processes by which the first tendency is realized are now familiar to us. In institutionalized (sacerdotal, aristocratic, feudal) systems, for example, we considered the mechanism of *ascription*. We saw that in this situation the economic and political sectors become subordinate to the social. The higher one's status (by birth, marriage, adoption, etc.), the greater the wealth one controls either immediately or potentially. And the presumption of political power or authority over other individuals or the system as a whole is also directly related to ascribed status. In a prototypical aristocratic system, for example, the elites have most of the prestige (status), most of the wealth and most of the political power. The point, however, is that here it is status that causes the concentration of economic and political prerogatives. Now, with the concentration of material benefits at the highest levels, the majority of the population has no real choice except to seek satisfaction through the exercise of spiritual or bio-emotional prerogatives. But to the extent that they are locked into a single corridor, they become susceptible to decreasing marginal satisfaction, from which the mixed benefit set of the elites may provide an escape. Much the same thing occurs within secularized systems, where power tends to concentrate, and with the concentration of power there is a clustering of status and material prerogatives. Here the clusteration or concentration may be even more severe than in institutionalized (e.g. aristocratic or feudalistic or sacerdotal) systems, and therefore the distance between the elites and the peasantry or proletariat reaches a maximum. In extreme cases, where a strong dictator or despot is operating, virtually all prerogatives may rest in the hands of a single individual. Here again it is necessary for the mass of the people to turn to either the spiritual or the social sector for their satisfaction, lacking effective access to the material.

To an expectedly lesser extent, rationalized systems (whether capitalist, socialist or quasi-communist) also suffer from concentration of prerogatives on the material dimension. We have already suggested that here prestige (status) tends to be positively correlated with wealth. And there is, of course, the fact that wealth may be used to purchase political influence, through subornation, bribery or simply electoral success due to the purchase of advertising, public relations, etc. Yet, within both capitalist and socialist economies, there is a much less severe concentration of material benefits, largely because of the tendency for such systems to use economic rewards as the major motivator for all forms of labor, but also because of the consensual political base. But, obviously, where material benefits predominate, and where there are just sufficient distributions to keep the lower classes from leaving the economic mainstream and turning to spiritual or social solace, we have problems of a different order. At any rate, it is clear that most members of socialist and capitalist systems feel that theirs would be a more satisfactory lot were material benefits to be less concentrated. We cannot argue with this logic, but it is a somewhat simplistic platform, for reasons which should now be clear.

In the first place, there might be no net gain in satisfaction were members of the lower economic strata simply to obtain higher economic or material benefits at the expense of spiritual or social benefits. But (and this is a critical point) in many cases such an exchange has already been made, historically and pervasively. For example, a majority of the population of the industrialized nations have already become victims of this tradeoff through the industrial revolution, which destroyed the extended family and the agrarian commune and weakened the church as a vehicle for transcendental reference. In many cases, the crisis in the work ethic, which is now seen as affecting most industrialized systems, becomes expressible in terms of this tradeoff. The increasing militance of workers—in both the private and public sectors—tends to be pointed in two directions, in accord with our analysis. First, there is the obvious move to gain a larger share of the national product and thus lessen the asymmetry in the distribution of material benefits. The positive socio-economic correlation with which most workers have been indoctrinated (initially to improve their productivity and inject assiduity, etc.), has now become a weapon to be turned against secular capitalism. This should have been predicted, for the relevant dynamic here is the titillation or treadmill model. Fundamentally, as the base of discretionary material benefits increases, the net satisfaction each wage hike brings is smaller (under the law of diminishing returns). Therefore, wage demands must accelerate just to keep the worker enjoying the gratification level he once realized at a lower wage.

We must remember that the concept of wage increases being geared to productivity means different things to management and to labor. Productivity to the manager means an incremental increase in real output. But productivity to the wage earner means incremental increase in satisfaction. Lacking a common concept of labor productivity, whether we employ the Marxian concept of "labor value" or the more algorithmic concepts of utility, etc., we are virtually bound to see wage demands get further and further ahead of real productivity. For labor does not work to produce a product, but to produce an increase in satisfaction. And the tendency toward decreasing marginal returns sets in to make each new wage advance fundamentally less valuable—in terms of net satisfaction—than the previous increment. Thus, the capitalist's attempt to despiritualize and desocialize society—in the name of efficiency and productivity—must eventually fail, because it was predicated on ignorance of the calculus of satisfaction.

But the social scientists have generated prescriptions that promise, in part, to repair the damage to worker motivation. Of particular interest are the trajectories shown in figure 6.4. As the reader will, I hope, rec-

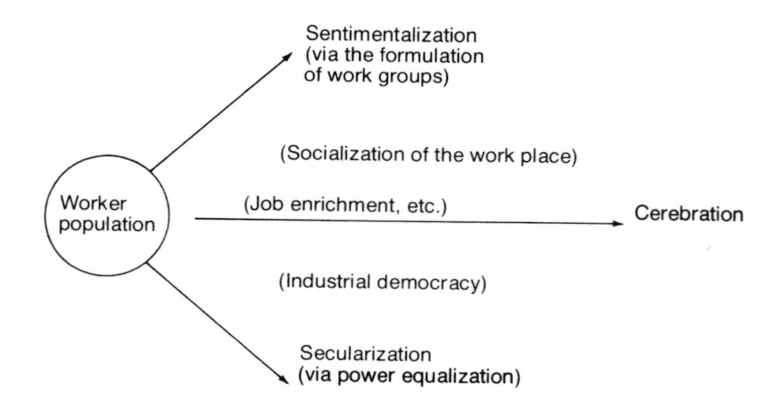

FIGURE 6.4: Prescriptive trajectories for industrial motivation.

ognize (and was the case with our analysis of the therapeutic trajectories available to the urban Negro in figure 6.1), there are both opportunities and perils here, in each of the three trajectories.

For example, sentimentalization is indeed inimical in many ways to material ambitions, and emotional relationships have a way of absorbing the individual. Indeed, we know that the benefits of emotional symbiosis may be directly related to the depth of absorption. Therefore, for the causes of industrial society to be served, there will have to be some constraints on the socialization process. However, as these constraints increase, there may be a perverse deepening of the intra-group relations, which may be amplified to the extent that the group now becomes the open enemy and adversary of the corporation. We must, then, always be careful of the lack of assiduity that attends socialization, and the tendency for emotionally symbiotic groups to increase their sastisfaction by *inviting* exogenous repression, or at least using it.

As for the trajectory that leads to the professionalization of the workplace, there is a potential problem here as well. It is in the process of professionalization—however ill founded—that we find the origin of the coercive union movements and the formalization of the adversary relationship between management and labor. In short, a man with the idea that he is not redundant becomes less tractable, reliable and manipulable, and so will contribute less on the whole to routine production. Thus, while professionalization may very well serve those organizations whose mission is to generate originative responses to unprecedented situations, it becomes dysfunctional when applied to more pedestrian occupations.

Finally, power equalization has two obvious problems. First, as we have already mentioned, it is possible for gratification through pursuit of power to replace gratificatio through material benefits as a societal rationale. In short, labor now takes on a distinctly political rather than economic significance, and the result is invariably the type of internecine conflict that one finds among states where industrial and material benefits have become subordinated to the constant-sum game of control and counter-control. Northern Ireland, Portugal, Lebanon, and to some extent the urban public sector in the United States have now become areas where politicization has entirely displaced interest in productivity, and the adversary relationship between labor and management has become most arrant. To a certain extent, politicization is due to the conjunction of power equalization (industrial democracy) and professionalization (e.g., job enrichment). These two trajectories reinforce each other, and may accelerate the disruption of the industrial and governmental base far more rapidly than either alone. It is no accident, then, that the potential for conflict is greater in Portugal or even England than in the Scandinavian countries, where job enlargement has been instituted largely without the rhetoric of power equalization. But when power equalization acts alone—without the complementation of moves toward professionalization—another phenomenon occurs that has great behavioral significance.

Power equalization, or industrial democracy, suffers from the same socio-psychological deficiency as political democracy: it reduces the individual's opportunity to gain placidity through resignation. For example, one of the avantages of being in a distinctly subordinate position in industry or government is that one is automatically relieved of the cognitive demands entailed by responsibility. It is readily evident to most workers—although the fact might surprise many social scientists—that their superiors carry greater burdens of anxiety than they, and to some extent they accept this as defense for the higher salaries of their managers. It is by no means clear that the worker will be happier (and therefore more productive) if he is given control over his own work habits or functions. While it is certainly true that most social scientists would be happier were this the case, it is a dangerous simplification to suggest that all individuals will accept anxiety for the sake of self-determination [13]. In short, the discipline and determinism of the workplace, up to a point, may carry satisfaction for the worker, and we eliminate it only at the peril of latent job dissatisfaction. Under some circumstances, then, the manager who seeks to delegate decision responsibility to his subordinates—in the name of egalitarianism or industrial democracy—may paradoxically earn the worker's enmity or resentment rather than his affection. And the organization that sponsors a sweeping removal of authoritarian constraints may find itself more the adversary of the worker than before.

Of course, we must finally comment on the emerging data base that seems to suggest that socialization, job enrichment and power equalization improve productivity. Like most academic exercises carried out under the guise of strict empiricism, the experiments producing the data tend to be less well controlled than the experimenters suggest. There are two problems. First, as any economist will confirm, it is very difficult to get an accurate measure of productivity in the first place, and the issue here is more theoretical than practical—that is, we have difficulty properly defining the notion, so that in many cases the measurement problems themselves are gratuitous. The second difficulty is in the problem of interpretation. For the most part, those who conduct the evaluation experiments are also those who profit from the experiment itself—that is, paid human-relations consultants. But there is another difficulty: the time frame. The experiments have hardly been conducted long enough to suggest whether supposed productivity increases are temporary or permanent. And with this, we have the conceptual quandary that perhaps supersedes all others. For, given the logic of our titillation model, it may be possible that *any* change whatsoever would produce temporarily favorable increases in measurable productivity. Thus, if changes are not introduced on a fairly regular basis, there is the possibility that incremental increases will be erased.

It is not the intention of these discussions to suggest that job enrichment, power equalization and industrial socialization are dangerous or insidious exercises, or that they are useless. When applied with moderation and caution, they point to a way out of the mechanical materialistic wilderness into which so many of our societal systems seem to have wandered. For, as can readily be seen, the human-relations concepts—often inadvertently, perhaps—do find the path to societal improvement along the trajectory of hybridization of benefits. As a last argument in this volume, I want to amplify this concept and extend it beyond the context of strictly material systems.

Throughout this section, we have argued that the satisfaction available from societal membership is less than it might be because of the tendency for systems to stress a specific benefit category, and for individuals within modern materialistic systems to confine themselves to a single corridor. The logic behind the assertion that societal satisfaction is suboptimal (or at least lower than is theoretically feasible) rests with the concept of the diminishing marginal satisfaction that sets in when the corridor elected is an exclusive one. Beyond some point, the ceiling of satisfaction begins to take effect, and the individual's behavior becomes adjusted in one of two ways:

1. He seeks a shift to some other benefit category, as in the worker's entering the political corridor for the sake of power equalization, or in the more recent phenomenon of sentimentalization disrupting industrial life (the advertising executive leaves Madison Avenue and moves to Vermont to grow cranberries).
2. Perhaps more commonly, treadmill behavior sets in: more and more demands are made in an effort to offset decreasing marginal satisfaction.

In earlier chapters, we suggested that the suboptimality of societal systems is also due to the faults of the prophetic process—its pendular and reactive properties—and in general to the social sciences' failure to provide meaningful, operational inputs to societal design. At this point, however, we can be more specific: the key to increasing the net satisfaction associated with membership in human society consists in systematically attempting to halt the pendular process through directed *syncretization* (hybridization). [14]. As things stand now, and as we have attempted to show through these many pages, societal systems tend mainly to redefine themselves rather than evolve *per se*. For example, the shift from institutionalization to sentimentalization mainly involved a substitution of the benefits of emotional symbiosis (identity) for the prior benefits of placidity. The amount of potential satisfaction may not have increased significantly. The same thing may be said of any of the other societal transformations we considered. The terms of the traditional tradeoff are such that we tend to exchange benefit sources without real amplification of satisfaction. We become rationalized and increase our material stock only at the expense of the spiritual and social significance of our system; we go to sentimentalization and gain emotional significance only at the direct expense of placidity or material welfare, etc. Thus, the key is to break the hold of the incomplete dialectic, the engine that finds us constantly substituting thesis for antithesis but never remaining at or defining a point of synthesis [15].

For this reason, the social sciences must be prepared to rise above the current cliches, which are also, as we have suggested, the ancient cliches as well. The now popular cry for a return to religion is too simplistic, for a return to a religion might merely imply a trading off of gratification for placidity. And when placidity became sufficiently entrenched, the decreasing marginal satisfaction would start another version of the Reformation, and that Reformation start another round of materialism, etc., to put us back where we started. By the same token, the siren song that the romantics are singing—asking for love rather than ideology or emotional rather than material significance—has been sung before, and merely resulted in the transition from institutionalized to sentimentalized systems. There we find the affectionate family, the spontaneously charming and guileless commune member. But there, also, we find rickets, penury, high death rates and the absence of any transcendent aim. And then there are the legions of development economists whose ambition is to take all the world's hunters, fishermen and subsistence farmers and turn them into accountants or lathe operators. Few of the reformational platforms popularly broadcast today escape from the bounds of pendularity, reactionism and pedestrianism. For the Marxian and his strange bedfellow, the developmental capitalist, all societal significance rests with the material. For the radical sociologist or human-relations consultant, all societal significance eventually gets back to love, in one or another of its forms (some quite maudlin). For the theologist and existentialist, the only domain of significance is the numinous, and the dominant survival strategy is a return to some sort of solipsism. But the challenge for the serious social sciences is not to *return* to anything, but for once to move ahead.

As a practical matter, the sophistication of our societal critera—and hence the improvement of the societies we reside in—depends on our becoming privy to the properties of the non-normative systems set out in Table 6.3. For, again, each of the hybrid states (with the exception of the "migrant" case) involves a situation where the expected benefits of societal membership are higher than in any of the three normative cases (where our seven ideal-type system referents are located). The logic here is by now evident: by having a hybridized benefit set, each of these systems has the opportunity to escape the crippling effects of decreasing marginal satisfaction. Therefore, the ceiling level of satisfaction is theoretically able to rise above that available to systems where only one benefit category is available. In such a situation, the individual is able to *optimize at the margin*: when a certain level of material benefits has been achieved, he can amplify his social benefits; when social benefits are marginally satisfied, there is spiritual dimension to exploit. In short, the high-quality societal system is simultaneously of spiritual, social and material significance, and the individual therein has adequate satisfaction on all three dimensions: placidity, identity and gratification. I shall defend this prescription no further here, as our task was to evolve a methodology for comprehending the way socio-behavioral systems operate, and not to suggest the way they *should* behave.

But before I complete this work, there are two caveats I would offer those who might want to extend the prescriptive inferences we've developed here—those whose ambition is to "manage" societal systems. First, as I have tried to suggest, the behavioral engination of most individuals will tend to lead them into one of the corridors at the expense of another. Therefore, it is not enough simply to argue for toleration of different benefit categories, or to try to force hybridization. For if the ultimate result is no net increase in satisfaction for individuals, then the fact of freedom does not absolve the social scientist of responsibility. Rather, we have an enormous educative task ahead of us: *the terms of the available tradeoffs must be articulated and broadcast* so that choices may be made intelligently. We must, as well, provide some rather specific directions as to *when a shift from one benefit category to another should be made*, and what types of configurations are likely to provide the highest probability of satisfaction. In short, the social scientist must, in part, assume the mantle of the proper prophet [16].

Now, the second caveat really derives from the first, for the popular prescription for benefit shifts derives directly from Maslow's theory of the hierarchy of needs [17]. That is, it is suggested that once basic survival needs are met, the individual then moves on to higher levels of satisfaction (e.g., security or self-actualization). There is much wisdom here, but there is also a difficulty. The sophisticated, high-quality societal system would most likely not be one where benefits become intelligible in terms of a simple hierarchy. Rather, there would be a constant interchange among benefit categories, given our knowledge of the tradeoffs. Thus, the individual would have to make the decision whether to try to get rich today, or to get drunk; whether to eat or seek sexual gratification, etc. In short, lower-level "needs" must be viewed as competing with higher-level "needs". So what one would get when mapping such a societal system is not a hierarchical configuration, but a complicated *reticular* one, where there is no *a priori* determination of what is a higher- or lower-level behavior, or what is a less or more gratifying experience. To live in such a system—where infinite choices are available—demands great analytical skill. Thus, the proper reference for truly hybridized societal systems would probably not be Maslow, but Epicurus.

In summary, then, we are not irrevocably condemned to be victims of the pendular engination we have historically obeyed—not, at least, once it is defined. But neither will the improvement of human society be a simple or straightforward task. However, we may take some comfort in knowing that many of the faults of society are merely faults of mind—and therefore reparable—rather than largely irredeemable faults of character. The dialectic gripping us is a dismal one, and has constrained us for all our ages. And yet the future is the place where all such constraints can disappear. This knowledge is perhaps cause enough for encouragement.

Notes

1. For the classic critique of such positions, see Jean-Paul Sartre's *Being and Nothingness* (New York: Philosophical Library, 1953), especially Part One, Chapter 2.
2. For example, see Mirceas Eliade's description of yoga as anti-human, in *Yoga, Immortality and Freedom* (Princeton: Princeton University Press, 1970).
3. The appreciation of the extent to which appearances or self-determined properties play a role in social processes was brilliantly set forth by Erving Goffman in *The Presentation of Self in Everyday Life* (Edinburgh: University of Edinburgh Press, 1956). The importance of Goffman's thesis to general social theory has been noted by Alvin Gouldner in Chapter 10 of *The Coming Crisis in Western Sociology* (New York: Basic Books, 1970).
4. The differences between a symbiotic process and a synergistic one have seldom been made explicit. In system terms, however, the effectiveness of any synergistic process will be related to the degree of difference between system components (both structurally and functionally).
5. With regard to the behavioral significance of social redundancy, see the remarkable paper by Arthur Chickering, "How Many Make Too Many" (in *The Case for Participatory Democracy* edited by Benello and Roussopoluos; New York: Grossman Publishers, 1971).
6. For some very insightful notes on the organization of agrarian enterprise, see Arthur Stinchcombe's "Agricultural Enterprise and Rural Class Relations"(in *Readings on Economic Sociology*, edited by Smelser; Englewood Cliffs: Prentice-Hall, 1965).
7. I think the best introduction to the cultural implications of existentialism is that given by Paul Tillich in *The Theology of Culture* (New York: Oxford University Press, 1959).
8. Indeed, Andras Angyal depicts human personality in terms of a dynamic balance between autonomy and homonomy; see his *Foundations for A Science of Personality* (New York: Viking Press, 1964).
9. This is a central point in the sociology of knowledge, especially as reformulated by P. Berger and T. Luckmann in *The Social Construction of Reality: a Treatise in the Sociology of Knowledge* (New York: Doubleday, 1966).
10. For an outline of this process, see Laing et al., *Interpersonal Perception: A*

Theory and Method of Research (New York: Harper and Row, 1966, pp. 3–48).

11. To some extent, the weakness of the original formulation of the concepts of inner- and other-directedness has been repaired, in passing, in D. Riesman's *Individualism Reconsidered* (Glencoe, Ill. Free Press, 1965). There is still the problem, however, of a limit to the beneficiality of inner-directedness. Beyond some point, intrinsic behavioral referents cease altogether to be functional, and therefore cease to be admirable.
12. An interesting variation on this theme is given by Ivan Illich in his intriguing *Tools for Conviviality* (New York: Harper and Row, 1973, pp. 54–73); for another slant on the issue, see Watzlawicki et al., *Change: Principles of Problem Formation and Problem Resolution* (New York: W. W. Norton, 1974).
13. The bias of most social researchers in the human-relations movement has been beautifully brought out by George Strauss in "Some Notes on Power Equalization" (in *The Social Science of Organizations*, edited by Leavitt; Englewood Cliffs: Prentice-Hall, 1963).
14. Though anthropomorphic, one argument for hybridization is presented by George Lockland in *Grow or Die: The Unifying Principle of Transformation* (New York: Random House, 1973).
15. It has been suggested that the Maoist movement in China represents an attempt to produce a hybrid, where material and social significance may be had simultaneously. But, in practice, what Communist China has done is largely to speed up the frequency of the pendular process. For a while, material-productive interests predominate, and economic development moves ahead. But at some point it is seen that economic development has achieved the price of weakening the spirit of egalitarian communism: that materialism is driving out the socio-spiritual factors. At this point, the ideological engine is exercised, and the economic sector depressed. The politico-economic history of Communist China has been largely of this kind, with dramatic and rather rapid swings between quasi-capitalism and ideological purity (via, for example, the Cultural Revolution). For more on this somewhat hysterical pendularity, see the essays in *China After the Cultural Revolution* (New York: Random House, 1969).
16. I have attempted elsewhere to describe what might be a methodology for proper (rational) prophecy. See, in this respect, my "Architecting the Future: A Delphi-Based Paradigm for Normative System-Building," (in *The Delphi Method: Techniques and Applications*, edited by Linstone and Turoff; Reading, Mass.: Addison-Wesley, 1975.
17. See, for example, his book *The Farther Reaches of Human Nature* (New York: Viking Press, 1971).

Glossary

ABULIA A clinical condition where the individual loses either the capacity or the will to make decisions. As used in this text, it refers to societal systems that tend to emphasize reflectivity over reactivity, and are hence characterized as speculative or indecisive.

ADHOCRATIC A term coined by Alvin Toffler in his *Future Shock* to describe a condition wherein an organization is structured with maximal plasticity, such that new organizational forms are defined only as a new need emerges, and then abandoned immediately after that need has been met. An adhocratic organization would then contrast, in terms of responsiveness, with bureaucratic systems.

AFFECTIVE BEHAVIOR (Predicate) Behavior which is determined by essentially emotional bases, such that cognitive or logical mediation is minimal.

ALGORITHMIC PROCESS Here used in a general way to describe any process that is defined in terms of a definite sequence of well defined procedures (as contrasted with heuristic processes).

ANGST A clinical situation where an individual displays feelings of anxiety and depression whose causes are obscure or unspecified.

ANOMIE A situation where an individual is deprived of purpose or ambition, or of the ability to associate with any broader societal system (the latter resulting in an effective loss of identity).

APODICTICAL Characterizes the property of models or theoretical devices which makes them susceptible to eventual validation via normal (empirical) scientific procedures. A nonapodictical construct is thus one that can be neither disproved or validated by any universal rules of evidence or procedure.

A POSTERIORI / A PRIORI EVENT As used in these pages, an a posteriori event is an action conditioned by (or taken in response to) a tangible, empirically accessible predicate; e.g., a force or stimulus operating in the individual's environment. An a priori event, on the other hand, is one that takes place in the absence of any tangible, external cause or evidence. In a more general sense, a priori behavior is thus essentially deductive or ab intra in origin, responding to empirically "transparent" predicates (e.g., ideological or affective stimuli); a posteriori behavior, in this sense, would thus be behavior of an accomodative sort, conditioned by "realities" to which the individual responds. In a procedural context, a priori information is that which is not derived (or induced) from factual predicates, but which is essentially theoretical or deductive in origin; a posteriori information, on the other hand, has its origins in the empirical world, and is often obtained using procedures of formal statistical inference, etc.

ASCRIPTION A situation where economic and political prerogatives flow to the individual on the basis of the status into which he was born, or in response to the social status he is able to demonstrate or acquire (through marriage into the aristocracy, etc.). In general, then, an ascriptive system is one where societal benefits are made available on the basis of "class" rather than individual distinctions.

ASSOCIATION The process where a cause and its effect(s)—or a stimulus and a response—are closely and directly connected in time or space (as contrasted, for example, with bisociation, where an effect or response is seen as the product of two disparate causes or stimuli). Associative behavior thus becomes comprehensible as a situation where, for every precedented event, there is one and only one probable response, this response being invoked automatically (i.e., without cognitive mediation or reflection).

AUTOCHTHONOUS EVENT An event (or response) whose cause (or stimulus) is distributed in the properties of the system itself. To the outside observer, then, it would look to be self-caused, or spontaneous in origin.

AXIOLOGICAL PREDICATE A determinant of individual (or collective) behavior which is a product of normative origin, specifying an "ought" or a "should", without offering any empirical (instrumental) or logical defense for the imperative or dictate. Axiological predicates are thus often associated with revelatory or moralistic sources.

AXIOMATIC PREDICATE A determinant of individual (or collective) behavior that has a logical (or quasilogical) defense, such as the "oughts" and "shoulds" that are derived from theoretical or discursive reasoning, and passed off to individuals or societal systems in the form of proper paradigms, sets of principles or rationalized protocols.

BAYESIAN PROCESS A generalized form of the statistical procedure which allows the formal admixture of a priori and a posteriori—subjective and objective—probabilities or information. In the general sense, a Bayesian process is an analytical approach which attempts to converge on some optimal or adequate solution to a problem by allowing the disciplined interconnection of judgemental and empirical data. Bayesian approaches thus seek to force a complementation between positivistic and deductive analysis.

COGNITIVE PROCESS A process that involves the application of directed (purposive) intellectual energy, and which thus becomes comprehensible as an exercise in either deduction or induction . . . as contrasted with, say, intuitive behavior.

DEDUCTIVE BEHAVIOR/PREDICATE The deductive process, formally defined, is the act of reasoning from the general to the specific. As used in these pages, deductive behavior refers therefore to activities that are a priori predicated, or driven by some theoretical, normative or other non-empirical determinant. More generally, deductive behavior is that for which no significantly probable empirical (real) stimulus can be found, or in operational terms, behavior which finds only a weak or insignificant correlation between stimulus and response. Creative or originative behaviors are thus special cases of deductive behavior. In process terms, a deductive system is indicated by a negligible relationship between the morphology (quality) of inputs and outputs; in short, as a process where inputs are considerably and significantly altered.

DISCURSIVE PROCESS The act of relating conclusions to premises by relying on a clear-cut, definable series of logical steps (as contrasted with intuitive behavior). As used in this text, discursive behaviors are defined generally as those where the reasoning process—allegorical or otherwise—plays the predominant role, and where empirical or factual data are absent, overridden or ignored.

ELLIPTICAL CONSTRUCT A model or theoretical system where there is a lack of specificity in the prior arguments, or where the terms employed are ambiguous, parabolic or otherwise ill-defined.

ENGINATION (of behavior) A clinical term used to describe the situation where a behavior (or response) has two or more causative agents, including the subject (individual) himself, and where we do not know the relative strength of the several determinants. In terms of our specific usage here, we speak of behavior being engined (as opposed to determined) when it is a product of the individual acting in concert with some axiomatic or axiological predicate. Axiomatic and axiological phenomena thus become behavioral "engines" in that they operate only through the individual, and not as independent agents.

ENTHYMEME/ENTHYMETIC CONSTRUCT As used here, an enthymeme is a theoretical system or cognitive construct where at least one of the premises is left unspecified or tacit, such that a conclusion (or prescription) is not irrevocably defensible given the prior arguments or defenses. In operational terms, an enthymetic construct is one where several different logical conclusions could have been derived from the prior premises or arguments, but where the author selected to pursue only one or a limited number.

ENVIRONMENT The system of tangible, empirical forces and properties surrounding an individual, extending to the point where the strength of interactions drops below some threshold of significance (contrasted with *milieu*, which contains both empirical and non-empirical properties, and with cultural context which, as used here, contains only axiomatic and/or axiological predicates).

EQUIFINAL PROCESS Implies a system or process which can reach the same conclusion along several different (and independent) causal trajectories. In operational terms, this means that simply by recognizing the system's concluding state, no deterministic inference can be made regarding the cause of that conclusion.

EXISTENTIALISM/EXISTENTIALISTIC SYSTEM As used in this volume, existentialism implies the situation where the interdependence among members of any social ordering passes below some threshold and becomes minimal or negligible. In short, an existentialistic sytem is one populated by essentially self-concerned, self-motivated individuals (and, by extension, by individuals whose behavior is determined largely ab intra and ab initio rather than in response to contextual or cultural predicates).

GESTALT A phenomenon in which the parts are effectively inseparable, such that the whole cannot be readily or meaningfully reduced to its constituents.

HEURISTIC PROCESS A search process where the final destination is unspecified, but where the individual or agency has two sources of discipline: (a) A set of a priori constraints which serve to artifically bound the search space (or to determine a first, tentative step); and (b) a set of criteria for distinguishing a favorable from an unfavorable (a promising from a sterile) trajectory. A trial-and-error process lacks both the above. In short, a heuristic process is employed when we know only vaguely where we want to go and are not sure how to get there. In a more extended form, a heuristic process is employed when we know that we want to get somewhere else, but we don't know where that might be till we actually strike it. In operational terms, then, a heuristic process is the procedure recommended for operations in an extremely complex, protean and rapidly changing environment.

HYPOSTATIZATION The process of attributing a reality or tangible essence to products of conceptual origin. As used here, hypostatization implies two types of activity: (a) A pathological behavior where the individual *hears* sounds, *sees* ghosts, etc., which are not empirical in nature; and (b) the process of reifying a mental construct such that it gains empirical significance (e.g., the translation of a concept into a working model).

HYPOTHETICO-DEDUCTIVE PROCESS An analytical scheme where an initial hypothesis is derived from deductive or a priori sources, and is used to initiate a normal cycle of scientific inquiry (as opposed to the inductive scientific process, where the initiating hypothesis is derived from empirical observations and subsequent generalization).

ICONOGRAPH A device which employs mathematical symbology but does not obey mathematical rules, used primarily to lend a graphical significance to essentially verbal or qualitative phenomena that would defy proper mathematical translation.

IDEAL-TYPE/TYPOLOGY A deliberately generalized or "idealized" model of some phenomenon or process, from which definite predictions are derived, and which hence has theoretical significance. An ideal-type thus serves to define a "class" of phenomena, but does so on the basis of theoretical or a priori reasoning rather than empirical-inductive process of normal science. A typology is an ordered collection of ideal-types.

IDEMPOTENCY A law peritnent to the algebra of propositions, which defines the case where, for example, $a \cup a = a$ and $a \cap a = a$.

IDEOLOGY A set of behavioral proscriptions or prescriptions which may be either axiomatic or axiological in origin, thus seeking to define terms of behavior that have a normative or logical rather than empirical (accomodative) basis.

INDUCTIVE PROCESS The act of reasoning from the specific to the general, usually characterizing the attempt by normal science to generate laws or classes by empirical observation of multiple individual phenomena, and then isolating and generalizing behavioral or structural consistencies. In an operational sense, as used in this volume, inductive behaviors are those whose original predication was on the basis of acquired knowledge stemming from trial and error learning (transmitted and codified as *lore*). From the standpoint of the individual, inductive behaviors are thus those whose predicate rests with the experience or observation of the individual, and whose rooting rests in the empirical (real) world. As a process of cognitive significance, inductive procedures are characterized by the situation where the outputs are closely correlated (in terms of quality or configurational properties) with inputs, and where the modifying that is done takes place mainly in terms of magnitudinal or quantitative changes. Thus, inductive behavior, in all its forms, always responds to properties of the individual's environment or experience.

INSTRUMENTALISM The act of treating ends as paramount, and means as strictly subordinate to ends. In a more general sense, instrumental behavior is that unregulated by any a priori (moral, ethical or normative) constraints, and which therefore responds merely to the volition or ambition of the individual or subject. By extension, instrumental science is concerned with the control of phenomena, not with explanation (except as useful for exploitation).

ISOMORPHISM A situation where two or more phenomena share certain configurational or structural similarities. In extension, isomorphism may also occur where two or more entities have functional (behavioral) properties that can be described in terms of the same essential set of dynamics or which otherwise become intelligible in terms of the same set of abstracted criteria.

PARADIGM/SUPRAPARADIGM A set of ordered suppositions of empirical and/or a posteriori origin which serves to predirect an investigator's attention into certain paths as opposed to others which might be available in the absence of the paradigm. As a rule, the paradigm itself is not subject to investigation, but is accepted as a given. In a more general sense, a paradigm becomes roughly the equivalent of a Weltanshauung or *mind-set*, which serves as a more or less coherent residence for those things an individual (or collectivity) thinks it already knows about the world. A supraparadigm is an artifice of qualitative analysis, and is a tentatively ordered collection of paradigms such that redundancies among the paradigmatic constituents have been removed.

PHENOMENOLOGICAL BEHAVIOR Refers to the conscious attempt to eliminate any a priori (axiomatic, axiological) predicates which might interfere with relations between subject and object. In an operational sense, phenomenological inquiry is thus that which attempts to proceed in a hypothesis-free context.

PSYCHOGENETIC BEHAVIOR/PREDICATE Psychogenetic behaviors are those postulated as deriving from the operations of the mind itself, and over which the individual has no control (and which thus become, in our terms, the lowest-order behaviors individuals may exhibit, being precognitive in origin). A psychogenetic predicate is thus a behavioral determinant which has no substance outside the individual mind itself (and which arises, as it were, ab intra and ab initio). Psychogenetic behaviors are often characterized as instances of instinct; thus, the psychogenetically driven individual becomes comprehensible as one driven by his own internally generated concepts or impulses, and one whose behavior is thus mediated neither by environmental or cognitive factors.

RHETORICAL CONSTRUCT A model, paradigm, etc., designed more to alter behavior than to serve as a source of comprehension or prediction of some phenomenon. More generally, a rhetorical construct is one that is tendentious rather than objective, and which employs terms that are affective in implication.

RETICULAR SYSTEM One which cannot be graphed in terms of a hierarchical structure, in that each and every component of the system has potentially the same degree of determinancy over every other. Thus, reticular systems are allegorized in terms of networks and become comprehensible as the ultimately democratic organizational modality.

SYNCRETIC CONSTRUCT A model which seeks to reconcile competitive paradigms by raising the subject of inquiry to a higher level of abstraction. In general, a syncretic process of inquiry is one that builds on restricted (parochial) perspectives and attempts to produce a broader paradigm (supraparadigm) by eliminating the a priori predicates that underlie the several competing constituents, thus isolating the grounds available for complementation.

TOPOLOGICAL CONSTRUCT As used in this volume, the term topological construct refers generally to any graphic representation that abstracts the properties of phenomena and distorts them for purposes of theory-building, but which does not violate the fundamental relationships of the phenomena (it is thus an extension of topology as the mathematical exercises studying the geometric properties of figures that remain essentially unchanged even when distorted).

Index

A

Absorption, 263–264
causality and, 264
in existentialistic states, 276
Abstraction, 95, 96, 99
Abulia, as collective pathology, 185
Accommodation, 43
Accommodation, transgenerational, 100
Accommodative system, 98
Acculturated ideal types, table of, 181
Acculturated systems, 98, 99
Action-research paradigms, 92
Adaptive behavior, 154, 156
complexity and, 169
modalities of, 97
sacerdotal systems and, 173–174
Adhocracy, 215
see also, Toffler, A.
Adhocratic elaborations, of prescriptive base, 167
Advancement associations, of Medieval Europe, 215
Adversary base, and competing axiomatic systems, 176–177
Affective behavior, 195–213
modalities of, 159
Affective rhetoric, 175
Affluent youth, 167, 188, 241
Africa, asymmetrical economic distribution in, 228
Alienation, 184
Algorithms, 93–94, 120, 132, 179, 214
heuristic, 123
in rationalized economic systems, 217
of resources and distributions, 248
of utility, 295
Allah, 270

Allegorical prophecy, 104
Allegorical reflexivity, 45
Allegorization, 95
Amos, 230, 240
Ampliative behavior, 57, 159–160
automatic engination in, 161
exploitation and, 169
gratification and, 168
secularization and, 161
Ampliative induction functions, 26–33
diagram of, 27
Ampliative variants, 154
Amplification, and adaptive behavior, 156
in repressive proscriptive bases, 163
socio-economic, 235, 237
Analogues, 147
Analytical skills, in hybridized systems, 302
Anarchic societies, 187
and compulsive behavior, 185
Anarchic socio-economics, 129
Anarchy, 193, 243, 264
Anomie, in a compulsive society, 186
Anthropomorphism, 263
Anti-heros, 188
age of, 242
Antithetical positions, 63–67
Annuitied set, and tenure, 287
Anxiety, 103, 162, 181, 185, 259
and materialism, 282
Aphoristic utterances, 109
Apodictical, constructs, 94, 153, 197
quality of models, 8
Arabian states, medieval, 202
Archearchy, 146
Aristocratic systems, 242
British and Indian, 292
Japanese, 244
Saxon, 244
wealth of, 223
Asceticism, 220, 241, 263–264
and symmetrical distribution, 228
by default, 271
Ascription, 219–220, 294
Europe's liberalization and, 237
imputational, 280
Ascriptive differentiation, of societal systems, 258
Ascriptive economics, 223–225
rationalized, 232
Assimilative function of cultures, 98
Association, and adaptive behavior, 156
Associative behavior, 160, 161, 182, 188
versus ampliative behavior, 169
Aspirations, definitions of, 81–82
legitimate, 81–82
related to achievements, 81
Asymmetry, socio-economic, 176, 228
Athletics, 273
Atomistic amalgams, 99
Attention of individual, preempted, 177
Aquinas, St. Thomas, 187, 234
Augustine, St., 181, 235, 279
Autarchic systems, 201
see also, Mechanistic systems
Authoritarian ideology, 141
Authoritarian industry, 110
Authority, transcendental replaced by rational, 192
Automaticity, and obedience, 191
in collective behavior, 169
ritualization and, 182, 188
Axiological base, 267
Axiological differentiation, 267
Axiological engines, 137
Axiological platforms, 192
Axiological predicates, 51, 56, 57, 97–99, 109, 118, 119, 120, 124, 130, 132, 134, 215, 256
absent in axiological systems, 227
definition of, 52–53
diagram of, 51
of Protestantism, 234
spiritual benefits and, 285
Axiological prophecy, 103, 104, 161
Axiological proscription, and leader-

ship, 166
and the status quo, 164
Axiological speculation, 100
Axiological systems, trajectory to goals, 167
Axiologies, 100, 147, 156
diagram of, 101
prescriptive, 105
proscriptive, 131, 157
Socratic, 166
Axiomatic base, 179, 192–193, 194
of capitalism, 252, 287
Axiomatic capitalism, 252
Axiomatic engines, 156
and material gratification, 168
in commerce, 134
socio-psychology of, 168
Axiomatic platform, 168, 192
and material corridor, 281
politico-economic order function of, 169
Axiomatic predicates, 51–58, 97–99, 109, 118, 119, 120, 124, 130, 132, 134, 136, 178, 195, 200, 229
absence of in secularized systems, 227
Axiomatic prophecy, 104, 108, 109
Axiomatic reasoning, 100
Axiomatic systems, child's education and, 177–178
contradictions in 180, 193
political paradigms and, 179
prophetic base and, 180
table of components, 110
Axiomatic proscription, subsumed by axiological predicates, 239

B

Bacon, R., 92, 93
Bargaining process, 257–258
Barth, 230
Basques, 285
Baye's theorem, 143
Bayesian convergence, 9
Bayesian decision making, 30
Bayesian investigation, 249
Bayesian process, 199
Beatific visions, and UFO's, 167
Behavior, idiosyncratic, 123
Behavioral bases, *ab intra*, 55–56
appealing phrasology of, 54
collective, 158
contextual, 100
data driven, 11
deductively predicated, 11
difficulty in changing, 254–255
indeterminacy of, 43
individual to collective, 45–46
inductively predicated, 11
intrinsic, 51, 52–53
need fulfilling, 53
simplicity of, 43
table of, 44, 55
Behavioral codes, 109
Behavioral engines, corridor logic and, 301
types of, 90
Behavioral expectations, as paragons, 180–181
Behavioral logic, totalitarian, 150
Behavior modalities, 56
adaptive and exegetical, 97
ampliative, 56
associative, 56
inductive, 56
Behavioral predictability, 120
Behaviorism, 11, 26, 32, 91, 78, 82
Behaviorists, 187
and homeostasis, 79
Bentham, J., 187
Buddhism, 233, 270
Tibetan, 175
Zen, 202
Buddhist model, 138–139
Bio-chemical bases, for deductive behaviors, 42
Bio-emotions, 294
relationships and, 275
symbiosis of, 270
Birthrights, 236

Birthrights, *continued*
and institutionalized systems, 217
Black Power, as white American instrumentality, 287
Black Pride, social benefits of, 288
Black revivalism, placidity via theonomy, 288
Blocs, defined, 215
Bolsheviks, 201
Bourgeoisie, 236, 248
Bribery, 281, 295
Bureaucracy, 243, 247
and repression, 167
Bureaucratization, 173

C

Calculus, compensatory, 53–54, 138, 162, 164, 194, 231, 235, 273
of satisfaction, 289
of socio-economics, table of, 230
of status, 291
of titillation, 261, 270
of zero-sum game, 218
pleasure-pain, 30
Calvin, J., 181
Capital-labor ratio, in primitive production, 226
Capitalism, 248, 281
axiological predicates of, 236
axiomatic base of, 287
axiomatic engination of, 167
axiomatic platform of, 194
Catholic church and, 200
proscriptive roots of, 168
Protestantism and, 229
supply and demand of, 217
Cartels, 218
as diseconomic, 228
and instrumentality, 218
Caste, versus socio-economic mobility, 238
Catholic church, 213, 229
capitalism and, 200
confession in, 135
decline of Holy Roman Empire and, 200
economic dogma in, 223
Catholicism, 233–234, 238, 278
medieval economic dogma of, 286
Causality, 140
absorption and, 263
in social science, 78
sequence and, 268
versus correlations, 211
Centripetal economy, 285
Cerebrated systems, 195
see also, Sapience
Cerebration, 193
eclectic implications of, 196
Certainty, 158, 165
Ceteris paribus assumption, 10
Change, management of, 255
sacerdotal systems and, 173–174
Chaos, political-economic, 200
Chomsky, N., 37
Christian activism, as a shift from secularization, 167
Christian, versus Buddhist model, 138–139
Class, versus socio-economic mobility, 238
Classical conditioning, 17
Cluster hierarchy, versus stratified hierarchy, 247
Clustering, 71, 73, 80, 99, 118, 120, 122, 211
congruent, 157
in conditioned systems, 183
in prescriptive systems, 166
of cultural predicates, 137
of societal system attributes, 257
of socio-economic properties, 244
of status and material benefits, 294
of theonomy, 263
producing communes, 214
symbiotic, 274
Coefficient estimation, qualitative analysis of, 7
Coefficient of cultural control, 118, 122

variables of, 119, 120
Coercion, 281
in socialism, 239
sacerdotalism and, 173
towards spiritual benefits, 277
unions and, 297
Cognitive behavior, 17–18
versus pre-cognitive behavior, 18–20
Cognitive dissonance, 260
Cognitive referents, 5
Collective behavior, diagram of bases, 158
mechanized behavior and, 168
Collective cynicism, 196
Collective lore, versus cultural base, 100
Collective pathology, abulia as, 185
Collectivities, 98–99
decreasing marginal satisfaction in, 292, 294
defined, 100
Collectivized agriculture, in the Soviet Union, 239
Collegial government, 246
Collegial system, emotional relationships in, 274
Collusion, diseconomics of, 228
Collusive counter-competitive industries, 194
Common denominator, in behavior modalities, 70
Communes, 201, 213
agrarian and the industrial revolution, 295
as a business, 223
as an artifice of production, 252
Christian, 286
intra-species contact and, 274–275
North American Chinese families as, 285
origins of, 214
Communism, Chinese, 202, 214, 239
simularity to medieval Catholic states, 239
Soviet, 213
Compensation, material and interpersonal security and, 214
Compensatory calculus, 53–54, 138, 162, 164, 194, 231, 235, 273
in prescriptive systems, 170
in proscriptive systems, 176–177
Competition, 138–139, 246, 269
as a rhetorical concept, 220
capitalism and, 218
Darwinism and, 193
model of, 184
objective performance and, 178
Complementation, 272–273
Complexification, 190–191
Complexity, and adaptivity, 169
Compromise, 26
Compulsive action, 185–186, 195
Computer resources, asynchronous remote participation of, 148
Comradeship, and intra-species contact, 274–275
Conceptual filters, and cultural dimensions, 98
Conditioned systems, as prototypical, 188
lore in, 182
name and form in, 272
transformed by complexification, 191
transformed into institutionalized systems, 191, 198
Conditioning, and church attendance, 159
Conferences, and sapience, 215
Configurations, of societal phenomena, 62
Confiscation, sanction of, 184
Congregation, defined, 213
in proscriptive protestant nations, 214
Congruence, 140, 192
Consensual political base, 295
Consensus, 243
Conspicuous consumption, 293
Constitutions, as an exegesis, 179
Constraint, 105

Constraint, *continued*
cultural, 97, 100, 135
Constructs, isolated descriptive, 96
Consuetude, in conditioned societal systems, 182
Contextual events, 52
diagram of, 51
Contradiction, human capacity for, 259
Control, and comprehension, 272
Convergence, 26
Copts, 233
Correlation, between social and economic benefits, 229–240
in the deductive process, 35
Correlation coefficient, 148
Corridor concept, 168
isomorphism of, 283
levels of, 291–293
societal trade-off and, 266
Covens, 215
Creative release, in collective behavior bases, 158
Creativity, defined, 185
satisfaction and, 169
Credentials, and recruitment in sapient systems, 185
Credit, as potential wealth, 216
Criminal activity, and prescriptive systems, 175
Crusades, 173
Cult, 142
word origin, 140
Cultural balance, 132
and set theory, 132
Cultural base, algorithm of, 120
diversity of, 136
intensity of, 134–135
heterogeneous, 119, 136
homogeneous, 119, 136
partitioned, 123
stratified, 137
Cultural conflict, 137–139
Cultural derivatives, 127
Cultural directivity, and set theory, 132
Cultural constraints, 82–84, 100
diagram of, 83
in the form of lore, 182
Cultural control, coefficient of, 118–124
Cultural dialectic, diagram of, 97
Cultural dimensions, 120–122
table of, 121
Cultural envelope, 97–98
Cultural evolution, 127
Cultural mediators, 98
Cultural predicates, 255
Cultural revolution, 127
Cultural set concepts, 124–139
Culturally conditioned behavior, 154–156
Culture, potency and set theory, 133
significance of, 83
word origin, 140
Cybernetics, 14–20, 77–78, 91, 187
Cyclical enviornment, 71
Cynicism, collective, 196

D

Data base, need in trial-and-error situations, 24
Data behavior, 14–20, 26, 28, 55
diagram of, 18
see also, Cybernetics
Darwin, C., 41–42, 226
Darwinism, 184, 193
Decentralized systems, 244
Deduction, 74, 153–154, 159, 211
in societal systems analysis, 92
Deductive behavior, 36–39
discursive, 39–44
diagram of, 38
exegetical, 38–39
diagram of, 38
heuristic, 40–41
idiosyncratic, 36–38
hypostatized, 36
interpretive, 36
principled, 40

Deductive inferences, 8
and incongruence, 179
Deductive process, 33–36
diagram of, 35
limitations of, 33–34
Deductive-rhetorical process, 94
Derivative hypothesis, 8
Delayed gratification, 162
Delphi process, 65
Democracy, decline of axiomatic base in, 287
Demographic factors, 137
Deprivation, sanction of, 183
Deprived system state, diagram of, 288
Derivative hypothesis, 8
Dialectical, defined, 55
Dialectical engines, 50–59
diagram of, 51
as a "switch", 154
Dialectical paradigm, 3
Dichotomies, heuristic potential of, 62
Dictators, 295
Didactic posture, and bureaucratic form, 247
Difference formulations, use of in non-linear constructs, 10
Differentiation, 191
elitism and in sentimental systems, 176
in conditioned systems, 218
in Maoist thought, 241
in totalitarianism, 245
of aspirations and abilities, 193
Dilettantism, 277
Discrimination, 111
Dogma, 105, 108, 160–161, 168, 192, 194, 195, 213, 220
Catholic compensatory calculus and, 231
Maoist, 247
Medieval Catholic, 286
prescriptive repression and, 269
shifts in, 233
Drug culture, 186, 202
cognitive mechanisms, 263–264
as a parabolic phenomenon, 271
Drugs, as dysfunctional absorption, 288
Dukhobors, 213
Duty, and cosmological mandates, 201
Dysfunctional surrogates, gangs, drugs, and muggings as, 288

E

Eclectics, and cerebration, 196
Ecological fallacy, 147
Economic centralization, 248
Economic deprivation, and proscriptive systems, 176
Economic engines, and distribution of wealth, 216
Economic mechanisms, table of properties, 221–222
Economic mobility, 232
Economists, developmental, 300–301
Economization, 173
Effeteness, 185
Efficiency criteria, of human behavior, 18
Egalitarianism, 166, 238
as a natural system, 256
drawbacks of, 298
economic prerogatives and, 223
proto-Communism and, 174
temporal appeal of, 175
Ego differentiation, and identity, 261
and social satisfaction, 297
Einstein, A., 41–42, 93, 149
Elegance, 170
Elites, 141, 164, 219, 223, 232, 233, 294
ascriptive systems and, 231
fanaticism rebounding on, 173
feudal, in Tibet and Africa, 286
oligarchic, 223
sentimentalized systems and, 176
spiritual benefits of, 269
paternal responsibility of, 244

Elites, *continued*
  reinforcement of artificial behavior, 167
  view of lower classes by, 270
Elliptical logic, 94
Emotional engine, calculus of, 274
Emotional symbiosis, 286–287
  absorption depth, 297
  in Negro community, 288
  shift to placidity, 300
  social satisfaction and, 291
Emotions, 20
Empirically transparent factors, 98
Empiricism, structures of, 91
Empiricist analytical engine, 92
Empiricist positivist, naivete of, 165
Engels, F., 230, 238
Engine of marginal satisfaction
  see Treadmill model
Environmental conditions, and societal evolution, 188
Envelopes, social, economic, and political, 103
Epicurus, 41, 187, 279, 302
Epistemology, 10
  dialectical, 59–62, 68
    field phenomena in, 60–62
    logical extensions of, 60
  dualistic, 233
  phenomenological bias and, 92–93
  predicates of, 59–60
Equifinality, 22
Equipotentiality, 22
  in human behavior, 19
Errors, in model building, 9
  predictive, 9
  projective, 9
Eskimos, 227
Evangelists, 270
Evolution, of societies, 192
  of civilizations, 198
  of societal systems, 255
  towards sapient societies, 196
  versus redefinition of societal systems, 300
Excommunication, 192, 232
Exegesis, 99, 107, 109, 161, 172, 247
  and the Constitution, 179
  modalities of, 154
  of the environment, 100
Exegetical analysis, revelatory, 233
Exegetical behavioral modalities, 97
Exegetical platforms, 112
Exile, 191
Existentialism, 194–196
  and absorptive spirituality, 264
  dissolution of, 195–196
  shift of sentimentalism to, 202
Existentialistic systems, 185, 193, 215
  and absorption, 271
Expatriated systems, 285
Exploitiveness, 169
  in instrumental systems, 166
Extended family, 213
  industrial revolution and, 295
  prescriptive movements and, 214

F

Fabians, the, 230, 238
Family, and intra-species contact, 274–275
Fanaticism, 172–173
Feedback loop, 15
  diagram of, 15
Feedback systems, 16–17
  compromise, 16–17
    diagram of, 16
  convergence, 16–17
    diagram of, 16
Feudal systems, 213, 242
  wealth in, 223
Finite state programs, 30
Finite state societal systems, 122
Freud, S., 13–14, 115
Freudian theory, 13–14, 19, 82, 134, 179
  synthesis with structuralists, 20
  see also Freud, S.
Flow diagrams, 6
Force diagrams, 61–62

Franchise system, 247
Franco, F., 201
Free enterprise, as a rhetorical concept, 220
Friendship, and intra-species contact, 274–275
Frustration, 167–168
Functionalists, 187
Functional differentiation, of societal systems, 256
Functional niche
  see, Complementation
Functions, in aprioristic base, 150
Fuzzy set theory, 7–8

G

Galbraith, J. K., 204
Game theory, 205
Gangs, as dysfunctional surrogates, 288
Generational transmission, 181, 182
Generic production function, 283
Geographical conditions, and conditioned behavior, 182
Georgescu-Roeyen, N., 84
Gerard, J., 41
Gestalt, 11, 75
Gestalten, 41, 77
  of right hemisphere of brain, 42
Gouldner, A., 78–79
Gratification, 293
  ampliative behavior and, 168
  categories of, 279
  ceiling of satisfaction and, 289
  delayed, 162
  generalized as titillation, 261
  power replacing material benefits in, 297
  sexual through pornography, 200
  threshold levels of, 266
Guilt, 204, 282
  process and principles of, 179
  repression and, 276
  shared, 134

H

Hari Krishna movement, 286
Harvey, 41–42
Hedonism, 169
  in instrumental systems, 166
  secularization and, 195
Hedonistic systems, 187
Hegelian dialectic, 106, 110
Hegel, 105
Heidegger, 187, 230
Herbst, 150
Heresy, 173
Hermits, 186, 215, 264
Heros, versus anti-heros, 190
Heuristics, 40–41, 55, 94, 161, 185
  algorithms, 123
  engination, 156
  in societal systems analysis, 91
  masks, 97
  potential of, 62
  utility in sapient systems, 196
Heuristic behavior, and reality feedback, 185–186
Hierarchalization, 93, 191
  ascriptive, 176
  asymmetry and, 247
  charm and, 283
  eased in sentimentalized system, 214
  Europe's liberalization and, 237
  of needs, 302
  opposed to membership credentials, 215
  people's contentment with, 259–260
  sacredotal, 269
Hippies, 202, 279
Hobbes, T., 181
Holy Roman Empire, decline of, 199–200
Homeostasis, 20, 190
  of physiological functions, 18, 20
  society and, 79
Homeostatic behavior, 14
Homogeneous clusters, 166

Horace, 238
Hosea, 230, 240
Hull, 13
Human judgment, inadequacy of, 149
Humanism, of St. Thomas Aquinas, 233
Human-relations consultants, 300–301
Hume, 13, 26, 32, 77, 187, 279
Hungarian repression, 178
Hunting societies, 188
Huxley, A., 187
Hybrid benefit mixes, 294
Hybrid states, 301
  versus normative states, 289
Hybridization, halting pendularity, 300
  of benefits, 299
Hybridized systems, 302
Hyden, 13, 41
Hypostatization, 185–186
Hypothesis-free research, 93
Hypothetico-deductive, 156
  distinguished from deductive-rhetorical, 94
  model building, 196
  process, 92–193
  real-world systems, 211

I

Iconographs, as topological constructs, 62–63
  clustering in, 64–65
  of dialectical behavior modalities, 56–58, 63–67, 74–77
    graphs of, 56, 57, 65, 76
Icons, 268
Id, 17, 275
  see also, Freudian theory
Id-level behavior, 55
Id-level variants, 154
Ideal types, 89, 96, 122
  acculturated, 172–181
  adaptive, 181–188
  analogous nature of, 283, 289
  diagram of, 69–70
  dogmatic, 172–174
  dynamics of, 187–197
  elaborative, 153
  formulation and manipulation of, 95–96
  hypothetical, 68
  interdisciplinary implications of, 210–211
  normative dynamics, diagram of, 189
  polarized, 98
  purpose of, 152
  rationalized, 177–181
  sentimentalized, 174–177
  societal, 5, 171
    table of, 171
  synoptic, 153–154, 169
    table of, 187
Identity, 294
  in cerebral systems, 277
  in sentimentalized systems, 275–277
  localized, 274
  of self, place, and purpose, 261
  status, wealth, power and, 282
  symbiotic reference to, 265
  synergistic reference to, 265
  through identification, 273
  through successive differentiation, 272
Ideologic methodologies, 148
Ideological defense systems, 149
Ideological motivators, 161
Ideological salience, 141
Ideologics, new science defined, 141
Ideology, 40, 88, 105
  cerebration and, 164
  definition of, 140
Idiosyncratic, 55
  behaviors, 123, 154, 156, 160, 184
  compulsion, 55
  formulation of prophets, 103
  hypostatization, 55
Idiot savant, 37–38

Imagination, and proscriptive loopholes, 163
Immanence, 261
Immigrants, in primitive systems, 182
Incontinent grouping, and covens, 215
Index of confidence, 199
Index of satisfaction, 133
Individual behavior, 154
Individuals, as media, 142
Individuation, 261
  ceiling of satisfaction and, 289
  dampened by fanaticism, 173
  emotional benefits of, 274
  heresy and, 191
  in sentimentalized systems, 272
  institutional and primitive systems compared with, 271
  in symbiotic synergistic contexts, 265
  threhold levels of, 266
  versus liberalism, 166
Indoctrination, 192, 217
  axiomatic predicates of, 177–178
  discontinuity of, 219
  in capitalism, 239
  socio-economics of, 295–296
Induction, 57, 74
  efficiency of, 91
  in societal systems analysis, 91
  interface with deduction, 211
Inductive engination, diagram of, 29
  examples of, 27–30
Inductive inference process, 160
Inductive modality, 25–33
  versus ampliative modality, 31–32
  versus associative modality, 30–31
Inductive transformation, 26–29
Inductive variants, 153
Industrial democracy, 110
  see also Power equalization
Industrial revolution, and work ethic, 295
Inertia, and conditioning, 159
Influence, and control, 268–269
Inherited privileges, 167
Inner-directed individuals, 277
Insinuation, and indoctrination, 178
Institutionalized systems, 105, 269–270, 284
  birthrights and, 217
  lessons of, 256
  shift to sentimentalized systems, 300
  transformed from conditioned systems, 190
Institution, 105
Institutionalization, 172–181, 220, 237, 255
Instrumental genius, of concept of afterlife, 54
Instrumental license, 169
Instrumental systems, 166
Instrumentalism, distinguished from capitalism, 168
Instrumentality, 10, 187, 193, 264
  blocs and, 215
  corporation growth and, 218
  elitist compensatory calculus and, 164
  material corridor and, 281
  secularized systems and, 265
  zero-sum games and, 227
Intention, quality of, 166
Interdependence, 215
  as symbiotic reference, 265
Interdisciplinary, imperative, 210–212
  interfaces, typology of, 249–251
  models, 94
  past failures, 95
Interfaces, axiomatic systems and, 180
  interdisciplinary, 94
    typology of, 249–251
  socio-economic, 229
  traditional social sciences and, 210
Internal drives, 20
Interrelationships, levels of, 211
Intra-species contact, 274–275
Intrinsic behavior, 100
Intrinsic predicates, 123, 180
Intuition, 94

IRA, 292
Irrational behavior
see, Pathological behavior
Isaiah, 240
Isomorphisms, 96–97, 136, 157, 255
and corridor concept, 283
and societal benefits, 260
Isomorphic analysis, 95

J

Jaspers, 187
Jehovah, 270
Jeremiah, 240
Jesus Christ, 103, 181, 194, 233, 234, 270
parables of, 104
Jesus "freaks", 202
Jewish communities, 213
Judeo-Christian ethics, 134
Jung, C. G., 204
Just price, 252
and artists, 223

K

Kant, E., 13, 187
Keynes, J., 109, 181, 248
Kierkegaard, S., 187, 230
Knowledge, creative nature of, 149
space, 147
theory and, 146
Koestler, A. S., 13, 41
Kuhn, 93

L

Labor strife, 178
Labor theory, of value, 217
under axiomatic socialism, 219
Labor unions, 297
Labor value, Marxian, 296
Laissez-faire, 218
Land, equity of, 233
Language, as hypostatized behavior, 37
Laplace, 93
Laplanders, 227
Law, 176
as codification of what people want, 102
assessed by prophets, 103
extenuating circumstances and, 174
letter versus spirit of, 174–175
punishment and, 166
Leadership, and axiological proscription, 167
Learning curves, 17, 290
graph of, 128–129
morphology of, 129
societal, 108
see also, Feedback systems
Learning loop, 8
diagram of, 6
Learning paradigms, 92
Lenin, N., 181, 230
Levi-Strauss, 37
Liberalism, versus individuation, 166
License, 105
and behavior latitude, 166
Limbic system, 14–15, 20
compared to inductive modalities, 25
diagram of, 15
Locke, J., 13, 26, 32, 77, 109, 187, 236
Logic, transcended, 148
Logical closure, 177
Logical preemption, 177
Lombardi, V., 183
Lore, 160
as a cultural constraint, 180
Love, 300–301
Luther, M., 181

M

MacLean, 13
Macro-relationships, 120, 124, 138
Magic, 268

Magnitude of relationships, in the coefficient level, 7
Magnitude of satisfaction, 135
Magnitudinal coefficients, 197
Mandation, 229
economic distribution of cerebral-sapient systems and, 220–222
Manipulation, 211
Mao, 103, 219, 241, 247
aphorisms of, 104
Maoism, ideo-economics of, 281
Markov process, 89, 153, 197, 205
Marshal, A., 109, 181, 194, 236
Marx, Karl, 39, 106, 107, 109, 137, 165, 181, 219, 226, 230, 238, 247, 250, 257, 270
Marxists, and capitalists, 301
and Karl Marx, 204, 217, 270
Masks, created by prophets, 102
heuristic, 97
historical, 106
Maslow, A., 21, 41, 109, 279, 301
Masses, 167
as prophet's target, 103
Material gratification, and axiomatic engines, 168
Materialism, 277–283
pendular swing of, 300
status, wealth, and power in, 280–281
Matrist societies, 175, 181, 234
Catholic church, decline and, 201
Holy Roman Empire, decline and, 199
Matrix, model, 198
table of, 197
transitional, 197–199
Maze, context, 23–25
environmental, 71
see also, Trial and error
Mechanistic systems, 182
Mechanisticity, and associative systems, 168
Mechanized behavior, 160
Medieval Europe advancement associations, 215
Membership credentials, as opposed to hierarchies, 215
Mental immaturity, of human species, 22
Merit, and salvation, 234
and self denial, 103
Meta-hypotheses, 156–157
ideal types as, 152
Methodology, managerial, 283–301
of social sciences, 3–85
of societal systems, 301
Middle class, 225
Migrant groups, 287
Milieu, emergent, 71
table of, 72
Mill, J. S., 187, 236
Minimum opportunity cost, 160
Models, 6
axiomatic, 53
deductive inference of, 53
elliptical structure of, 53
correlation in, 6
levels of analysis of, 5
coefficient, 7
diagram of, 6
parametric, 7
relational, 5–6
state-variable, 5
matrix, 198
of conflict, 138–139
of cultural conflict, 147
predictive and descriptive, 95
predictive value of, 8
static, 8
vector of congruence, 249
versus quantitative methods, 6–7
Model building, as abbreviation of complex realities, 95
morphological transformations and, 96
Modular authority, 246
Mohammed, E., 103, 173, 181
Monarchy, 202
Monopolies, 194, 228, 258
and instrumentalities, 218
Moon, Rev. S. M., 279

Moral authority, 169
Moral force, in capitalism, 239
Moral predicates, 112
Moral suasion, 174, 177
More, T., 230, 238
Mormons, 213, 286
Morphological change, in deductive process, 35–36
  diagram of, 35
Morphology, 157
Mortal sin, 158
Moses, 199
Motivation, industrial, diagram of, 297
M test, 142
Multinational corporations, 252
Mystics, 187
Mythology, as hypostatized behavior, 37

N

Napoleonic regime, 200
National socialism, 39, 200, 213, 278
Natural law, 236
Nazism, see National socialism
Needs, hierarchy of, 221
Nepotism, and economic distribution, 228
Nestorians, 233
Networks, versus pyramids, 215
Newton, I., 41, 93
  algorithms for mazes, 23
Nobel prize, 229
Noise, filtering of interpersonal and rhetorical, 148
Nomadic existence, and associative predominence, 189
Nomadic system, man-hour productivity in, 226
Nonconstant-sum game, contrasted with zero-sum game, 205
Non-normative systems
  see, Normative systems
Non-parochial social science, 94
Non-redundancy, of the individual, 272–273
Normative, 104
  behaviors, 129
  component of prophetic platform, 104
  dictates, 106
  dynamics of ideal types, 188
    diagram of, 188
  states, versus hybrid, 289
  strictures, 168
  systems, 284
Nuclear family, 252
  and axiomatic systems, 214
Null calculus, 234
Null hypothesis, of behaviorists, 69
  of psychology, 14
Numinosity, 195
  of national movements, 200
Numinous authority, 213

O

Obedience, and automaticity, 191
Objective performance, and competition, 177
Oligarchies, 242
Open society
  see, Proto-rational system
Open-loop, 36
Open-loop process, 21
Operant conditioning, 30, 32, 97, 99, 100
Opportunistic behavior, 169
Organized labor, 217–218
Original sin, and civil government, 238
Originative behavior, and elegance, 169
Originative systems, weakened by cerebration, 184

P

Pantheism, 263, 269
Parabolic logic, 94
Parabolic prophecy, 104
Parabolic revelation, 245

Parabolic utterances, 109
Pardigms, 92–93
paradigmatic camps and, 211
Parametric change, 106
Paranoia, collective, 185
Parish, 213
Parochialism, among social sicentists, 3–4
Parsees, 285
Partial derivatives, in non-linear constructs, 10
Participatory democracy, 243–244, 246, 253
Pathological behavior, 18–19, 22–23, 26–37, 53–54, 62, 70, 72–73, 77, 156, 160, 161, 169, 184
Pathology, collective, 180
see also, Paranoia, collective
Patriotism, 176
Patrist societies, 175, 181
Pavlovian conditioning
see, Classical conditioning
Pendularity, 102, 105–108, 191–192, 195, 302
counter-correlation of, 241
history of society and, 74
prophetic process and, 300
Penfield, 41–42
Penury, 176, 213
and economic base level, 227
Permissiveness, curtailed by associations, 166
Persuasion, as vehicle of prophet, 104
Philosophy, and ideologics, 146
Physio-mechanical bases
see, Biochemical bases
Piaget, 13
assimilative theory of, 21
Pilgrims, 213, 286
Placidity, 260–261, 282, 291, 293
ceiling of satisfaction and, 289
cognitive, 268
God-ordained, 286
milieu properties of, 70–71
societal benefits and, 277
threshold levels of, 266
Plato, 230, 238
Platonic societal system, 243, 273
Pleasure-pain calculus, 30
Poetic prophecy, 104
Polar opposites, and proscriptive-prescriptive attraction, 191
Polarity, of opposing economic systems, 226
Polarized ideal types, 98–99
Polarized value systems, of prophets, 103
Polarization, 104, 107, 172
Polanyi, 187
Popper, 187
Pornography, 200
Positivists, 187
Positivist-empiricists, 160
Potential wealth, as credit, 216
Potlatch, 252
Power equalization, 297–298
Predestination, 235
and poverty, 236
Predicates, word origin, 140
Predicate sets, diagram of, 99
Predictive-descriptive models, 95
Predictive specificity, 6
Prejudice, 32
Preemption, 211–225, 281
behavior control and, 183
totalitarian power and, 245
zero-sum game and, 218
Preludial conceptual stage, of ideologies, 141
Prescribed behavior, 130
Prescriptions, 84
a priori, 110
set of, 100
Prescriptive axiology, 133, 135, 159, 191, 195, 235, 274
Prescriptive behavior, graph of, 128
set theory and, 128–129
Prescriptive platform, 108
Prescriptive predicates, 132
and liberality, 165
Prescriptive system, 161
communes and, 214
compensatory calculus of, 176
sentimentality and, 173–174

Priesthood, 172
Primitive food gathering, 189
Primitive production, capital to labor ratio of, 226
Primitive religion, 172
Primitive systems, 182, 186–187, 263
Principle and practice, contradiction in axiomatic systems, 193
  qualitative difference between, 103
Probability, transformational, 198
  vectors, 197
Processual relationships, 186
Prodigal genius, 37
Productivity, measurement of, 298
  workers versus management, 295–296
Professionalism, 281
  in sapient systems, 184
Professionalization, and coercive union movements, 297
  collective phenomena of, 273
Professionalized systems, 287
Programmation, 180
Proletariat, 110, 164, 270
  and American politicians, 291–292
Prophecy, 191
  axiological, 106, 161
  axiomatic, 108–109
  configurations of social tradeoffs and, 302
  pendularity of, 300
  prescriptive, 104–105
  proscriptive, 105
Prophetic base, 130
Prophetic platform, 177
Prophetic process, 98–99, 103, 127
  as reactive, 102
  simplicity and implementability of, 102
Prophets, 41, 93, 101, 213, 240
  as managers of society, 255
  in the material arena, 172
  prescriptive, 174
  role of, 102
Proscriptions, 84, 158–159, 193
  from prescriptions, 236
  set of, 100
Proscriptive axiology, 131, 133, 138, 157, 164, 170–171, 191
  axiomatic systems and, 179
  elites and, 165
  material corridor, 281
  psychological benefits of, 270
  quasi-legality of, 163
  repression and amplification in, 163–164
Proscriptive behavior, 130
  and set theory, 125–126, 130–132
  graph of, 125, 126, 131
Proscriptive platform, 108
  differentiation and, 166
  psychological utility and, 162
Proscriptive predicates, 132
  compensatory calculus of, 170
  passing as axiomatic system, 219
  switch to prescriptive system, 167
Protean societal phenomena, 94, 108
Protestant ethic, 84, 177
  and capitalism, 229
Protestant mystics, 271
Protestant Reformation, 199
Proto-capitalism, and purpositivity, 167–168
Proto-capitalist theory, 215
Proto-Christian, sentimentalistic base of, 174
Proto-Communism, sentimentalistic base of, 174
Proto-rational systems, 187
Proto-socialism, income and effort in, 217
Prototypical societies, dissolution of, 190
Psychogenetics, 194
  and covens, 216
Psychogenetic engines,absorption by, 272
Pseudo-science, 141
Psychological utility, 162
  of Catholic church, 173
  of certainty, 165
Psychology, cliches in, 81
  dialectical presence in, 73–74
  differentiation and, 258, 277

identification problem in, 20–23
schools of, 11, 13, 61
syncretic analysis applied to, 4–5
Psycho-physiology, 161
Psychotic, cognitive placidity of, 271
see also, Pathological behavior
Public investments, 223
Pulitzer Prize, 229
Punishment, in prescriptive systems, 175
Puritanism, 137
Purposivity, in socialism, 168
as a benefit of axiomatic platforms, 167
Pyramids, versus networks, 215

Q

Quakers, 135, 137, 213
emotional symbiosis proscribed, 286
Qualitative analysis, 88–89, 91, 153
Qualitative constructs, 5
Quasi-legislative proscriptive axiologies, 163

R

Range estimators, in the coefficient level of analysis, 7
Rappoport, A., 96
Rational behavior base, 169
Rationalism, 217, 255
of Anglo-Saxon nations, 199
Rationality, 57
Rationalized systems, 177–181, 294
consistent with axiomatic bases, 179
lessons of, 256
material benefits of, 281
Reactive, constructs, 106
versus creative in prophetic process, 102–103
Realism, 245
cognitive demands of, 260
Realistic behavior
see, Instrumentality
Reality comprehension, 150
Reality feedback and compulsive behavior, 185
Realpolitik, 200
Real wealth, 226
opposed to potential wealth, 216
Real world precepts, 162
Real world processes, as surrogates in education, 178
Real world systems, 284
clustering and, 211
lack of variety in, 99
Reciprocity, in proscriptive systems, 166
Recruitment, and credentials, 184
Reductionism, 9
Reflexive model, 157
Reformation, the, 232–233, 300
as dogmatic regression from humanism, 234
leading to secularization, 167
release from repression and, 238
Refugee groups, 284–285, 287
Regression, 58, 181, 205
logic, 123
statistical, 30
Regression coefficients, 7
Requisitional economic systems, 223–225
graph of, 224
Requisitions, in sentimentalized systems, 217
Reification, 36, 52, 150
in the U.S. Constitution, 179
Reinforcement, 96
Religion, as opiate of the masses, 270
idiosyncratic, 271
Space Age, 165
Remission syndrome, 175
Repression, 132, 135, 173, 197
and amplification, 163–164
benefits of, 163
bureaucratic elaboration and, 167
calculus of titillation and, 261
increasing satisfaction and, 297
Reformational axiologies and, 238
resignation and, 103, 261

Resonance, and franchise system, 247
Responsibility, by default, 103
resignation opposed to, 103
Reticular configurations, 302
opposed to hierarchies, 257
Reticular referents, 244
Revelation, 94
Revolution, versus evolution, 61
Revolving of societal systems, 301
Rhetorical engination of prophets, 103
Rhetorical process, 106
Rhetorical science, 93–94
Ricardo, 236
Ritual, in primitive-subsistence systems, 245
Ritualization, 159, 188
and automaticity, 182–190
Romanticism, 187–188
Rote, conditioning, 159
engination, 168
transmission of primitive societies, 268
Rousseau, 181, 226, 229, 241, 256–257, 267
Russell, Lord, 23

S

Sacerdotal coercion, 161
Sacerdotal systems, 169–294
Sacerdotalism, 94, 106, 107, 173, 190, 213, 269
Sacrifice, and Protestant ethic, 236
Salazar, 201
Salem witch trials, 172
Salvation, man's autonomy over, 233–234
Sapience, 141, 160–161, 186, 196, 215
Sapient behaviors, and speculative science, 169
Sapient-cerebral system, 196
Sapient systems, 201
credentials and recruitment in, 184
economic distribution in, 211
Janus-faced nature of, 184
within traditional systems, 185
Sartre, J. P., 187
Satisfaction, 231
calculus of, 289
ceiling of, 289
graphs of, 289
increased by exogenous repression, 297
productivity and, 295–296
socio-economic level of, 232
see also, Gratification
Savonarola, 240
Scarcity, 220, 269, 279
Scholarly study, and transformational probabilities, 198
Schoolmen, 229, 232
Science, and philosophy, 94
superstructure of, 96
Scientific method, 33
Seadler, S. E., 149
Secular capitalism, 295
Secularization, 161, 166–169, 183–186, 192–195, 200, 219, 234
Secularized systems, economic base in, 227
instrumental behavior in, 265
titillation and gratification in, 276
Security, and continuity, 173
Segmentation, 188, 218, 226, 256, 258
Self-actualization, 41, 81, 192–193, 302
see also, Maslow
Self-determination, drawbacks of, 298
in conditioned contexts, 269
Selfishness, opposed to transcendental aim, 103
Seneca, 238, 279
Sense datum, 28
Sense driven variants, 154
Sensory perception, from political perspective, 151
Sentient man, 20, 169, 191–192, 195, 199, 202, 217, 234
Sentimentalized system, 213–214
conflict of conscience in, 275

elitists and, 176
prescriptive systems and, 173–174
set of benefits of, 276–277
symbiotic clustering in, 274
Sentimentality, as origin of socialism, 168
Sentimental-matrist system, 176, 297
Sermon on the Mount, 175
similarity to romantic socialism, 217
Set concepts, cultural, 124–139
Set space constructs, 6
Set theory, 149
Sex, and capitalism, 279
Sexually abberant experimentation, 194
Shame, 204
and axiological predicates, 195
Shankara, 103
Shared values, 79
Shares, and proto-Communism, 227
Simplicity, 96, 102, 120, 180, 182, 190
attempts at, in industrial societies, 189
change and, 106
complexity of societal management and, 210
Sin, and guilt, non-concative, 174
Skinner, B. F., 13, 77
Slavery, 178
Smith, A., 168, 181, 194, 225, 236, 238, 248, 279
and his "invisible hand", 101, 225, 238
Socialism, 110, 174
as axiomatic engine, 167
in axiological utopian base, 238
individual's purpose in, 168
Marxian, 129
materialistic, 215
Social psychology, individual versus society in, 22
Social science, criticism of, 77–84
criteria for constructs of, 5–11
Social scientist, as a prophet, 302
Societal benefits, table of, 262
Societal dynamics, components of, 105
pendularity of, 102, 106
Societal form, archetypical, 266
Societal goods, 217
Societal learning curve, 107
Societal management, 208–302
Societal referents, ideal properties of, 170–171
Societal systems, 165
analysis, 92
change in, 254–255
differing from organizations, 90
finite state, 122
high quality criteria of, 301
management of, 301
options available to improve, 293–302
reflexive nature of, 211
synergy of, 250
table of, 216
tribes as prototypical form, 215
Societal trade-offs, 302
thesis and antithesis, 300
Societal transformation, 188
Society, definition of, 80
Sociobiology, 277
emotion as instinct, 274
see also, Bio-emotions
Socio-economic calculi, table of, 230
Socio-economics, 229–242
Socio-psychology, of axiomatic engination, 167
of synoptic ideal types, 262
Sociology, origins of, 79
Socrates, 103, 181, 226, 241, 256–257
Socratic axiology, 166
Socratic schemes, 174
Solipsism, and collective cynicism, 195
pathological behavior and, 169
reflected differentiation and, 271
secularization and, 194
Solipsists, 185, 215
Southeast Asia, asymmetric economic distribution in, 228

Soviet Union, 178
communism in, 219
Spanish Inquisition, 172
Specialization, 190
arena of, 274
conditioned systems and, 218
in cerebral systems, 272
psycho-social utility of, 273
Speculation, 94
Speculative science, and sapient behavior, 169
Spirituality, 193
Spontaneity, 103, 166
and anxiety, 176
constrained, 269
in collective behavior bases, 158
prescriptive axiology and, 274
Spontaneous theory of societal form, 102
Status, 291
calculus of, 292
economic engines of, 216
identity and, 282–283
material benefits and, 280
wealth and power, 291
Stratification, and real world systems, 211
unit of primary social significance and, 213
Stratified cultural base, 120
Stratified hierarchy, versus cluster hierarchy, 247
Stochastic, 105
behavior in competitive subcultures, 123
distribution, 23
graph of stimulus-response, 122
Stochasticity, 93
Strategic reasoning, in emergent milieu, 72
Subcultures, 123, 138–139
hippie and beatnik, 202
symbiotic or competitive, 120
Suboptimability, pathology of, 184
Subornation, diseconomics of, 228
Subsistence cultures, 189, 301
Sufis, 187, 271
Superego, 178
see also, Freudian theory
Supply and demand, 252
distribution of economic benefits and, 225
in capitalism, 217
Surplus, as prerequisite for societal elaboration, 267
Surrogates, and qualitative operators, 7
money as, 237
Surrogate values, in measuring magnitude of changes, 67
Survival, 302
conditions, 100
in secularized system, 184
Survival value, choreographed, 182–184
Symbiosis, 261
emotional, 278
Symbiotic association, in harsh conditions, 277
Symbiotic relations, emotions in, 272
Symbiotic subcultures, 120
Symmetrical order, 93
Symmetry of advantage, 177
Syncretic, alternatives, 99
analysis, 3
constructs, 45
paradigms, 4–46, 77
epistemological predicates of, 13
procedures, 61
process, 11–14
diagram of, 12
Syncretic value, 107
Syncretism, 106
Synergistic associations, from complementation, 277
Synergy, 250, 261
potential in complementary strategy, 273
symbiosis and, 293
Synoptic ideal types, 153–154, 156–157, 169–170
socio-psychological implications of, 162–169
Synthesis, opposed to thesis and an-

tithesis, 300
Synthetic suprapardigm, 11
diagram of, 12
see also, Behavior bases
System states, 186

T

Tabula rasa, 25, 32
Tactical reasoning, 31
Tantricism, 240
Taoism, 240
Taylor, G. R., 175
Technocracy, 248
Teleconferencing, 148
Teleogy, 93
in collective behaviors, 158
prophecy and, 104
Temporal world, and axiomatic platforms, 167
Teutonic feudalism, 244
Theocracy, 244
Theocratic systems, wealth in, 223
Theonomy, and Black revivalism, 288
anthropomorphic creator in, 261
Thoreau, H. D., 181, 229, 241
Titillation, 194, 200
as calculus of sentient license, 270
calculus of, 282
in secular capitalist treadmill model, 296
productivity and, 299
ratchet function of, 282
Toffler, A., 215
Toleration, as a qualitative difference, 103
human capacity for, 259
Topology, ideologic, 145–148
Topological constructs, 6, 64, 67
iconographs as, 62–63
Totalitarianism, 243
appeal of, 201
as secularized-instrumentalized system, 245
behavioral logic of, 150
Tractability, 121, 123
Transcendental authority, and pretention, 172
Transgenerational, accommodations, 100
differentiation, 258
Transmogrification, of capitalist-socialist prescriptions into constraints, 238
Treadmill model, 293
of satisfaction, 291
Trial and error, as a real world response, 24–25
heuristic modalities as, 69
hysteric modalities as, 70
scientific method and, 93
variants, 154
Tribes, as prototypical societal systems, 215
Turkish feudalism, 244
Typology, of generic benefits, 284–289
of ideal-type constructs, 249
of interdisciplinary correlatives, 210, 215
table of, 212

U

U.F.O.'s
see Unidentified flying objects
UNESCO, charter, 149
Unidentified flying objects, and Carl Jung, 204
as technological religion, 166
Unidimensional understanding, and totalitarian logic, 150
Unification Church, 279
United States, Constitution of, 180
public sector of, 298
society rationalized into secularized system, 287
Urban Negroes, 287
Utilitarian ends, 159–160
Utilitarians, 187
Utility, 252
of order, decline of, 167
Utopia, 238, 281–288

V

Validation, 5
 see also, Apodictical
Value dependent imperative, 101
Value free science, 148–149
Variables, exogenous, 10
 homogeneity as, 80
 in factoring ideologics, 143
 in parametric models, 8
 needs and resources as, 241
Vietnamese War, 178, 201
Violence, to reality comprehension, 150
Voltaire, 42

W

War, 139
 see also, Vietnamese War
Watson, J. B., 13
Weber, M., 109, 150
Weimar Republic, 200
Whitehead, A. N., 255
Work ethic, 295
Wycliffe, J., 230, 238

Y

Youth, 241
 affluent, 166
Yoga, as shift from secularization, 166

Z

Zen Buddhism, 202
Zero-sum game, 56, 110, 183
 calculus of, 183
 contrasted with non-constant sum game, 205
 instrumentality and, 227, 246
 internecine conflict in industrial states, 297
 preemption and, 218